Adoption, Usage, and Global Impact of Broadband Technologies:

Diffusion, Practice and Policy

Yogesh K. Dwivedi
Swansea University, UK

Information Science REFERENCE

INFORMATION SCIENCE REFERENCE

Hershey · New York

Director of Editorial Content: Kristin Klinger
Director of Book Publications: Julia Mosemann
Acquisitions Editor: Lindsay Johnston
Development Editor: Dave DeRicco
Typesetter: Casey Conapitski
Production Editor: Jamie Snavely
Cover Design: Lisa Tosheff

Published in the United States of America by
Information Science Reference (an imprint of IGI Global)
701 E. Chocolate Avenue
Hershey PA 17033
Tel: 717-533-8845
Fax: 717-533-8661
E-mail: cust@igi-global.com
Web site: http://www.igi-global.com

Library of Congress Cataloging-in-Publication Data

Adoption, usage, and global impact of broadband technologies : diffusion, practice, and policy / Yogesh K. Dwivedi, editor.
p. cm.
Includes bibliographical references and index.
Summary: "This book provides a comprehensive coverage of broadband deployment, diffusion, adoption, usage, and policies as they have been realized by research in many countries around the world"--Provided by publisher.
ISBN 978-1-60960-011-2 (hardcover) -- ISBN 978-1-60960-013-6 (ebook) 1. Broadband communication systems. 2. Broadband communication systems--Government policy. 3. Telecommunication systems. 4. Internet service providers. 5. Globalization--Social aspects. I. Dwivedi, Yogesh Kumar.
TK5103.4.A36 2010
384.3'3--dc22
2010017215

British Cataloguing in Publication Data
A Cataloguing in Publication record for this book is available from the British Library.

To my loving daughters 'Shagun' and 'Saanvi'

Table of Contents

Detailed Table of Contents

Chapter 1 titled "*Still in Pursuit of the Fast Lane: The Crawl to Broadband Freedom*" by Justin Henley Beneke from University of Cape Town, South Africa begins by asserting that South Africa may be renowned for its natural attractions, warm climate and fine wine, but certainly not for high quality broadband. The country has fallen behind its international peers – in both developing and developed markets – in the race to rollout internet connectivity. In fact, even within the African continent, it is neither a leader nor progressive in comparison to its North African counterparts. This chapter aims to provide a chronology of the major developments in the provision of broadband internet services in South Africa, as well as touching on the challenges faced in bringing this phenomenon into the mainstream. Reasons for the lack of diffusion and adoption of such services point to the high-end user costs of the service, a limited geographical footprint of both fixed-line and mobile broadband infrastructure, as well as a lack of computer literacy and understanding of what broadband is able to offer. The chapter continues to look at possible solutions including introducing a greater degree of competition into the market to facilitate downward pressure on prices, provisioning further international bandwidth through undersea fibre optic cables, and the unbundling of the local loop, to further this objective.

Chapter 2, titled "*Strategic Interaction under Asymmetric Regulation: The Case of New Zealand*" by Bronwyn Howell from New Zealand Institute for the Study of Competition and Regulation aims to examine the specific consequences of the asymmetric tariff obligations and ensuing strategic interaction amongst sector participants on sector development – namely the effect of universal service retail prices and the allocation of the ensuing costs in determining the ongoing regulatory agenda; the role of a 'free local calling' obligation on the evolution of New Zealand's broadband market; and the consequent application of further asymmetric legislative obligations on the incumbent to address apparent 'problems' for which the asymmetric tariffs and rivals' strategic choices provide more credible explanations than the incumbent's exertion of its dominant position.

Chapter 3, titled "*Examining the Influence of Government Policy on Broadband Internet Access: The Case of China*" is authored by Qiuyan Fan from University of Western Sydney. The primary goal of this chapter is to examine how various policy factors interplay to affect broadband deployment and performance in the case of China. The findings from this chapter suggest that the Chinese approach to broadband development has not produced desirable results due to several factors. The most important factor is the lack of a detailed national broadband plan with clearly states goals, performance targets and action plans for implementation. China's regulatory arrangement and approaches to regulating the broadband industry have failed to establish an effective and competitive market and to drive high levels of broadband performance, investment and innovation.

Chapter 4 of this section is co-authored by John B. Meisel, John C. Navin and Timothy S. Sullivan from Southern Illinois University Edwardsville, USA and discusses the Federal Communication Commission's National Broadband Plan. The American Recovery and Reinvestment Act of 2009 charged the US Federal Communications Commission (FCC) to develop and deliver to Congress a national broadband plan by February 17, 2010. The FCC formally commenced the process of developing the plan by issuing a Notice of Inquiry (NOI) on April 8, 2009. The NOI identified broadband issues and critical questions and asked stakeholders to respond to these issues and questions with data and analysis. The purpose of this chapter is to analyse the written documents generated by stakeholders' responses concerning the specific issues of open networks and competition and to make recommendations to the FCC in its formulation of federal policy as to the position that makes the most economic sense on these issues. The chapter suggests that many of the arguments and concerns of stakeholders are dependent upon predictions regarding the competitiveness of ISP markets. Based on the discussion made within the chapter it predicts with confidence that technological innovations are likely to make many legal arguments (on all sides) obsolete in the near future.

Chapter 5 focuses on broadband and structural separation from the perspective of transaction cost economics authored by Hidenori Fuke from Komazawa University, Japan. Conduct regulation and structural separation are often discussed in industrial organisation studies as options to prevent the abuse of market power by vertically integrated firms toward the downstream market. Both the structural separation of NTT and conduct regulation have been discussed in the Japanese telecommunications industry since the introduction of competition in 1985 and the issue is still being discussed, although the industry is going through a transition from POTS (Plain Old Telephone Service) to the broadband internet. Past discussions have been inclined toward elimination of the harmful effects of vertical integration. However, there is a benefit of vertical integration in the sense that it will promote the efficient administration of the firm concerned. This chapter presents a new contention that it is important to employ a balanced analysis of costs and benefits of vertical integration based on transaction cost theory. Structural separation in the broadband market entails significant transaction costs between a carrier with access facilities and firms offering broadband services by renting these facilities as input. These kinds of transaction cost are comparatively negligible in POTS. The chapter attempts to make clear that the balance analysis of the costs and benefits of structural separation has become more important in broadband than in POTS, based on the actual differences in network structure.

The last chapter (Chapter 6) of this section compares mobile phones with other ICTs and presents the case of Greece and its adoption of mobile phones from a socio-cultural perspective. This chapter is authored by Panayiota Tsatsou from Swansea University, UK. The chapter explores mobile phones and how they have been received in juxtaposition with the Internet and in close association with the socio-cultural contexts of life. By examining the Greek case and its particularities, the chapter provides some sense of why different Information and Communication Technologies (ICTs), such as mobile phones and the Internet, might be received and adopted differently by people living in the same socio-cultural context (at the national level). In the case of Greece, statistical and historical data confirm the contrasting receptions of mobile phones and the Internet but there is a lack of empirical evidence to explain the exceptionally high adoption rates of mobile phones in the country. Thus, the chapter reports on original empirical evidence obtained in elite actors' interviews and focus groups of ordinary people to explain the contrasting ways that mobile phones and the Internet have been received in the country. On the basis of the empirical findings, the chapter suggests that certain socio-cultural contexts, such as that of Greece, favour mobile phones more than the Internet, thus making mobile telephony a distinctive case of ICT.

Section 2
Cross-Country Analysis and Global Perspectives on Policies and Strategies

Elizabeth Fife and Francis Pereira from the University of Southern California, USA contributed Chapter 7 on the effectiveness of Government policies in broadband deployment in Singapore, Hong Kong SAR and South Korea. While the potential economic and social benefits of broadband internet use are significant, adoption levels vary greatly between countries around the world. Many governments, particularly those in Southeast Asia, have adopted aggressive policies to deploy broadband networks and to encourage the use of applications. Governments are motivated to promote broadband adoption in order to realise both economic and social benefits. This chapter argues that the generally higher levels of broadband adoption rates witnessed in many Asian economies is attributable in part to the aggressive policies pursued by these governments. There is some evidence to suggest that these governmental policies have been successful in achieving their stated goals.

The three Greek authors, namely Christos Bouras (from the University of Patras), Apostolos Gkamas (from Research Academic Computer Technology Institute) and Thrasyvoulos Tsiatsos (from Aristotle University of Thessaloniki) co-authored Chapter 8 titled "Best Practices and Strategies for Broadband Deployment: Lessons learned from around the World". This chapter begins by asserting that broadband deployment is a necessity nowadays. It could help each country, municipality and region to grow and offer better quality of life to the citizens. Today, the emphasis on the development of broadband networks is on fixed Fibre to the Home solutions. The lessons learned from countries that are leaders in broadband penetration and Fibre to the Home deployment could prove very useful for under-served communities, regions and countries where the broadband penetration is low. Therefore, this chapter summarises the lessons learned from implementing (a) country-wide strategies formulated at the national level, and (b) local strategies formulated by the municipalities. Concerning the role of national and local governments, it should be noted that nowadays it is very urgent in terms of the involvement of government in the development of broadband infrastructure. Proposed noteworthy remarkable cases are Japan, South Korea and Singapore.

The last chapter (Chapter 9) of Section 2 titled "*The European Research and Education Networks: Ensuring Europe's Leadership in e-Science*" is co-authored by Navonil Mustafee from Swansea University, UK and Simon JE Taylor from Brunel University, UK. This chapter begins by asserting that e-Science is the future of Science. It necessitates the use of powerful computing resources, massive data sets, remote instruments, scientific/visualisation software and expertise that are distributed around the world. e-Infrastructures refer to the underlying computing technologies that facilitate e-Science. To weave these distributed resources into a cohesive entity, and thereby enabling such large-scale science to be conducted in highly distributed environments, needs the utilisation of high-speed communication networks. The European Research and Education Networks provide the connectivity required to ensure Europe's leadership in e-Science. These networks not only provide the communication resources required to conduct e-Science, but they also offer high-speed internet access to universities and educational institutes to facilitate teaching and learning. The objective of this chapter is to inform the readers of the organisation of the underlying European networks that are used by millions of researchers, academics and students. The importance of these networks is highlighted by presenting six e-Science and e-Infrastructure projects that are being funded by the European Commission.

Section 3
Impact and Social Consequences of Broadband Diffusion

The opening chapter (Chapter 10) of this section was contributed by Peter Stenberg and Mitchell Morehart from Economic Research Service, USDA, USA and attempts to build an understanding of U.S. rural-urban differences in broadband internet adoption and use. The chapter begins by asserting that the Internet has become embedded in the U.S. economy over the past 15 years. During this period, access to and use of the Internet has increased for all regions of the United States, most types of households and work places, and all income groups. This chapter explores how access to technologies may affect household online activity patterns and addresses some of the aspects that differentiate urban and rural household use of the Internet. Rural households are less likely than urban households to have broadband internet access but this varies regionally across the country. The results suggest that broadband internet access is no longer perceived as a luxury but as a necessity, and that there is pent-up demand for broadband internet access in rural areas.

Chapter 11 is titled "*Information Communication Technology (ICT) for Rural Women's Life in Bangladeshi Villages*" and is co-authored by four authors, namely Md. Mahfuz Ashraf from the University of Dhaka, Bangladesh, Helena Grunfeld from Victoria University, Australia, Syeda Rownak Afza from BRAC University, Bangladesh and Bushra Tahseen Malik from Brainstorm Bangladesh. The chapter begins by recognising that Information Communication Technology (ICT) has the potential to contribute to development, especially in rural areas of developing countries. But the mechanisms through which ICT can be combined with development agendas and an understanding of the actual development process and impacts of ICT are less well understood or properly defined in the academic literature. This research is an attempt to contribute to understanding this process by analysing the impact of two ICT initiatives in Bangladesh, aimed at improving the lives of rural women. An interpretive approach in the qualitative research tradition was adopted to identify emergent themes in this study. Findings presented in this chapter indicate that these ICT projects have significantly improved the socio-economic opportunities of many women. This paper will be useful for those academics, practitioners and policy-makers who wish to enhance their understanding of ICT projects in rural areas of developing countries.

Justin Henley Beneke from the University of Cape Town, South Africa contributed Chapter 12 titled "*The Application of Social Network Websites as a Marketing Platform to the Youth: An Emerging Market Perspective*". The chapter begins by stating that social networking is often touted as being a prominent application responsible for driving the adoption of residential broadband services. The growth of social networks is phenomenal – in many cases more than doubling in size on an annual basis. This study considers how social networking may be utilised for commercial purposes to spread word-of-mouth communication. The chapter therefore considers the characteristics of young adult social network users and how they behave and interact with other users on such platforms, as well as the manner in which marketers can make the most of this platform without experiencing a consumer backlash. The research suggests that if a symbiotic relationship does exist between broadband proliferation and the adoption of social networking, both have a vested interest in each other's continued success.

The last chapter (Chapter 13) of Section 3 is titled "*Consumer Usage of Broadband Internet Services: An Analysis of the Case of Portugal*" co-authored by Janice A. Hauge from the University of North Texas, USA, and Mark A. Jamison and Mircea I. Marcu, both from the University of Florida, USA. The chapter aims to analyse the intensity and patterns of use of fixed and mobile broadband consumers in Portugal. If usage across types of consumers is similar after controlling individual characteristics identified to be important drivers of adoption, then it is more likely that consumers will view mobile and fixed broadband as somewhat substitutable. Such a result is important for studies of broadband impacts; specifically, for discerning whether mobile broadband services will have a similar level of impact upon social and economic development as fixed broadband services have had. The results presented in this chapter indicate that broadband uses are similar across fixed and mobile users, suggesting that the technologies are somewhat substitutable from customers' perspectives and raising the possibility of limited differential effects on innovation and other social goals. Results of interest include the characteristics of internet users by technology, and differences of usage patterns reflected by individual characteristics.

Section 4
Technological Advances and its Impact on Emerging Applications

The opening chapter (Chapter 14) of the last section is contributed by Banani Nandi and Ganesh K. Subramaniam based at AT& T Shannon Laboratories, USA. This chapter is titled "*Evolution in Broadband Technology and Future of Wireless Broadband*". It begins by asserting that in the past two decades, the availability of worldwide high speed internet networks together with progressive deployment and adoption of broadband connections to the Internet has significantly enhanced the ability to transmit video, audio and voice through the same channel with high performance. In addition to that, since the beginning of this millennium, there has been a rapid growth in diffusion of wireless broadband. In this respect, mobile broadband is gaining popularity lately in both developing as well as developed countries. The main reason for this growth in wireless broadband lies in its ability to serve remote areas at a relatively lower cost than landline and its potential to transmit information seamlessly from anywhere in the world. The purpose of this chapter is to examine this changing diffusion behaviour of fixed versus mobile broadband technology and highlight the future possible path of adoption for these technologies by users around the world.

Nabyla Daidj and Pierre Vialle from the Telecom Business School, France co-authored Chapter 15 on the strategic moves related to broadband diffusion on the French VoD and TV market. The chapter begins by asserting that the launch of triple-play services, including IPTV, can be considered as a "critical event" that has had a strong structural impact on the market. This paper analyses the evolution of the VoD supply in France, and in particular the effect of the launch of IPTV systems. Firstly, the launch of attractive triple-play offers has induced a strong development of the French broadband internet access market, which in turn has resulted in the entry of numerous players in the VoD market. Secondly, the adoption of an IPTV system has changed the competitive and strategic positioning of main players in the VoD value chain, thus inducing or amplifying strategic moves along this value chain. In particular, the chapter shows how the diffusion of IPTV systems has induced a switch from complementary convergence to competitive convergence.

Finally, Chapter 16 is titled "*Media Platform Competition: The Displacement Effect of IPTV on Cable TV in Taiwan*" by Yu-li Liu & Wen-yi Hsu from National Chengchi University, Taiwan. The chapter begins by outlining three major TV platforms (terrestrial TV, cable TV, and IPTV) in the context of Taiwan. When faced with various TV platforms, the consumers in Taiwan will be able to choose from many broadcasts. When a new medium emerges, there are always concerns for its displacement effects on existing media. Many studies have explored the displacement effects of newly-emerging media, and have come up with a variety of sometimes conflicting findings. This chapter aims to analyse the media platform competition, namely, the displacement effect of IPTV on cable TV. The research methods used include a literature review and a telephone survey. With regard to the telephone survey, 612 of CHT's IPTV users were selected systematically based on the name lists provided by the service provider. In this study, while the platform displacement effect of IPTV on cable TV is clear, the statistics reveal a different result for the time displacement effect.

Foreword

When Dr. Dwivedi told me about his plan to edit a book on adoption, usage, and global impact of broadband technologies, I was not sure if broadband technologies had come that far to necessitate editing a book on their adoption and usage. Then one of my dear friends told me about an information revolution that is taking place in remote villages in developing countries like India, thanks to Internet and broadband technologies. The *information revolution* my friend was referring to is individual empowerment where a fisherman in a remote village in India can access market information in real time at an Internet kiosk before deciding where and at what price to sell his produce. While growing up in a developing country, I could not have fathomed something like this which seemed like a dream then to turn into a reality so fast. This made me realize that possibly, I am not up to date on something powerful which is reconfiguring the information landscapes in the remotest parts of the world. During my brief exploration of broadband technologies, I felt the idea of a book on adoption and usage of such technologies was a great one. I applaud Dr. Dwivedi to undertake this initiative to put together a compilation of practices and experiences about broadband technologies across the world.

We all have seen broadband rapidly emerge as the building block for the global economy. Already, we are observing widespread deployment of broadband networks as an essential component of infrastructure for global information economy, providing businesses and consumers with on demand access to Internet- based services, content, and applications. Recently, broadband technology has become more reliable, readily available, cost effective as well as easy to deploy. Despite significant growth in broadband networks and accelerating rate of broadband adoption, significant differences still exist between developed nations and developing nations in terms of existing infrastructure, regulatory environments, and urban-rural divide. Such differences point to the need of sharing best practices or experiences that facilitate even more rapid and more cost-effective deployment and adoption of broadband technologies across different countries.

The major challenge to widespread broadband adoption is the ability of policymakers to ensure that broadband supply meets the demand while keeping the costs low. The first two sections of the book provide rich insights into impact of government policies across countries on diffusion of broadband technologies. Documenting the experiences of broadband development and adoption across a wide range of countries from developing countries like South Africa to highly developed countries like USA, this book provides outline for the steps that other countries can follow. This book provides guidance to legislators, policy makers, and regulators to compare development of broadband in their respective countries to that in other countries and identify the key issues in regulation and spectrum management. A number of key issues surrounding market competition, national policies, regulatory challenges, and socio-cultural impacts of broadband technologies are highlighted in this section. The case of countries such as China, as presented in the first section, where regulatory arrangement and approaches are hindering the establishment of well functioning and competitive market, could be insightful for practitioners from other countries. On the other side, the lessons learned from the leaders in broadband deployment, as documented in the book, could be equally valuable for the countries that are just starting to catch up with others in broadband deployment.

The nationwide impact of broadband begins in communities and individual businesses, where research has repeatedly identified positive effects including greater productivity and rapid employment growth. Section 3 of the book explores such impact and social consequences of broadband diffusion. With experiences as diverse as the countries they are derived from such as USA and Bangladesh, this section covers a spectrum of impacts broadband adoption and usage is having on urban and rural communities. With broadband technologies empowering individuals either through "Village Phone" in Bangladesh or through social networking in South Africa, it underlines the depth of broadband penetration. Such experiences underscore the opportunities that lie ahead for broadband vendors, telecom executives, and community organizations to empower individuals. This is where the future is and this book provides a good overview of social impacts of broadband technologies. Finally, the book provides sneak peek into technological advances and its impact on emerging applications in Section 4. This is particularly important for technology vendors to chart out their future strategies.

In essence, I am convinced this book is an excellent collection of experiences and best practices in broadband adoption and usage across countries that can set the stage for broadband future. It is time now to empower someone, somewhere, in some remote part of the world to experience the power of information through broadband. It is time to provide guidance to vendors and influence policies of regulators across countries to ensure widespread reach of broadband, free from regulatory shackles. This book is a right step in that direction.

Vikas Jain
University of Tampa, USA

Vikas Jain *holds a Ph.D. in Information Systems and Decision Systems from The George Washington University, Washington DC. He completed his masters in information systems from IIT, New Delhi, India and a Bachelor of Technology degree in Electronics Engineering from Delhi Institute of Technology, Delhi, India. He has worked for nearly 12 years in IT industry in India in different areas ranging from software development, IT market research and consulting, and e-learning. He has worked on projects for clients like IBM, Microsoft, HP, Prudential Insurance among many others. Currently, Dr. Jain is an assistant professor of information and technology management at the University of Tampa, Tampa, FL and conducts research in the areas of enterprise resource planning systems, electronic government, and IT business value. He has served as guest editor for number of journals including Electronic Government: An International Journal, Journal of Global Information Technology Management (JGITM), and International Journal of Organization and End User Computing. He also serves as an editorial review board member for International Journal of Electronic Government Research.*

Foreword

The potential and already realized benefits of broadband have resulted in an upheaval of excitement about the opportunities of this range of technologies. I remember doing my first video-conference with other universities in 1998. We had connections with 6 persons at 6 distinctive locations and were really impressed by the quality of voice and video, which was far better than the standard television at that time (not high-density). This simple application saved us travelling time and still we were able to view each other's reactions in the discussion we had. At that time only some of the Dutch universities were connected by a very high-speed network. Nowadays the investments necessary for realizing broadband and other thresholds to create broadband connections have significantly lowered. However, despite all advances the availability of broadband is not at the level we desire. Broadband connections are only available to a small part of the world's population. In this respect the topic has not lost any of its importance in the past years. Instead, further development of broadband might have a huge impact on our society.

In some countries, like the Netherlands, broadband penetration is high and many people have the luxury of fibre-to-the-home, like I do. Many of the students in Delft have broadband to their livings which enables them to view on-demand video-streaming of lectures. They don't have to wake up early to be in the lecture hall and can simultaneously view the lecturer, the slides and (book) material. In this way broadband enables all kinds of new and innovative applications that can advance economic growth and contribute to the creation (enhancement) of the information society. The potential benefits offered by the widespread diffusion of broadband are astronomic. Yet, we are only just at the start of this take-off at a world scale.

The Dutch government played a pivotal role in the adoption and diffusion of broad band in the Netherlands. The Ministry of Economic Affairs recognized the potential benefits to economic growth. Regional and local cooperations saw the benefits that could be realized by empowering the social subclass by providing access. The well-educated part of the population realized the potential and many companies started to develop applications and providing services to take advantages of availability of broadband. This also demonstrates a classical chicken and egg situation; applications and services taking advantage of broadband should be available to stimulate broadband development by the market, whereas broadband should be available to stimulate the development of new application and services by companies.

Over time, our knowledge has advanced in many areas of the broadband playing field. Nevertheless there still is a lack of knowledge in several areas within this field. Therefore there is a clear need to expand our research to gain a better understanding in the mechanisms at work. Due to the relevance of the work' presented in this book to both practice and theory it is my pleasure to write this foreword. Much can be learned from cross-case and cross–country comparisons. Policies that work, those that might work and those that do not work are important to deal with the social impact of broadband.

I hope this book will contribute to achieving higher broadband penetration and will help to outline policies and strategies to help to address societal issues at stake. In my view this book gives knowledge, but also inspiration to develop ideas to overcome the digital and social divide, stimulate rural development and think about new societal forms.

Marijn Janssen
Delft University of Technology, The Netherlands

***Marijn Janssen** is an Associate Professor at the Information and Communication Technology section and Director of the interdisciplinary SEPAM Master program of the Faculty of Technology, Policy and Management at Delft University of Technology. He has conducted and managed several simulation and gaming projects, and published over 150 refereed publications and serves on several editorial boards and conferences in the area of e-government. For more information, visit www.tbm.tudelft.nl/marijnj.*

Foreword

It has been almost a decade now since we were talking about 'fast Internet' and 'second generation Internet' as a wishful thought while now we live this reality. Broadband has been a great revolution for Internet users all over the world and it has now been characterised as 'a citizens' fundamental right'. So what were the factors that lead to this wide spread of broadband Internet? The most obvious is content, watching TV shows and films and playing interactive games is an everyday reality of people in every age today. Social networking and the media capabilities they offer is another factor making people demand and use Internet which is fast and reliable. Low prices and network infrastructure have also been heavily contributed to giving broadband Internet the commodity nature it has today.

But is this the case in every part of the world? Apparently not as the different country analysis reports in this book show that there is still a large part of the human population who has never used the Internet due to lack of available infrastructure and training. A recent report from BBC world showed the striking difference between families in South Korea who were made to go off line for a week, creating withdraw Internet symptoms to them with families in rural areas in Nigeria who have never used or heard of the Internet before. So this is where the government policies come into place to iron out these inequalities and work not only on narrow national specific settings but go out to reach into a global environment of free information sharing and communication.

So, what is the future for broadband now? The answer comes immediately to mind is mobile broadband, is it already a reality and how much we need to do in terms of infrastructure development and content to reach the levels of broadband Internet? The indications are that this won't be too far from now. What about convergence with mobile, Internet and TV? Is that something we will experience soon? This book gives answers to all these questions and includes a comprehensive account of where broadband in today and where is going in the future. Have a good reading.

Natasha Papazafeiropoulou
Brunel University, UK

***Anastasia Papazafeiropoulou** is a lecturer in the Information Systems and Computing department at Brunel University, UK. Her PhD was awarded by Brunel University and she also holds a first degree in Informatics and a M.Sc. in Information Systems both from the Athens University of Economics and Business, Greece. She has been involved in a number of European and UK funded research projects with emphasis on electronic commerce and small and medium size enterprises (SMEs). She teaches information systems management and business integration at the postgraduate level. She also supervises researchers in the field of technology adoption by organisations with special interest in developing countries. She has 10 years of research experience on the study of diffusion and adoption of electronic commerce, broadband Internet, Enterprise Recourse Planning Systems (ERPs), Customer Relationship Management systems (CRMs), IP-telephony and mobile TV.*

Foreword

Recent studies stress the crucial role of broadband diffusion to enhance economic growth and performance. Therefore, the analysis of the factors shaping its adoption, diffusion and impact becomes a matter of special importance. In spite of the steady growth in broadband diffusion, many countries are still in the early stages of broadband deployment and are assessing policies to promote faster adoption. As rates of broadband access continue to rise across nations, understanding the principal causes of such diffusion is an important yet under-analyzed topic.

Broadband connectivity is considered a significant driver in exploiting the full potential of electronic commerce within an information economy, which in turn is resulting in significant gains of international competitiveness at a national level. Therefore, encouraging businesses and homeowners to adopt broadband connectivity and thereby increasing penetration rates have been a prime policy-issue in many countries (Dwivedi and Irani, 2009).

An interesting yet currently unanswered empirical question is, What is the underlying cause of this diffusion? If the increase stems from higher levels of infrastructure developing across the nation, policies to further expand such infrastructure may find additional public support. Alternatively, shifting returns to household characteristics such as education, income, or rural status through the normal process of innovation diffusion may be the dominant source of the increase (Whitacre, 2008). If this premise is correct, policies to increase future rates of access may focus less on promoting infrastructure, and instead, center on educating individuals with low propensities for adoption about the benefits of broadband access (Whitacre, 2008).

This book provides exhaustive coverage on a particular research issue that will provide a positive contribution to adoption, diffusion, and impact of broadband from a global perspective. The aim of this book is to provide an understanding of various factors and policies affecting the adoption of broadband. It also aims to understand the usage of broadband and its impact upon consumers. The book is likely to contribute towards theory, practice and Policy.

Roya Gholami
Aston Business School, UK

REFERENCES

Dwivedi, Y., & Irani, Z. (2009). Understanding the Adopters and Nonadopters of Broadband. *Communication of the ACM, 52*(1).

Whitacre, B.E. (2008). Factors influencing the temporal diffusion of broadband adoption: evidence from Oklahoma. *Ann Reg Sci*, 42, 661-679.

***Roya Gholami** is a lecturer in Operations and Information Management Group, Aston Business School, Birmingham, UK. Her current research interests are IT business value, Green IT, IT adoption and diffusion, IT and Development. She has published in IEEE Transactions on Engineering Management, Information & Management, Journal of Global Information Management, Information Resource Management Journal, World Economy, Technovation and Electronic Journal of Information Systems in Developing Countries.*

Preface

Several advantages of broadband for both the public and private sector organisations have been outlined and discussed in many previous studies. These benefits include cost savings, efficiency and competitiveness (Lee et al, 2003; Oh et al, 2003; Sawyer et al, 2003). Furthermore, high uptake of broadband is perceived to play a vital role in the growth and development of the emerging internet-based applications including electronic commerce, electronic government, and electronic health. It promises several societal benefits including telecommuting (Suomi & Pekkola, 1998), which in turn offers benefits in many ways such as contributing to flexibility of lifestyle for individuals, permitting space savings for organisations, and helping to decrease carbon footprints by reducing travelling between the home and the workplace (Dwivedi et al, 2010). Broadband allows implementation of telemedicine-enabled health service delivery, particularly for older people and for dispersed populations located in remote areas. Another societal use of the technology is support for e-Learning and distant learning. e-Learning both complements face-to-face classroom education with computer-aided teaching (so-called *blended learning*) and enables those who are unable to attend a formal educational environment (e.g., people in remote areas) to learn and gain knowledge by accessing online resources through an internet connection (distance learning). Many governments have realised the potential of broadband technologies and have made available an increasing number of government services for citizens to access online (Dwivedi et al, 2010).

Considering the aforementioned benefits, governments in a number of countries, including South Korea, Japan, Hong Kong, Sweden, Australia, Canada, the UK and the USA, have made large investments in developing a broadband infrastructure to deliver high-speed internet access to end users, including household consumers and Small and Medium Enterprises (OECD, 2001; Oh et al, 2003; Sawyer et al, 2003). Despite the early efforts by the governments of several countries, its demand has not increased in line with expectations (Dwivedi et al, 2010). In order to accelerate the growth momentum further, recently some of these broadband pioneers have rejuvenated their efforts to provide broadband access to all citizens. For example, the USA has formulated the 'National Broadband Plan' (FCC, 2009[1]). Similarly, the Australian Prime Minister has announced plans to build Australia's new national broadband network that will aim to reach 90% of Australian households at a cost of $US30 billion (Radio Australia News[2]). Such initiatives indicate that the development, deployment and diffusion of broadband infrastructure and technologies require continued and long-term planning and strategic thinking (Dwivedi et al, 2010).

Since broadband internet has the potential to profoundly impact science, business and society – and transform almost every aspect of everyday life – it is appropriate and timely to understand the deployment and adoption of broadband technologies (Dwivedi et al, 2008). Numerous researchers around the world have realised the importance of studying this area and have focused upon accumulating relevant knowledge of it. Broadband research has been prolific for a phenomenon that is quite young. However, an analysis of the current literature on broadband suggests that the available body of knowledge is fragmented with some studies looking at adoption or usage patterns, and others at the impact of broadband on existing or new internet applications (Dwivedi et al, 2008). Recognising this as an opportunity to contribute to the emerging research on broadband, IGI Global has published an authored book (Dwivedi, 2007) on *Consumer Adoption and Usage of Broadband* and a two-volume handbook (Dwivedi et al, 2008) titled *Handbook of Research on Global Diffusion of Broadband Data Transmission*. These two publications made initial efforts in synthesising existing research results in order to offer an overall picture and comprehensive understanding of exploratory issues related to the deployment, diffusion, adoption, usage and impact of broadband technology from a global perspective. However, the issues, problems and suggested solutions reported in these two IGI publications may have evolved, which may suggest a need for fresh thinking and strategies to advance the knowledge and practice in this area.

In line with the aim of the previous two publications (Dwivedi, 2007; Dwivedi et al, 2008) to access the latest research and provide an outlet to researchers in the field of broadband, the editor decided to launch this book where researchers with varied geographical, educational and cultural backgrounds assist in providing the necessary coverage of possible research issues within the area. Hence the primary objective of this edited volume was to assemble as much research coverage as possible related to the recent policy- and practice-oriented advances regarding deployment, diffusion, adoption, use and impact on emerging applications from studies conducted in various geographical settings. Additionally the book helps the reader to understand the differences in the adoption of broadband in different countries and examine comparative policy issues at national and international levels.

In order to provide the most balanced coverage of concepts and issues related to the selected topics of this book, researchers from around the world were asked to submit proposals describing their planned coverage and the contribution of such coverage to the book. All proposals were carefully reviewed by the editor in light of their suitability, the researcher's records of similar work in the area of the proposed topics, and the best proposal for topics with multiple proposals. The goal was to assemble the preeminent studies into broadband from all over the world to contribute entries to the book. Upon the receipt of full entry submissions, each submission was forwarded to at least two expert external reviewers on a double-blind peer review basis. Only submissions with strong and favourable reviews were chosen as entries for this book. As a result, this book includes 16 entries highlighting aspects of deployment, diffusion, adoption and use of broadband in various geographical settings. All entries were written by knowledgeable, distinguished scholars from many prominent research institutions around the world.

The extended and comprehensive coverage of broadband research in this distinctive book will contribute towards theory, practice and policy. The theoretical contribution of this collection of studies is that it synthesises the appropriate literature in order to enhance knowledge of broadband deployment, diffusion, adoption, usage and impact from the global perspective. In line with previous publications (Dwivedi, 2007; Dwivedi et al, 2008) this book contributes to theories and models from different areas such as Information Systems, Management, Marketing, Economics and other Social Sciences disciplines. Considering the relatively slow, heterogeneous and stagnant adoption of broadband, it can be learnt that the policy makers and providers of the innovation – in this case the telecommunications industry – hold a specific interest in the findings of this book. Policy makers in various countries, mainly from the developing world, are currently investigating how to increase the diffusion of broadband within their own countries, and so information on other countries' experiences could prove useful. Additionally, the telecommunications industry is interested in determining how to improve their current strategies. Therefore, for both policy and practice, this book offers an understanding of the broadband diffusion strategies at both macro and micro levels. Understanding the usage and impact of broadband will be helpful for content-developing organisations to integrate compelling content with new generation broadband, as well as for broadband service providers seeking to improve their services.

In order to cater to the information needs of a diverse spectrum of readers and at the same time to effectively present this global but complex topic, this book is structured into four sections, with each section including a number of chapters. A brief description of each section and chapter is provided below.

Section 1, titled "*Country Analysis and National Policies*," examines both the factors and policies shaping broadband deployment and diffusion in various countries including Greece, Japan, New Zealand, South Africa, and The United States of America. A number of pertinent factors and issues including national policy, market competition, structural changes, regulatory challenges, and socio-cultural perspective were discussed within this section. A brief account of the seven chapters included in this section is provided below.

Chapter 1, titled "*Still in Pursuit of the Fast Lane: The Crawl to Broadband Freedom*," by Justin Henley Beneke begins by asserting that South Africa may be renowned for its natural attractions, warm climate and fine wine, but certainly not for high quality broadband. The country has fallen behind its international peers – in both developing and developed markets – in the race to rollout internet connectivity. In fact, even within the African continent, it is neither a leader nor progressive in comparison to its North African counterparts. This chapter aims to provide a chronology of the major developments in the provision of broadband internet services in South Africa, as well as touching on the challenges faced in bringing this phenomenon into the mainstream. Reasons for the lack of diffusion and adoption of such services point to the high-end user costs of the service, a limited geographical footprint of both fixed-line and mobile broadband infrastructure, as well as a lack of computer literacy and understanding of what broadband is able to offer. The chapter continues to look at possible solutions including introducing a greater degree of competition into the market to facilitate downward pressure on prices, provisioning further international bandwidth through undersea fibre optic cables, and the unbundling of the local loop, to further this objective.

Chapter 2, titled "*Strategic Interaction under Asymmetric Regulation: the Case of New Zealand*," by Bronwyn Howell aims to examine the specific consequences of the asymmetric tariff obligations and ensuing strategic interaction amongst sector participants on sector development – namely the effect of universal service retail prices and the allocation of the ensuing costs in determining the ongoing regulatory agenda; the role of a 'free local calling' obligation on the evolution of New Zealand's broadband market; and the consequent application of further asymmetric legislative obligations on the incumbent to address apparent 'problems' for which the asymmetric tariffs and rivals' strategic choices provide more credible explanations than the incumbent's exertion of its dominant position.

Chapter 3, titled "*Examining the Influence of Government Policy on Broadband Internet Access: the Case of China*," is authored by Qiuyan Fan. The primary goal of this chapter is to examine how various policy factors interplay to affect broadband deployment and performance in the case of China. The findings from this chapter suggest that the Chinese approach to broadband development has not produced desirable results due to several factors. The most important factor is the lack of a detailed national broadband plan with clearly states goals, performance targets and action plans for implementation. China's regulatory arrangement and approaches to regulating the broadband industry have failed to establish an effective and competitive market and to drive high levels of broadband performance, investment and innovation.

Chapter 4 of this section is co-authored by John B. Meisel, John C. Navin and Timothy S. Sullivan and discusses the Federal Communication Commission's National Broadband Plan. The American Recovery and Reinvestment Act of 2009 charged the US Federal Communications Commission (FCC) to develop and deliver to Congress a national broadband plan by February 17, 2010. The FCC formally commenced the process of developing the plan by issuing a Notice of Inquiry (NOI) on April 8, 2009. The NOI identified broadband issues and critical questions and asked stakeholders to respond to these issues and questions with data and analysis. The purpose of this chapter is to analyse the written documents generated by stakeholders' responses concerning the specific issues of open networks and competition and to make recommendations to the FCC in its formulation of federal policy as to the position that makes the most economic sense on these issues. The chapter suggests that many of the arguments and concerns of stakeholders are dependent upon predictions regarding the competitiveness of ISP markets. Based on the discussion made within the chapter it predicts with confidence that technological innovations are likely to make many legal arguments (on all sides) obsolete in the near future.

Chapter 5 focuses on broadband and structural separation from the perspective of transaction cost economics authored by Hidenori Fuke. Conduct regulation and structural separation are often discussed in industrial organisation studies as options to prevent the abuse of market power by vertically integrated firms toward the downstream market. Both the structural separation of NTT and conduct regulation have been discussed in the Japanese telecommunications industry since the introduction of competition in 1985 and the issue is still being discussed, although the industry is going through a transition from POTS (Plain Old Telephone Service) to the broadband internet. Past discussions have been inclined toward elimination of the harmful effects of vertical integration. However, there is a benefit of vertical integration in the sense that it will promote the efficient administration of the firm concerned. This chapter presents a new contention that it is important to employ a balanced analysis of costs and benefits of vertical integration based on transaction cost theory. Structural separation in the broadband market entails significant transaction costs between a carrier with access facilities and firms offering broadband services by renting these facilities as input. These kinds of transaction cost are comparatively negligible in POTS. The chapter attempts to make clear that the balance analysis of the costs and benefits of structural separation has become more important in broadband than in POTS, based on the actual differences in network structure.

The last chapter (Chapter 6) of this section compares mobile phones with other ICTs and presents the case of Greece and its adoption of mobile phones from a socio-cultural perspective. This chapter is authored by Panayiota Tsatsou. The chapter explores mobile phones and how they have been received in juxtaposition with the Internet and in close association with the socio-cultural contexts of life. By examining the Greek case and its particularities, the chapter provides some sense of why different Information and Communication Technologies (ICTs), such as mobile phones and the Internet, might be received and adopted differently by people living in the same socio-cultural context (at the national level). In the case of Greece, statistical and historical data confirm the contrasting receptions of mobile phones and the Internet but there is a lack of empirical evidence to explain the exceptionally high adoption rates of mobile phones in the country. Thus, the chapter reports on original empirical evidence obtained in elite actors' interviews and focus groups of ordinary people to explain the contrasting ways that mobile phones and the Internet have been received in the country. On the basis of the empirical findings, the chapter suggests that certain socio-cultural contexts, such as that of Greece, favour mobile phones more than the Internet, thus making mobile telephony a distinctive case of ICT.

Section 2, titled "*Cross-Country Analysis and Global Perspectives on Policies and Strategies*," examines and compares policies, strategies and best practices in relation to broadband deployment, diffusion, adoption, and usage in various countries. A brief account of the three chapters included in this section is provided below.

Elizabeth Fife and Francis Pereira contributed Chapter 7 on the effectiveness of Government policies in broadband deployment in Singapore, Hong Kong SAR and South Korea. While the potential economic and social benefits of broadband internet use are significant, adoption levels vary greatly between countries around the world. Many governments, particularly those in Southeast Asia, have adopted aggressive policies to deploy broadband networks and to encourage the use of applications. Governments are motivated to promote broadband adoption in order to realise both economic and social benefits. This chapter argues that the generally higher levels of broadband adoption rates witnessed in many Asian economies is attributable in part to the aggressive policies pursued by these governments. There is some evidence to suggest that these governmental policies have been successful in achieving their stated goals.

The three Greek authors, namely Christos Bouras, Apostolos Gkamas and Thrasyvoulos Tsiatsos, co-authored Chapter 8, titled "*Best Practices and Strategies for Broadband Deployment: Lessons Learned from Around the World.*" This chapter begins by asserting that broadband deployment is a necessity nowadays. It could help each country, municipality and region to grow and offer better quality of life to the citizens. Today, the emphasis on the development of broadband networks is on fixed Fibre to the Home solutions. The lessons learned from countries that are leaders in broadband penetration and Fibre to the Home deployment could prove very useful for under-served communities, regions and countries where the broadband penetration is low. Therefore, this chapter summarises the lessons learned from implementing (a) country-wide strategies formulated at the national level, and (b) local strategies formulated by the municipalities. Concerning the role of national and local governments, it should be noted that nowadays it is very urgent in terms of the involvement of government in the development of broadband infrastructure. Proposed noteworthy remarkable cases are Japan, South Korea and Singapore.

The last chapter (Chapter 9) of Section 2, titled "*The European Research and Education Networks: Ensuring Europe's Leadership in e-Science*," is co-authored by Navonil Mustafee and Simon JE Taylor. This chapter begins by asserting that e-Science is the future of Science. It necessitates the use of powerful computing resources, massive data sets, remote instruments, scientific/visualisation software and expertise that are distributed around the world. e-Infrastructures refer to the underlying computing technologies that facilitate e-Science. To weave these distributed resources into a cohesive entity, and thereby enabling such large-scale science to be conducted in highly distributed environments, needs the utilisation of high-speed communication networks. The European Research and Education Networks provide the connectivity required to ensure Europe's leadership in e-Science. These networks not only provide the communication resources required to conduct e-Science, but they also offer high-speed internet access to universities and educational institutes to facilitate teaching and learning. The objective of this chapter is to inform the readers of the organisation of the underlying European networks that are used by millions of researchers, academics and students. The importance of these networks is highlighted by presenting six e-Science and e-Infrastructure projects that are being funded by the European Commission.

Section 3 examines the impact and social consequences of broadband diffusion. The chapters included in this section provide an in-depth discussion on the impact of broadband on societal issues such as digital divide, rural development and emergence of new societal forms. A brief account of the four chapters included in this section is provided below.

The opening chapter (Chapter 10) of this section was contributed by Peter Stenberg and Mitchell Morehart and attempts to build an understanding of U.S. rural-urban differences in broadband internet adoption and use. The chapter begins by asserting that the Internet has become embedded in the U.S. economy over the past 15 years. During this period, access to and use of the Internet has increased for all regions of the United States, most types of households and work places, and all income groups. This chapter explores how access to technologies may affect household online activity patterns and addresses some of the aspects that differentiate urban and rural household use of the Internet. Rural households are less likely than urban households to have broadband internet access but this varies regionally across the country. The results suggest that broadband internet access is no longer perceived as a luxury but as a necessity, and that there is pent-up demand for broadband internet access in rural areas.

Chapter 11 is titled "*Information Communication Technology (ICT) for Rural Women's Life in Bangladeshi Villages*" and is co-authored by four authors, namely Md. Mahfuz Ashraf, Helena Grunfeld, Syeda Rownak Afza and Bushra Tahseen Malik. The chapter begins by recognising that Information Communication Technology (ICT) has the potential to contribute to development, especially in rural areas of developing countries. But the mechanisms through which ICT can be combined with development agendas and an understanding of the actual development process and impacts of ICT are less well understood or properly defined in the academic literature. This research is an attempt to contribute to understanding this process by analysing the impact of two ICT initiatives in Bangladesh, aimed at improving the lives of rural women. An interpretive approach in the qualitative research tradition was adopted to identify emergent themes in this study. Findings presented in this chapter indicate that these ICT projects have significantly improved the socio-economic opportunities of many women. This paper will be useful for those academics, practitioners and policymakers who wish to enhance their understanding of ICT projects in rural areas of developing countries.

Justin Henley Beneke contributed Chapter 12, titled "*The Application of Social Network Websites as a Marketing Platform to the Youth: An Emerging Market Perspective.*" The chapter begins by stating that social networking is often touted as being a prominent application responsible for driving the adoption of residential broadband services. The growth of social networks is phenomenal – in many cases more than doubling in size on an annual basis. This study considers how social networking may be utilised for commercial purposes to spread word-of-mouth communication. The chapter therefore considers the characteristics of young adult social network users and how they behave and interact with other users on such platforms, as well as the manner in which marketers can make the most of this platform without experiencing a consumer backlash. The research suggests that if a symbiotic relationship does exist between broadband proliferation and the adoption of social networking, both have a vested interest in each other's continued success.

The last chapter (Chapter 13) of Section 3 titled "*Consumer Usage of Broadband Internet Services: An Analysis of the Case of Portugal*" co-authored by Janice Hauge, Mark Jamison and Mircea Marcu. The chapter aims to analyse the intensity and patterns of use of fixed and mobile broadband consumers in Portugal. If usage across types of consumers is similar after controlling individual characteristics identified to be important drivers of adoption, then it is more likely that consumers will view mobile and fixed broadband as somewhat substitutable. Such a result is important for studies of broadband impacts; specifically, for discerning whether mobile broadband services will have a similar level of impact upon social and economic development as fixed broadband services have had. The results presented in this chapter indicate that broadband uses are similar across fixed and mobile users, suggesting that the technologies are somewhat substitutable from customers' perspectives and raising the possibility of limited differential effects on innovation and other social goals. Results of interest include the characteristics of internet users by technology, and differences of usage patterns reflected by individual characteristics.

Section 4, titled "*Technological Advances and its Impact on Emerging Applications*," explores the advances in the area of broadband technologies and examines the impact of broadband on emerging ICT applications and business models. The chapters included in this section provide in-depth discussion on the evolution of broadband technologies and their impact on the delivery of entertainment services to consumers and citizens.

The opening chapter (Chapter 14) of the last section is contributed by Banani Nandi and Ganesh K. Subramaniam. This chapter is titled "*Evolution in Broadband Technology and Future of Wireless Broadband*." It begins by asserting that in the past two decades, the availability of worldwide high speed internet networks together with progressive deployment and adoption of broadband connections to the Internet has significantly enhanced the ability to transmit video, audio and voice through the same channel with high performance. In addition to that, since the beginning of this millennium, there has been a rapid growth in diffusion of wireless broadband. In this respect, mobile broadband is gaining popularity lately in both developing as well as developed countries. The main reason for this growth in wireless broadband lies in its ability to serve remote areas at a relatively lower cost than landline and its potential to transmit information seamlessly from anywhere in the world. The purpose of this chapter is to examine this changing diffusion behaviour of fixed versus mobile broadband technology and highlight the future possible path of adoption for these technologies by users around the world.

Nabyla Daidj and Pierre Vialle co-authored Chapter 15 on the strategic moves related to broadband diffusion on the French VoD and TV market. The chapter begins by asserting that the launch of triple-play services, including IPTV, can be considered as a "critical event" that has had a strong structural impact on the market. This paper analyses the evolution of the VoD supply in France, and in particular the effect of the launch of IPTV systems. Firstly, the launch of attractive triple-play offers has induced a strong development of the French broadband internet access market, which in turn has resulted in the entry of numerous players in the VoD market. Secondly, the adoption of an IPTV system has changed the competitive and strategic positioning of main players in the VoD value chain, thus inducing or amplifying strategic moves along this value chain. In particular, the chapter shows how the diffusion of IPTV systems has induced a switch from complementary convergence to competitive convergence.

Finally, Chapter 16 is titled "*Media Platform Competition: The Displacement Effect of IPTV on Cable TV in Taiwan*" by Yu-li Liu & Wen-yi Hsu. The chapter begins by outlining three major TV platforms (terrestrial TV, cable TV, and IPTV) in the context of Taiwan. When faced with various TV platforms, the consumers in Taiwan will be able to choose from many broadcasts. When a new medium emerges, there are always concerns for its displacement effects on existing media. Many studies have explored the displacement effects of newly-emerging media, and have come up with a variety of sometimes conflicting findings. This chapter aims to analyse the media platform competition, namely, the displacement effect of IPTV on cable TV. The research methods used include a literature review and a telephone survey. With regard to the telephone survey, 612 of CHT's IPTV users were selected systematically based on the name lists provided by the service provider. In this study, while the platform displacement effect of IPTV on cable TV is clear, the statistics reveal a different result for the time displacement effect.

As reflected by the brief accounts of the 16 chapters presented above, this book provides a comprehensive coverage of broadband deployment, diffusion, adoption, usage and policies as they have been realised by research in many countries around the world. Therefore, the editor believes that this book will provide a positive contribution to the area of Information Systems in general and, more specifically, to the adoption and diffusion of broadband.

Yogesh K. Dwivedi
Swansea, UK
March 17th, 2010

REFERENCES

Dwivedi, Y.K. (2007). *Consumer Adoption and Usage of Broadband.* Hershey, PA: IGI Global.

Dwivedi, Y.K., Papazafeiropoulou, A., & Choudrie, J. (2008). *Handbook of Research on Global Diffusion of Broadband Data Transmission*. Hershey, PA: IGI Global.

Lee, H., O'Keefe, B., & Yun, K. (2003). The Growth of Broadband and Electronic Commerce in South Korea: Contributing Factors. *The Information Society,* 19, 81-93.

OECD (2001). Working party on telecommunication and information services policies: the development of broadband access in OECD countries, OECD, Paris. Retrieved March 12, 2003 from http://www.oecd.org/dataoecd/48/33/2475737.pdf

Oh, S., Ahn, J., & Kim, B. (2003). Adoption of Broadband Internet in Korea: the Role of Experience in Building Attitude. *Journal of Information Technology, 18*(4), 267-280.

Sawyer, S., Allen, J. P., & Heejin, L. (2003). Broadband and Mobile Opportunities: A Socio-technical Perspective. *Journal of Information Technology, 18*(4), 121-136.

Suomi, R., & Pekkola, J. (1998). Inhibitors and Motivators for Tele-work: Some Finnish Experiences. *European Journal of Information Systems, 7*(4), 221-231.

ENDNOTES

1 http://www.broadband.gov/

2 http://www.radioaustralianews.net.au/stories/200904/2538028.htm

Acknowledgment

A number of people have extended their help in the development of the ideas presented in this book. I take this opportunity to convey my regards and thanks to those who have helped and supported me at various stages in completion of this project.

This book would not have been possible without the cooperation and assistance of the authors, reviewers, editorial advisory board, our colleagues and the staff at IGI Global Publishing. The editor would like to thank the people at IGI Global, namely: Kristin M. Klinger for handling the book proposal, Jan Travers for managing the contract, and Dave DeRicco for managing this project especially for answering queries and keeping the project on schedule.

A special word of thanks goes to the members of the Editorial Advisory Board for their continuous support in the editorial processes. A special word of thanks also goes to reviewers for their useful and constructive comments that have been incorporated in the final versions of the chapters. I am greatly obliged to Dr. Vikas Jain, Dr. Marijn Janssen, Dr. Anastasia Papazafeiropoulou, and Dr. Roya Gholami for providing the forewords. Last but not least my sincere gratitude to the chapter authors for their valuable contributions to the book.

Yogesh K. Dwivedi
Editor

Section 1
Country Analysis and National Policies

Chapter 1
Still in Pursuit of the Fast Lane:
The Crawl to Broadband Freedom

Justin Henley Beneke
University of Cape Town, South Africa

ABSTRACT

South Africa may be renowned for its natural attractions, warm climate and fine wine. But certainly not for high quality broadband. The country has fallen behind its international peers – both developing and developed markets – in the race to rollout Internet connectivity. In fact, even within the African continent, it is neither a leader nor progressive in comparison to its North African counterparts. This chapter aims to provide a chronology of the major developments in the provision of broadband Internet services in South Africa, as well as touching on the challenges faced in bringing this phenomenon into the mainstream. Reasons for the lack of diffusion and adoption of such services point to high end user costs of the service, a limited geographical footprint of both fixed-line and mobile broadband infrastructure, as well as a lack of computer literacy and understanding of what broadband is able to offer. The chapter continues to look at possible solutions including introducing a greater degree of competition into the market to facilitate downward pressure on prices, provisioning further international bandwidth through undersea fibre optic cables, as well as the unbundling of the local loop, to further this objective.

INTRODUCTION

South Africa, situated at the southern most tip of Africa, has an abundance of good fortunes – natural beauty, mineral wealth, a warm climate and good wine. Yet, the digital divide is omnipresent on these shores – the penetration of Internet access is truly pitiful. The pace of development is even more staggering. Internet penetration stood at five percent of the population at the turn of the century and, a decade later, this statistic has just ventured into double digits. This effectively means that nine out of tens residents are on the wrong side of the digital divide. In contrast, leading broadband nations experience an inverse situation whereby a small fraction of the country's population isn't

DOI: 10.4018/978-1-60960-011-2.ch001

connected – and that's usually by choice. Alas, the dream of affordable, reliable, high speed Internet access still appears to be some way off. The crawl towards broadband freedom continues.

This chapter aims to provide a historical account of the provision of broadband services in South Africa through scrutinising the milestones achieved, and pivotal challenges faced, in bringing this technology into the mainstream.

A BRIEF HISTORY OF INTERNET ACCESS DEPLOYMENT IN SOUTH AFRICA

Telkom, the state controlled telecommunications company, was awarded a monopoly on fixed-line telecommunications services between 1997 and 2002. As part of its obligations, the company was mandated to implement PSTN services in rural and semi-urban areas around the country, many of which would otherwise not have proved commercially viable to operators in a purely competitive environment. Although initially successful in its endeavours to connect the masses, the gravity of the situation soon became apparent as the company was forced to disconnect over two million telephone lines owing to non-payment.

Telecommunications prices remained high due to the need for cross-subsidisation (i.e. affluent customers needing to finance network rollout to the poor), recouping of losses due to extensive copper cable theft, and monopoly inefficiencies created by the absence of any competition in the fixed-line sector. Value-added telecommunications services were also rather limited in the late nineties and early part of the current decade. Consumers were relatively complacent as they had no alternative and little knowledge of what level and variety of service they should expect to receive.

Due to the fact that Telkom had invested significant capital in its Integrated Services Digital Network (ISDN) platform, it was reluctant to cannibalise on this and install ADSL lines. The incumbent was also very weary of introducing a new service to market which would negatively impact its metered-driven call revenues. Nonetheless, fearing competition shortly after the expiration of its monopoly, the company announced the provisioning of broadband services in 2002 and launched its first fixed-line consumer broadband offering in August of that year. This was an always-on connection to the Internet that operated at a downlink speed of 512 kilobits per second and an uplink speed of 256 kilobits per second. Previously, consumers were restricted to dial-up connections operating at 56 kilobits per second (analogue), 64 kilobits per second (singe band ISDN), or 128 kilobits per second (dual band ISDN). Apart from being inherently slower, the dial-up service was billed on a per-second basis, with the result that users accessing the Internet through this technology were typically watching the clock to ensure that their online sessions were kept as brief as possible. Their exposure to the Internet was thus minimised as lengthy online sessions equated to a hefty telephone bill at the end of the month.

Table 1 lists the cost of spending ten, twenty and forty hours online per month during each year between the period 1993 and 2003. The Telkom costs are calculated using the local call rate during peak call-times (i.e. business hours) and factoring in the mandatory line rental.

Figure 1 reveals the escalating cost of spending time online. It is particularly worthwhile noting the sharp increase between 2001 and 2003since the year 2001.

However, whilst ADSL promised users a greater degree of freedom, the service was launched with a three gigabyte monthly data 'soft cap', meaning that consumers could only use the service at its maximum capability until this amount of data had been sent and/or received. After this cap was reached, the service would be severely restricted in terms of international transmission speeds. This resulted in the access to international servers being throttled, whilst local web

Table 1. Before broadband (Internet access between 1993 and 2003)

Year	1993	1994	1995	1996	1997	1998	1999	2000	2001	2002	2003
Average ISP costs	80.00	80.00	80.00	80.00	80.00	80.00	88.00	89.00	90.00	94.00	100.00
Year on year increase		0.00%	0.00%	0.00%	0.00%	0.00%	10.00%	1.14%	1.12%	4.44%	6.38%
Telekom (10 hours)	60.06	66.67	88.01	97.69	111.38	141.53	168.62	184.93	201.24	287.61	323.71
Telekom (20 hours)	85.91	94.57	131.78	145.79	173.17	218.82	264.62	295.93	327.24	485.25	546.19
Telekom (40 hours)	137.62	150.39	219.33	241.99	296.74	373.41	456.62	517.93	579.24	880.53	991.15
Year on year increase		9.89%	40.91%	10.55%	19.74%	26.23%	21.27%	12.24%	10.91%	49.26%	12.56%
Total (10 hours)	140.06	146.67	168.01	177.69	191.38	221.53	256.62	273.93	291.24	381.61	423.71
Total (20 hours)	165.91	174.57	211.78	225.79	253.17	298.82	352.62	384.93	417.24	579.25	646.19
Total (40 hours)	217.62	230.39	299.33	321.99	376.74	453.41	544.62	606.93	669.24	974.53	1091.15
Year on year increase		5.36%	23.11%	6.82%	13.21%	18.57%	18.49%	9.70%	8.84%	40.48%	11.66%

Source: Internet.org.za (2003)

sites remained unaffected. Telkom have subsequently reversed this policy and applied a 'hard cap' meaning that all Internet connectivity is terminated once the cap has been reached. Local bandwidth is no longer unlimited and is charged on a per gigabyte basis after the pre-assigned

Figure 1. A graphical representation of Internet access costs between 1993 and 2003 (Source Internet.org.za (2003))

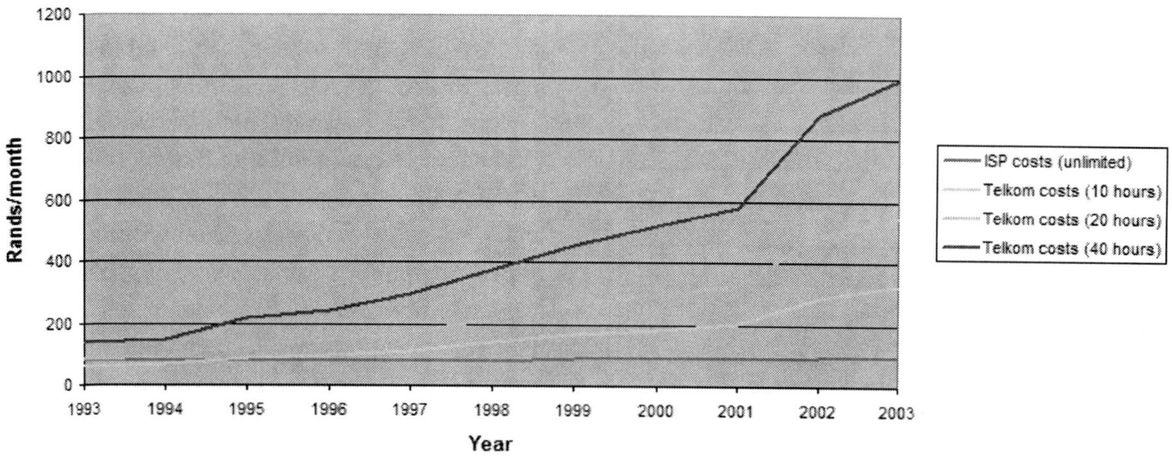

threshold. Whilst this proved largely unproblematic for Internet users with basic online needs, power users were less than impressed with this product revision. Nonetheless, despite the limitations, the deployment of broadband heralded an era where consumers were granted a much greater degree of Internet freedom and didn't need to be as time-driven in their online behaviour as was previously the case. Hence, broadband was deemed by the South African Internet community to be a significant step in the right direction.

The table below illustrates the migration from dial-up to broadband. It is evident that dial-up subscribers have increased year-on-year until 2004, but have dramatically declined since this time.

Early in 2004, Sentech (the government-owned national television signal carrier) launched their MyWireless service. This allowed Internet users the ability to experience broadband connectivity through a wireless medium. The service used advanced radio technology to send/receive data between the PC and Internet. However, this was not a true mobile solution in that the computer needed to be in a fixed-location (i.e. remain stationary) whilst an Internet session was in progress. Furthermore, Sentech's limited network presence meant that the service was only available in a small selection of metropolitan areas. The service was launched without a data cap, instead relying on fair use by its customers. This attracted many would-be ADSL customers who weren't enthusiastic about using a service which essentially limited them to only a few hours of usage per day. Unfortunately, the MyWireless service attracted a considerable amount of negative press due to teething network problems and their lack of support in dealing with customers' frustration. This resulted in a small group of disgruntled customers becoming incredibly vocal about their experiences, even going so far as to establish a hate site (www.SentechSucks.co.za). The company was also forced to reconsider its "all you can eat" philosophy in terms of usage and thereafter imposed absolute restrictions on uploads and downloads. Eventually, in 2008, as financial loses mounted and broadband competition intensified from much larger plays, the company pulled the plug on the MyWireless service and began dismantling the transmission equipment.

Table 2. Growth and decline of dial-up subscribers in South Africa

Year	Subscribers (growth/decline)
1994	15 000
1995	33 600 (124%)
1996	79 700 (137%)
1997	196 620 (147%)
1998	366 235 (86%)
1999	560 000 (53%)
2000	782 000 (40%)
2001	960 000 (23%)
2002	1 008 000 (5%)
2003	1 048 320 (4%)
2004	1 088 000 (4%)
2005	1 080 000 (-1%)
2006	1 030 000 (-5%)
2007	908 000 (-12%)
2008	700 000 (-23%)

Source: Goldstuck (2009)

Wireless Business Solutions (WBS), a private operator tasked with the responsibility of running the network of National Lottery point-of-sale terminals, commercially launched their iBurst service in early 2005. This was very similar in nature to Sentech's offering, although it featured a higher theoretical download transmission speed of 1 Mbps. It did, however, include a monthly three gigabyte data cap, similar to that of Telkom. The iBurst team appeared to have learnt from Sentech's MyWireless failures and, combined with improved funding for network expansion, now boasts an impressive infrastructure covering the major centres and growing on a monthly basis. The appetite for its services seems encouraging, as it operates in areas where Telkom's ADSL service is unavailable. It also capitalises on Telkom's tardy customer response times. The company promises a sign-up window of a day rather than weeks. Its major hindrance is obtaining permission to erect new transmission towers. These base stations are often delayed due to onerous requirements such as protracted Environmental Impact Assessments, objections from residents concerned about radiation leaking from transmission towers, and aggressive bargaining on the part of property owners where transmission equipment is installed. In 2008, iBurst partnered with Vodacom to roll out a WiMax nework, targeting high end residential and business customers. This technology allowed for more robust, higher capacity connection to the Internet, as demanded by this market segment.

The most recent entrant into the broadband space is Neotel, touted as the country's second network operator, the official competition to Telkom. Neotel was founded in 2006 and shortly after being awarded its licence, set about to establish a nationwide network. The principle of convergence shaped the formation of its infrastructure, in that all telecommunications services were designed to run over a single IP-based network. This meant that all data-driven services could be integrated simultaneously on its network. In 2007, it launched its portfolio of corporate services and, a year later, its consumer offering. The company uses CDMA2000 for its access network, connecting consumers. The average customer therefore experiences a throughput in the range of 250kbps to 750 kbps. Whilst this is not the fastest connection on the market, the company reckons it is adequate for most users. It has recently launched a suite of new products, including a multi-user platform (NeoFlex) which is able to reach peak speeds in the region of 3Mbps. Neotel has been ambitious in terms of its pricing and offers a rate of R 0.08 (one US cent) per megabyte of data transferred. This is considerably lower than any other wireless operator and is within the realm of fixed-line broadband. The major stumbling block has been coverage. Whilst the company boasts a metropolitan and national network of colossal proportions, extending the reach to end users is stifled by a limited number of transmission towers. Neotel has suffered many of the same problems as iBurst in this respect, particularly in the Cape Town region where local authorities apply a strict stance on base station erection.

The above broadband services require users to remain in a fixed location whilst using the Internet. However, the cellular networks' offerings (i.e. those of MTN and Vodacom) were designed to be more flexible. This allows a user to be in motion, for example travelling on a train or bus, whilst keeping the connection active.

Prior to Christmas in 2004, Vodacom fulfilled their promise of launching 3G services (operating at a speed of 384 kilobits per second) in the local market. MTN followed suit in mid 2005. Due to the cellular networks' extensive coverage throughout the country, the area in which wireless broadband could be obtained was significantly expanded. However, the high speed service was initially launched on a limited scale – within metropolitan areas – and is still constrained to urban and CBD areas where Internet usage is rife. Furthermore, data charges (i.e. cost per megabyte transferred) were originally considerably higher than those offered by fixed-location broadband providers. This resulted in potential 3G users being offered a one gigabyte package for a similar price to that being charged for a three gigabyte (or larger) account on the other networks. This immediately signalled that users would be made to pay for true mobility and extended network coverage. In 2006, MTN and Vodacom optimised their 3G service with the introduction of High Speed Downlink Packet Access (HSDPA). This increased download speeds to a maximum theoretical limit of 1.8 Mbps, providing the fastest wireless form of broadband. This has subsequently been ramped up to 3.6Mbps and even 7.2Mbps in selected coverage areas. This development was seen as a major step forward for customers who required multimedia services (particularly Web 2.0 applications such

Figure 2. Broadband market share by access technology

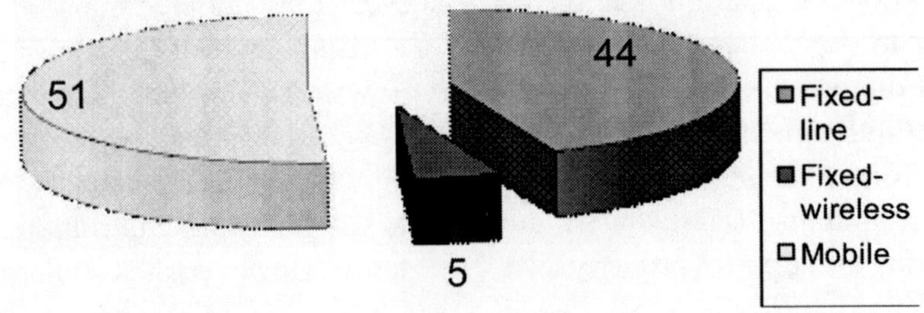

as YouTube, MySpace, etc) which are bandwidth intensive by nature.

Figure 2 illustrates the current split of broadband users between fixed-line (Telkom), fixed-wireless (iBurst) and mobile (MTN & Vodacom) providers. It is evident that the mobile broadband providers occupy the lion's share (51%) of the market, followed by Telkom's ADSL service (44%) and lastly iBurst's 5% market share. This is something of an anomaly in an international context. In contrast, Point-Topic (2009) reveals that, globally, fixed-line providers dominate the residential broadband market (with a two thirds majority), primarily owing to considerably lower cost structures and enhanced economies of scale. This is further discussed in the following section of this chapter.

Table 3 provides a year-on-year tally of the number of broadband users in South Africa.

The decreasing growth rate doesn't necessarily provide cause for concern. The low base somewhat flatters growth in the early years, with the size of the broadband base at least doubling on an annual basis until 2008. That said, whilst the market is experiencing a promising growth rate in percentage terms at the time of writing, it should be acknowledged that the growth is based on a very small pool of users. By taking into account the shrinking base of dial-up users discussed above, it may be safely inferred that a migration is occurring. However, the gains to the broadband user base outweigh the losses to the dial-up user base, suggesting that more South African are connecting to the Internet as they see the advantageous of broadband connectivity.

Table 3. Growth of broadband subscribers

Year	Subscribers (growth rate)
2003	15 700
2004	51 500 (228%)
2005	175 200 (240%)
2006	392 000 (123%)
2007	818 000 (108%)
2008	1 352 000 (65%)

Source: Goldstuck (2009)

Double counting remains a problem though as a particular user may have fixed-line broadband at home and a 3G account for mobile broadband whilst travelling. Likewise, the number of broadband subscribers is unlikely to equate exactly to the number of broadband users as an entire family might share a single ADSL connection. To a limited extent, these two effects do counteract one another.

According to the latest statistics by Goldstuck (2009), of the 48 million South African residents, only 4.5 million (9.4%) have Internet access and fewer than four people per hundred have access to broadband services. The following section now compares South Africa's performance relative to the rest of the world - considering both developed and developing nations.

PENETRATION OF BROADBAND IN AN INTERNATIONAL CONTEXT

The Organisation for Economic Co-Operation and Development conducts an annual broadband penetration survey. The latest findings reveal that Scandinavian and European countries dominate in this respect (Table 4).

Point-Topic, a commercial supplier of broadband statistics, reports that the world's largest broadband markets remain China and the United States, with both markets adding the highest

Table 4. Broadband subscribers per 100 inhabitants (June 2009)

Rank		DSL	Cable	Fibre/LAN	Other	Total	Total subscribers
1	Netherlands	22.5	13.7	1.1	0.8	38.1	6 262 500
2	Denmark	22.4	9.9	3.9	0.9	37.0	2 031 000
3	Norway	22.7	7.7	3.5	0.7	34.5	1 645 619
4	Switzerland	23.3	10.0	0.2	0.3	33.8	2 603 400
5	Korea	7.2	10.5	15.1	0.0	32.8	15 938 529
6	Iceland	30.7	0.0	1.3	0.7	32.8	104 604
7	Sweden	18.5	6.3	6.7	0.1	31.6	2 915 000
8	Luxembourg	26.0	5.3	0.0	0.0	31.3	153 172
9	Finland	24.9	4.1	0.0	0.8	29.7	1 579 600
10	Canada	13.2	15.2	0.0	1.3	29.7	9 916 217
11	Germany	26.7	2.4	0.1	0.1	29.3	24 043 000
12	France	27.5	1.6	0.1	0.0	29.1	18 675 000
13	United Kingdom	22.8	6.1	0.0	0.1	28.9	17 742 676
14	Belgium	16.3	11.8	0.0	0.2	28.4	3 041 311
15	United States	10.3	13.8	1.6	0.9	26.7	81 170 428
16	Australia	19.4	4.3	0.0	1.2	24.9	5 356 000
17	Japan	8.5	3.3	12.4	0.0	24.2	30 927 003
18	New Zealand	20.4	1.4	0.0	1.0	22.8	980 649
19	Austria	14.5	6.8	0.1	0.5	21.8	1 821 000
20	Ireland	15.5	2.8	0.1	3.0	21.4	950 082
21	Spain	16.6	4.0	0.1	0.2	20.8	9 477 901
22	Italy	19.2	0.0	0.5	0.1	19.8	11 878 000
23	Czech Republic	7.0	3.9	0.9	6.3	18.1	1 891 958
24	Portugal	10.0	6.7	0.1	0.2	17.0	1 809 354
25	Greece	17.0	0.0	0.0	0.0	17.0	1 908 000
26	Hungary	8.2	7.6	1.0	0.0	16.8	1 688 414
27	Slovak Republic	6.6	1.3	2.7	2.0	12.6	680 351
28	Poland	7.4	3.7	0.1	0.1	11.3	4 307 992
29	Turkey	8.5	0.1	0.0	0.0	8.7	6 188 676
30	Mexico	6.3	1.9	0.0	0.2	8.4	8 959 426
	OECD	13.7	6.6	2.1	0.5	22.8	271 134 392

Source: OECD (2010)

number of new subscribers over the past year. The Middle East and Africa comprise a mere 5% of the world's broadband connections, despite boasting over a billion residents in the region, punching far under their weight in relative terms. North Amerca, Western Europe and South East Asian feature just short of seventy percent of all broadband connections on the planet (Point-Topic, 2009).

As per their calculations, the leading countries in penetration of broadband lines are depicted in Figure 3 below. This largely mirrors the results of the OECD study, with the notable exception of the principalities of Liechtenstein and Monaco. It is surmised that their extremely high level of affluence, coupled with the miniature amount of geographical space occupied, resulted in them topping the chart.

Although South Africa doesn't feature in either of these reports, comparisons between it and the leading broadband nations are immediately apparent. South Africa's broadband penetration, at less than four percent (Goldstuck, 2009), is outclassed by emerging markets such as Mexico and Turkey and dwarfed by developed countries such as those depicted above, most of which feature ten times the level of penetration achieved within South Africa.

A study by Cisco and the University of Oviedo (BBCNews, 2009) recently clustered countries, based on the quality of their broadband services, into four categories, namely 'ready for tomorrow', 'comfortable for today', 'meeting needs for today' and 'below needs for today' (Table 5).

The leaders are considered to be Korea, Japan, Sweden, Lithuania, Bulgaria, Latvia, Netherlands, Romania and Denmark. Somewhat surprisingly, the United States (categorised as 'comfortable for today') and the United Kingdom (categorised as 'meeting needs for today') don't appear significantly future proof. South Africa, alongside China, is categorised as being 'below needs for today'.

DRIVERS AND INHIBITORS OF LOCAL BROADBAND DIFFUSION AND ADOPTION

Whilst the penetration of broadband in South Africa may be serious cause for concern, there are a number of factors stimulating development. Goldstuck (2007) identifies three key ongoing drivers of broadband adoption in the country.

Lifestyle-Oriented Demand: The new generation of online gamers, file-swapping enthusiast

Figure 3. Broadband lines per 100 population (Source Point-Topic (2009))

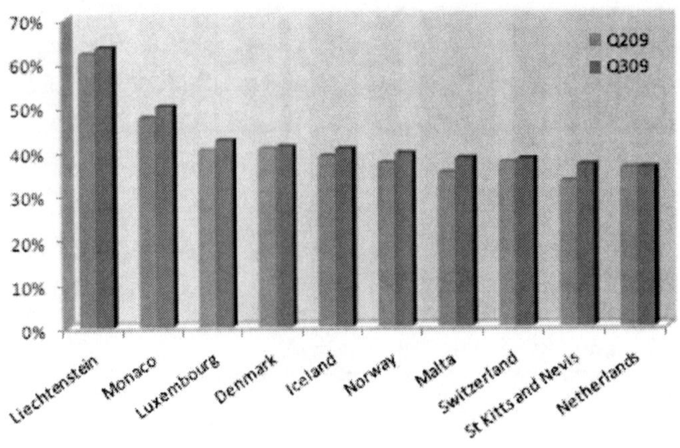

Table 5. Future proof status of broadband services in selected countries

Ready for tomorrow	Comfortable for today	Meeting needs for today	Below needs for today
Korea	Switzerland	Iceland	Malta
Japan	Czech Republic	Estonia	Luxembourg
Sweden	Norway	Greece	Chile
Lithuania	United States	Singapore	China
Bulgaria	Slovakia	Canada	Qatar
Latvia	Portugal	United Kingdom	Brazil
Netherlands	Finland	Australia	Argentina
Romania	France	Spain	Saudi Arabia
Denmark	Germany	Poland	Cyprus
	Hungary	New Zealand	Costa Rica
	Russia	Ukraine	Bahrain
	Belgium	Turkey	Thailand
	Slovenia	Ireland	Tunisia
	Taiwan	Italy	Mexico
	Austria		Philippines
	Hong Kong		UAE
			Malaysia
			Pakistan
			Colombia
			Morocco
			Vietnam
			South Africa
			Indonesia

Source: BBCNews (2009)

and social networkers demand speed and quality of access. For the gamers, speed and quality are paramount – even more so than for the business market. For the file-sharing market, which uses applications that accelerate data usage through multiple upload streams from a typical user's machine, a data cap of any size is highly resented. Social networking site (e.g. MySpace, Facebook, Twitter) and Web 2.0 applications (e.g. YouTube) have brought about a shift in mindsets, whereby consumers are socializing and sharing content online to a much greater extent than they did before. This has even prompted some employers to restrict access to social networking and such sites in the workplace in order to minimize the negative effect on productivity. Nonetheless, the Internet (and particularly broadband) is definitely seen as a lifestyle enhancement platform.

SME Demand: ADSL has found a ready and enthusiastic market among small businesses. Such users are not as price-sensitive as home users, nor are they as focused on capacity issues like contention ratios and upload/download restrictions. Here, the solution represents a significant cost saving by preventing the business having to dial-up on a regular basis to send/retrieve e-mail and has also had the positive spin-off of allowing the enterprise to embrace online communications. Thus, the SOHO market, micro-enterprises and small business alike have realised instant cost benefits as a result, and their experience has become an excellent word of mouth marketing tool for broadband service providers

Pent-Up Consumer Demand: Home users who have been on the Internet since the mid or late 1990s have become frustrated with the limitations and cost of dial-up, and have been ready to spend a higher upfront monthly amount on a better form of connectivity. Many such individuals have been exposed to the merits of broadband on their travels overseas or have heard about the benefits of the medium through communication with their friends, colleagues and families settled overseas. Whilst it is true that consumers have resisted the price levels set by Telkom and others, the need for speed is ever present. The World Wide Web of the current decade is fundamentally different from that of the nineties, as discussed above, whereby web designers usually took account of slow (typically, dial-up) connections by building sites that could download quickly. However, in a broadband-dominated developed world, where most web sites are hosted, this priority is fading into the background. Bandwidth intensive sites have resulted in dial-up connections becoming ill equipped to handle the flow of data. Therefore, lengthy downloads result in higher cost to dial-up users, and timeouts mean wasted time, money and energy… over and above simple frustration.

The 3CC framework summarises the drivers and inhibitors of growth in broadband usage within a South African context. The updated version (based on that initially published by Beneke (2008) is represented in Table 6.

A WARPED VIEW OF 'BROADBAND'

The perception of broadband in South Africa is somewhat distorted compared to how the technology is perceived in many developed countries. In this respect, there appear to be significant disparities between the characteristics of local broadband services and international broadband services, reinforcing the notion that South Africans are obtaining a raw deal in contrast to users overseas. Many broadband providers in South Africa restrict the usage of their service in a number of noteworthy respects. These issues are discussed below.

Firstly, whilst most standard broadband services in developed countries effectively aren't capped, ADSL accounts in South Africa typically feature data caps ranging between two and five gigabytes. The mobile broadband networks, strictly speaking, don't cap individuals, but charge users in a manner which results in the service being prohibitably expensive for bandwidth intensive applications. This means that using a broadband connection in South Africa is not feasible for many individuals wishing to download music albums, full length movies or even stream live television. In this respect, either the data cap will be rapidly exhausted or, if the user is subscribed on a pay-per-usage basis, the end-of-month account is likely to be exorbitant.

Although a number of uncapped packages have been launched into the market by innovative broadband providers, these are typically labelled with a disclaimer purporting 'fair usage'. Whilst fair usage is seldom quantified, this is usually deemed to be in the region of 30 gigabytes per individual account per month. However, on a 4Mbps ADSL line, this amount of data could conceivably be downloaded in two days, leaving the user effectively limited, or even blocked outright, for the remainder of the month. Until recently, Telkom, the largest provider of broadband connectivity in South Africa, considered downloading large amounts of data through an ADSL line to be 'abuse' of the service. This despite marketing its 'Do Broadband' service explicitly claiming the package may be used for online gaming, downloading music, and the like.

Secondly, many broadband service providers utilize 'protocol shaping'. This technology gives network preference to certain applications over other applications. In order to comprehend

Table 6. The 3CC Framework

Factors	Drivers of Growth	Inhibitors of Growth
Costs	Since introducing its ADSL service, Telkom has reduced subscription rates on a number of occasions between 2005 and 2007. The company has also increased the value proposition by upgrading (free of charge) 192kbps customers to 384kbps and upgrading 1Mbps customers to a 4Mbps line speed. Mobile providers have decreased their per megabyte data rates from R 25.00 in 2006 to a mere R 2.00 at the time of writing. Per megabyte rates within large data bundles have been reduced phenomenally further.	South Africa's broadband providers (particularly those offering fixed-line services) are constantly lambasted for high tariffs. As per the 2007 Africa Competitiveness Report (World Economic Forum, 2007), South Africa's relative broadband value was ranked behind countries such as Morocco, Egypt, Madagascar, Senegal, Botswana and Mauritius.
Coverage	Telkom has realised the potential of data revenue and is rapidly equipping its telephone exchanges to accommodate ADSL lines. The phenomenal growth of subscribers between 2004 and 2009 bears testimony to this policy. The two leading cellular networks are competing, head-to-head, to acquire the more lucrative business customers who require advanced services such as mobile broadband access. For this reason, both 3G/HSDPA mobile broadband provides have ubiquitous coverage of CBD areas and many suburban areas. The networks have recently been upgraded to 3.6Mbps and can even reach a throughput of 7.2Mbps in some parts. This has resulted in the performance of mobile broadband actually outshining the performance of fixed-line broadband. Newcomers iBurst and Neotel, both fixed-wireless broadband providers, are also engaging in extensive network expansion which presently includes all major cities and most large towns in the country.	Broadband coverage outside metropolitan areas in South Africa is virtually non-existent. Copper theft in rural and peri-urban areas is rife, therefore limiting rollout of fixed-line broadband services in these regions. Furthermore, the demand for 3G/HSDPA mobile broadband services is limited to urban and peri-urban areas and the operators have therefore focused on increasing capacity in these areas at the expense of expanding such networks countrywide. The state's signal broadcaster, Sentech (the previous operator of MyWireless services), has now been tasked with establishing broadband infrastructure in rural areas as a government sponsored initiative.
Competence	Internet access is beginning to reach the "man in the street". In part, this is due to computer education in the workplace, as well as in the primary, secondary and/or tertiary education environment. A number of companies and non-profit organisations, including the Internet Service Providers Association and Telkom Foundation, have been proactive in donating hardware and software to educational institutions and providing training to educators so that they can impart these skills to their students.	Broadband appears to be an unknown quantity for many South Africans, the vast majority of whom aren't computer literate and are unlikely to have ever accessed the Internet before. There are estimated to be between five and ten million Personal Computers and laptops in the home (as suggested by Laschinger & Goldstuck, 2006), predominantly in formal settlements such as suburban areas. Many consumers in informal settlements (i.e. township areas) have therefore had no exposure to the Internet of any sort.

'protocol shaping', it should be understood that most consumer broadband accounts are 'contended'. This means that a 512 kbps connection, for example, isn't guaranteed the full line speed at all times. Instead, a number of users essentially share the same Internet pipeline, meaning that if all users are concurrently online, each user is likely to achieve a mere fraction of the theoretical throughput. Protocol shaping takes this a step further. The principle here is that certain applications should be given more than their equitable share of bandwidth, whilst others should be given less than their fair share. In this manner, certain applications are given low priority and thus their use is discouraged. For example, if the network detects that the user is transferring files through a peer-to-peer application such as Limewire, Kazaa or BitTorrent, his/her performance is likely to be intentionally degraded and thus bandwidth removed from the connection and donated to other 'network-friendly' applications such as general web browsing. Another example is mobile broadband providers downgrading the Voice over Internet Protocol to protect their voice stream revenue which attracts significantly higher profits. Hence, broadband users might experience limited throughput and high latency, translating into broken speech patterns and abnormal delays when holding a conversation through an application such as Skype.

To this end, it is argued that potential abusers of the network are throttled so that performance is not degraded for the entire broadband user base. Unsurprisingly, such individuals complain that they are being unfairly victimised. Herein lies the fundamental problem: broadband promises a multimedia rich environment whereby users can experience the true benefits of the information age without prejudice in terms of application. However, by labelling certain users as rogue customers or certain activities as abuse, the potential for broadband to revolutionise the way in which the Internet is used is severely stifled. Further unhappiness is caused by the fact that this is a relatively unique stance foisted upon consumers in the South African market.

THE 'RIPOFF' RATIONALE

The high costs of telecommunications services, particularly broadband connectivity, provides much fuel for ongoing debate. A number of reasons have been put forward as to why South Africa is out of synch with other emerging markets apropos telecommunications costs. These are addressed below.

Firstly, it is claimed that the South African broadband subscriber base is still relatively limited in size. Thus, the country has been unable to benefit from the economies-of-scale experienced elsewhere in the world. However, it is reasoned that as the number of broadband subscribers continues to grow, subscription rates should naturally decrease over time. Fortunately, there appears to be some merit in this argument as there have already been a number of price reductions in the market. The price tag of many other broadband services has actually stayed static and has not, as expected, increased with inflation. This is, in effect, a price decrease in real terms. It is expected that as broadband infrastructure is built, capacity creation (i.e. supply of bandwidth) will increase and prices will be pushed downwards. Progress, however, remains slow. Prices, likewise, remain sticky as there is little incentive for companies to drastically slash prices.

Secondly, telecom incumbents claim that international bandwidth is exceptionally expensive in South Africa due to the geographical distance from the main Internet centres (i.e. United States of America, Western Europe, South East Asia, etc.). Until mid 2009, the SAT-3/SAFE undersea fibre optic cable was the exclusive fixed connection integrating South Africa with the Internet backbone. Whilst satellite connectivity has been in operation for over a decade, the exceptionally high costs and poor latency meant that this capacity was typically reserved for backup purposes. Telkom, being South Africa's only shareholder in the undersea cable and holding exclusively landing rights, exercised its market power and held wholesale connectivity prices artificially high. This situation has subsequently been rectified with open-access to the SAT-3/SAFE cable and the landing of the SEACOM cable.

Thirdly, Telkom claims that establishing their ADSL network has been extremely capital intensive. The incumbent argues that it has had to convert previously analogue exchanges into digital format, and then equip these to handle fixed-line broadband. It therefore maintains that it needs to recoup these costs and that further investment in infrastructure, without the required financial return, wouldn't make good business sense. In this respect, the company has even threatened to pull the plug on its broadband network if it were to be forced to slash the price tag of its broadband offerings.

At present, most fixed-line broadband providers merely resell white-labelled (core) Telkom products. These products are facilitated by Telkom and make use of its own network – the alternative providers merely operate their own billing systems, support desks, marketing campaigns, and the like. Telkom does offer a means through which providers can provision their own bandwidth and hence merely use Telkom to connect the customer to the closest telephone exchange. However, the platform (called IPConnect) is sold in such a manner that the margin between the wholesale and retail tariffs is negligible. This means that only large scale ISPs, with very high volumes of Internet traffic, would consider this a feasible business model. For this reason, a few large players (such as Internet Solutions and WebAfrica) have to date utilized this platform to provision residential broadband services. Nonetheless, as the volume of broadband traffic grows, other providers are conducting feasibility studies into the prospect of migrating their user base onto their own backbone using Telkom's IPConnect.

Fourthly, South Africa's teledensity is lower than that of many broadband leaders. In the centre of London, for example, residents live in close proximity to each other and less cable is thus required to reach a given number of residents. In South Africa, due to high levels of dispersion of the population and a country five times the size of the United Kingdom, more cable is required to build a comprehensive network. Another key point in this regard is that a negative correlation exist between the length of the local loop and the theoretical limit of the ADSL data throughput rate. Currently, the local loop is between four to five kilometres in length. Telkom feel this should be closer to two kilometres. In order for ADSL2+ speeds to be possible, it will be necessary to redevelop parts of the network in order to implement shorter local loops. Again, the infrastructure costs may not prove economically viable unless favourable returns can be generated on the service.

Fortunately for telecoms operators in South Africa, urban sprawl has been reigned it through escalating property costs. There now appears to be a trend towards living in clustered accommodation. This is advantageous for the purpose of increasing teledensity as security complexes and blocks of flats, often referred to as gated communities, are now being wired internally for high speed networks. In fact, new clustered housing developers are ensuring that their projects are cabled for high speed connectivity from the outset, therefore preventing retrofitting such infrastructure. This has meant that such communities are able to benefit from fibre deployment, lending itself to triple play services such as cable television, broadband connectivity and telephonic services from a single service provider.

GOLDSTUCK'S INTERNET HIERARCHY FRAMEWORK

Arthur Goldstuck, a prominent Internet historian and researcher in South Africa, considered Maslow's hierarchy, grounded in human behaviour theory, and developed a parallel hierarchy relating to consumers' online needs. The model encompasses levels for access, quality, utility, community and strategy (in ascending order) and is depicted in Figure 4.

The individual layers are detailed in Table 7 below.

Unfortunately, it appears that whilst broadband users elsewhere in the world have ascended to the top of the pyramid (i.e. the community or strategy level), problems at the quality level have prevented South African broadband users from obtaining maximum benefit from the Internet, limiting most to the utility and community layers.

Examining the situation from an industry perspective, Goldstuck developed a further iteration of the model based on the hierarchy of needs depicted above. As can be seen, the left-hand side of the triangle is superimposed over a quadrant-based model that reveals the impact of various strategies on ISPs and their customers.

The Necessary Evil / Unconditional Need axis (x axis) indicates the perspective of the corporate

Figure 4. Goldstuck's Hierarchy of Internet Needs (Source Goldstuck (2007 82))

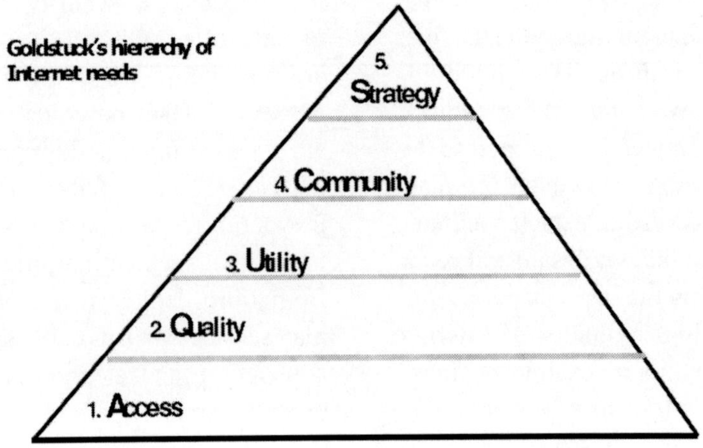

Table 7. Five Layers in Goldstuck's Hierarchy of Internet Needs

Access	This infrastructure level equates to connectivity and merely entails establishing a point-to-point connection into the Internet.
Quality	This infrastructure level equates to the nature of the connection – e.g. narrowband (dial-up), broadband (ADSL/3G/HSDPA), etc. In essence, the more mature and affordable the technology (that is, great reliability and increased throughput rates at lower costs), the better for the user.
Utility	This value added services layer equates to the spread of applications that the consumer is able to use on the Internet. In essence, the more online activities the user can pursue (e.g. transferring music, online banking, reading breaking news, etc), the more favourable the experience.
Community	This value added services layer equates to a social dimension and entails the consumer reaching out and experiencing the world through computer-aided human interaction. Examples include using instant messaging, joining online discussion forums, etc.
Strategy	This value added services layer relates to the phenomenon whereby the user feels that the online world is an extension of his/her life in the physical world. Hence, almost everything achieved offline may actually be replicated online. This is evident in applications such as Second Life, whereby your Avatar (character) represents you in a cyber world. This layer may also entail operating a business online, pursuing relationships on a dating portal, publishing a blog, etc. Therefore, this layer encompasses all of the above by taking the consumer into a seamless virtual realm.

client. Much like the end user, the client progresses through the access, quality, utility, community and strategy layers as the Internet becomes more and more fundamental to the business. The final layers suggest that the lack of such services would effectively cripple the business as the Internet becomes an unconditional need, and is tightly integrated into overall business processes.

The Infrastructure Focus / People Focus (y axis) represents a similar evolution for the ISP, which must gear up its services in accordance with the needs of its customers (corporate users) The ISP is challenged to provide the right mix of technology and intellectual capital to serve the needs of its client base in the most cost-effective way possible.

The two axes, then, represent the respective upgrading of requirements by the corporate user and of the services rendered by the ISP. Inevitably, both of these processes tend to be associated with increasing cost.

Four quadrants in the model are identified by considering the ISP's value proposition, whereby the

Figure 5. Goldstuck's model of Internet industry structure (Source Goldstuck (2007 83))

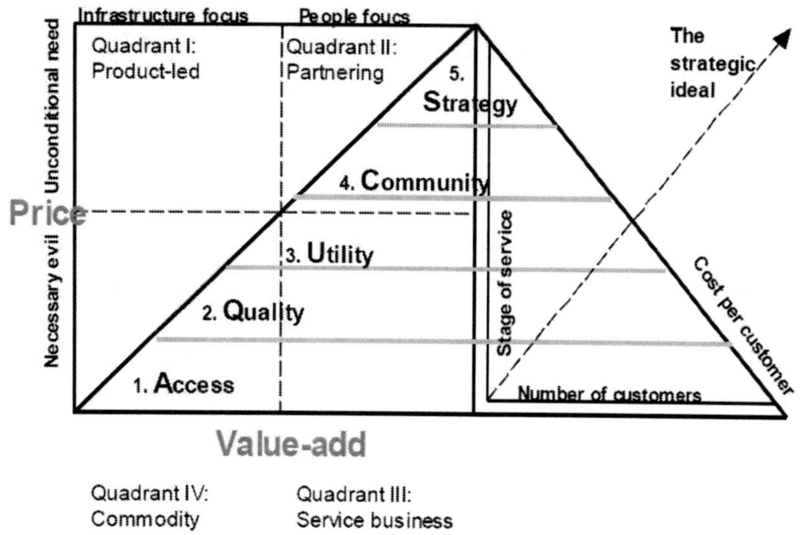

x axis indicates increasing price levels and the y-axis represents increasing value-added services on offer. These are categorised as product-led (the ISP attempts to cater for business strategic business needs through an extensive product suite of products), partnering (the ISP focuses on partnering with a client in order to meet its needs), service (the ISP provides basic strategic needs) and commodity (the ISP provides core access needs).

On the right-hand side of the triangle, the impact of these business models on the revenue streams of the ISP is demonstrated. Typically, an ISP operating in the access arena (i.e. a commodity business) will possess a large amount of clients, but relatively low average revenue per client. An ISP operating in the strategic arena (i.e. tailoring offerings as per client expectations) may have fewer clients of a higher average revenue each. The strategic ideal is to migrate clients to as high a level as possible.

LIGHT AT THE END OF A DARK (UNLIT) TUNNEL

Reducing the cost structure of broadband connectivity is an imperative. However, the cost components in offering a broadband service are somewhat complex. Admin related expenses such as technical support, billing and general back office functions need to be catered for. Technical expenses such as local bandwidth provisioning, international bandwidth provisioning, network fault resolution and even the construction of a 'last mile' access network (connecting end users to the network backbone) are highly pertinent to end user pricing. The latter areas, in particular, have spent much time in the media spotlight of late. This is surmised to be an area wherein improvements are likely to have a material effect on reducing broadband pricing in South Africa.

Until recently, most ISPs were dependent on a small group of national carriers such as Telkom to relay their traffic between cities. Somewhat surprisingly, a sector of national (local) bandwidth may be more expensive than a sector of international bandwidth. To this end, ISPs have reported that Telkom charges a higher price for data transit between Johannesburg and Cape Town than it does between Cape Town and London. This lucrative market has sparked investment in fibre optic cables criss-crossing the country. Vodacom and MTN (the operators which control about 80% of the mobile market) are in the process of building their own national fibre-optic networks, having recently completed a number of metropolitan fibre rings. Infraco (a government initiative) and Dark Fibre Africa (a private company) are, likewise, aggressively rolling out fibre-based networks of their own in a bid to sell capacity to smaller service providers. However, even small operators now have the right to set up their own national network, should they choose to do so. In 2008, the Department of Communications was taken to court over the matter and was forced to issue network operator licences (labelled Electronic Communications Network Services, or ECNS, licences) to service providers wanting to establish their own national, or regional, network infrastructure.

Another stumbling block has been frequency allocation. In order to operate a wireless network of any sort, operators require access to radio frequency for data transmission purposes. This is a nationally controlled resource and, as such, operators need to apply for spectrum. This ensures that no two operators are using the same frequency in the same geographic area, which would otherwise lead to interference and, ultimately, network failure. Policing spectrum also results in rogue operators being punishable by law. Unfortunately, incumbents such as Sentech and Telkom are, at present, hogging spectrum, meaning that whilst they have the sole right to use it, they choose not do, and no other operator therefore has the opportunity to use this finite resource. This has the net effect of crippling competition in the wireless arena. The industry regulator, Independent Communication Authority of South Africa (ICASA), is presently drafting a policy whereby both prospective and current wireless operators will have easier access to spectrum and all operators will face a 'use it or lose it' scenario.

A key cost saving measure is local peering, brought about by ISPs interconnecting with each other within the borders of the country. This prevents data from being routed internationally and is obviously a much cheaper mechanism. The Johannesburg Internet Exchange (JINX) is the peering point through which most large players (with the notable exceptions of Telkom and Vodacom) exchange data. The Cape Town Internet Exchange (CINX) has recently been re-commissioned and provides a similar facility for ISPs which are based in the Western Cape region. For ISPs who wish to bypass JINX and CINX completely, direct linkages are the alternative. For example, Telkom and Internet Solutions (a major supplier of corporate bandwidth) operate direct peering links between themselves.

International bandwidth is still in short supply. SEACOM, a privately funded undersea cable connecting the east coast of Africa, was brought into operation in July 2009. With a capacity of 1.2 Terabits per second, this promises to pose serious competition to SAT-3/SAFE bandwidth. Kenya, in particular, has recorded a surge in bandwidth consumption due to the cheaper SEACOM bandwidth, with some operators reportedly increasing their data allowance at no extra cost to consumers. The effect has yet to filter through to South African consumers in a meaningful context. Unfortunately, the SEACOM cable is rumoured to have some limitations in that it has no redundancy should the cable break. Furthermore, backhaul (terrestrial) links have proved somewhat temperamental in bringing this bandwidth onto the national grid, heightening fears about unreliability.

Around the launch of SEACOM, Telkom announced it was increasing the data cap on its retail accounts (doubling it in most cases), but that bandwidth is provisioned through the SAT-3/SAFE cable. Telkom has also announced that the capacity on the SAT-3/SAFE cable is being radically increased (from 120Gbps to 340 Gbps) in order to accommodate escalating demand. It seems apparent that that the increased international capacity has renewed competition is beginning to drive prices downwards.

There are a number of other cables in the pipeline which aim to have a presence in South Africa during the coming years. These are depicted in the image to the right. The East African Submarine System (EASSY) cable is due to be launched in mid 2010 with the West African Cable System (WACS) set to be active the following year. This should radically reduce Africa's dependence on satellite connectivity.

In a study conducted by Balancingact-Africa (Southwood, 2006), it was revealed that some 29 of 55 African countries obtain 80% of their total international Internet bandwidth by satellite, an inherently more expensive medium than submarine fibre optic cable. Source: manypossibilities (2008)

Local Loop Unbundling (LLU) has been proposed as a solution to give other players in the market direct access to households and businesses. The creation of the fixed-line telephone network was established over a number of decades, using tax payers' money, and this is neither easily nor cheaply replicated by competitors. Some commentators therefore see this is a strategic national asset which should not be held in private hands under these circumstances. During its period of exclusivity, Telkom was given the exclusive right to provide telecommunications infrastructure connecting commercial and private establishments in the country. This meant that no two parties could establish a connection between themselves, even if a company wished to connect two regional offices together. This also had the unfortunate effect of cellular operators, Internet Service Providers, and the like needing to use the incumbent to connect their inter-company nodes if not located within the same premises. These artificially raised telecommunications costs and kept internet penetration to modest levels (as highlighted earlier in the chapter).

Local loop unbundling would allow the second network operator (Neotel) and other pivotal players the opportunity to provide direct telecommunications services to consumers and businesses without 'piggy backing' on the Telkom network.

Figure 6. Forthcoming cables connecting Africa to the global Internet backbone

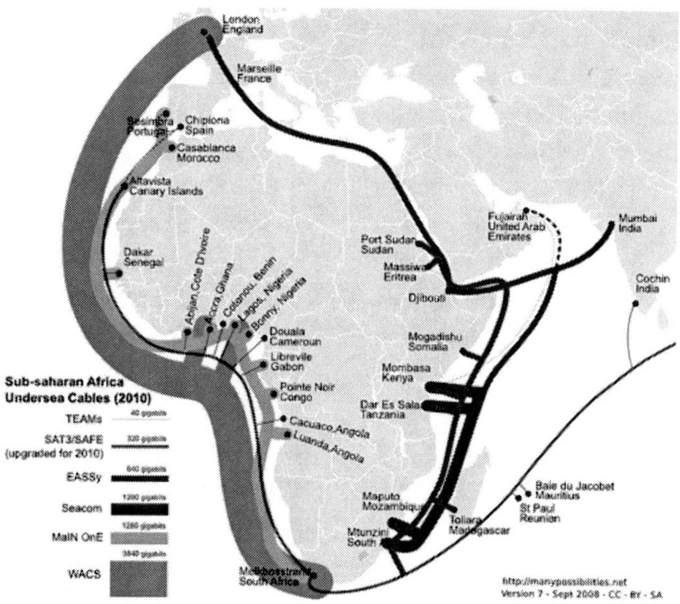

This was first mooted in 2006. The Minister of Communications at the time, Ivy Matsepe-Cassabury, took a policy decision that the unbundling process should be implemented and completed by 2011. Although little progress has been made to date, this is once again on the industry regulator's agenda and has been highlighted as a strategic objective in their recently released 'Programme of Action'.

CONCLUSION

The emergence and growth of broadband players in the market in certainly cause for celebration. Convergence also appears to be taking hold. Traditional voice-oriented players such as Telkom, MTN and Vodacom are shifting their focus to broadband services, whilst traditional data-oriented players (such as Internet Solution) are marketing VoIP offerings. Competition is therefore intensifying as these companies congregate on the broadband space.

Backhaul (national) and international connectivity, as well as the local loop, remains troubling. Two companies, Neotel and Telkom, control access to the SEACOM and SAT-3/SAFE international gateways, as well as being responsible for the bulk of intercity backhaul traffic in South Africa. Telkom maintains ultimate control of all fixed-line broadband connections through its ADSL service, whether the end-user is a direct customer of Telkom or not. MTN and Vodacom sew up the remainder of the broadband market through provision of their 3G/HSDPA services. Wholesale prices still remain artificially high as a small set of players (effectively reflecting an oligopoly scenario) has tight control over these electronic pipelines.

The political will is required to further liberate the market or to forcibly drive down prices, at the wholesale level, through legislation. Unfortunately, the Department of Communications (responsible for overseeing the telecommunications and broadcast sectors) and the Independent Communications Authority of South Africa (the body tasked with regulating the sector) have a chequered past in this respect. ICASA is widely regarded as being ineffective and 'toothless' – the body frequently backs down in the face of legal action if an operator happens to disagree with a directive or policy stance.

The Department of Communications, until very recently, had preferred a 'managed liberalisation' approach which under the guidance of the late Minister of Communications, Ivy Matsepe-Cassabury, saw the status quo held untouched for several years.

The frustrations of local users are evident in their thousands and are vented on a daily basis on activist sites such as www.MyBroadBand.co.za and www.Hellkom.co.za (a pun intended on combining the words 'hell and 'Telkom'). Yet, consumers' voices are still largely ignored and prices remain out of synch with international standards.

Hope may arise in the form of local municipal networks and wireless user groups, intent on providing residents with low cost or even free Internet access. These fledgling networks are starting to surface and are already in existence in Knysna (a small coastal town) and within the Gauteng (Johannesburg and Pretoria) region. Both models are based on Wireless Fidelity (WiFi) mesh networks such as those deployed in Philadelphia and San Francisco. The municipalities of Cape Town, Durban and Johannesburg are thought to be in favour of such developments and are planning similar projects of their own at the time of writing.

As the power engine of the sub-Saharan economy in Africa, broadband may prove a key ingredient to economic growth and prosperity within the country. Research clearly indicates a strong correlation between broadband adoption and GDP growth (OECD, 2009). Yet, there are other national priorities – namely, fighting HIV/AIDS, constructing housing, improving primary and secondary education systems, and equitably distributing social grants – to contend with. Broadband is still perceived to be an elitist service and therefore not within reach of the average South African. Unless pricing, coverage, computer literacy and nonchalant mindsets are altered, the country is unlikely to experience a fundamentally different landscape in the next ten years. The digital divide will persist and the (very) long journey to broadband freedom will continue, albeit at a snails pace.

ACKNOWLEDGMENT

The 3CC framework was developed in consultation with Rudolph Muller, founder of MyADSL and former Information Systems lecturer at the University of Johannesburg.

REFERENCES

BBCNews. (2009). UK broadband 'not fit' for future. Retrieved October 1, 2009 from http://news.bbc.co.uk/2/hi/tech nology/8282839.stm

Beneke, J. (2008). Long walk to broadband freedom. In Y. Dwivedi, A. Papazafeiropoulou & J. Choudrie (Eds.), *Handbook of Research on Global Diffusion of Broadband Data Transmission* (pp. 13-28). IGI Global: Information Science Reference.

Bridges.org. (2001). South Africa telecommunications overview, commentary, and statistics (policy brief). Retrieved May 2, 2001 from http://www.bridges.org/publications/123 Christian, C. (2005). Telkom's ADSL now officially over 1000% more expensive than other countries. Retrieved November 11, 2005 from http://www.mybroadband.co.za /nephp/?m=show&id=1167

du Plessis, M. (2005). Tit for tat. Retrieved September 15, 2005 from http://www.iweek.co.za/View Story.asp?StoryID=155165

E-Smith. J. (2009). Internet future in SA. Retrieved November 5, 2009 from http://mybroadband.co.za/news/Internet/10332.html

Franz, L. (2005). Pushing the city limits. iWeek. Retrieved December 1, 2005 from http://www.iweek.co.za/View Story.asp?StoryID=157791

Goldstuck, A. (2008). Internet Access in South Africa 2007. World Wide Worx: Johannesburg. Retrieved June 2, 2006 from http://free.financialmail.co.za/06/ 0602/technology/atech.htm

Internet.org.za. (2003). Cost of Internet access in South Africa (1993-2003). Retrieved October 5, 2006 from http://www.internet.org.za/costs.html

Laschinger, K., & Goldstuck, A. (2006). *PC users in South Africa 2006*. Johannesburg: World Wide Worx.

ManyPossibilities. (2009). African Undersea Cables. Retrieved July 28, 2009 from http://manypossibilities.net/ african-undersea-cables

Mawson, N. (2006). Telkom moves to shorten local loop in high-demand areas. *Engineering News*. Retrieved May 19, 2006 from http://www.engineeringnews.co.za/ eng/news/today/?show=86554

McLeod, D. (2006). *Ivy's Extra Mile* [online]. Financial Mail.

Monteiro, A. (2006). Education, lack of engineers hamper SA IT [online]. Moneyweb. Retrieved March 29, 2006 from http://www.moneyweb.co.za/s hares/ict_sector/995758.htm - 39k

Muller, R. (2006). SA slipping further behind in broadband rankings [online]. Moneyweb. Retrieved April 20, 2006 from http://moneyweb.iac.iafrica.com/ shares/ict_sector/252408.htm

MyBroadband. (2009a). JINX and CINX traffic growth. Retrieved October 7, 2009 from http://mybroadband.co.za/new s/Internet/9909.html

MyBroadband. (2009b). Reducing ADSL and broadband costs. Retrieved August 4, 2009 from http://mybroadband.co.za/n ews/Broadband/9073.html

MyBroadband.co.za. (2006a). Telkom 1000 times more expensive. Retrieved May 8, 2006 from http://www.mybroadband.co.za / nephp/?m=show&id=2680

MyBroadband.co.za. (2006b). iBurst coverage expansion moving forward. Retrieved May 29, 2006 from http://www.mybroadband.co. za/ nephp/?m=show&id=2943

Mzolo, S. (2005). Lower telecom costs 'only talk' [online]. Fin24.co.za. Retrieved December 20, 2005 from http://www.fin24.co.za/ articles/default/ display_article.asp?Article ID=1518-1786_1853631. Dated

OECD. (2009). OECD Broadband Statistics: December 2008. Organisation for Economic Cooperation and Development. Retrieved May 20, 2009 http://www.oecd.org/document/54/0,334 3,en_2649_34225_38690102_1_1_1_1,00.html

Point-Topic. (2009). World Broadband Statistics: Q1 2009. Retrievbed June 1, 2009 from http:// point-topic.com/contentDownload/operator-source/dsl reports/world%20broadband%20statistics%20q1%202009.pdf

Rens, A. (2008). What does an ECNS licence amount to anyway? Retrieved November 7, 2008 from http://www.shuttleworthfoundation. org/our-w ork/blogs/what-does-ecns-licence-amount-anyway

Simons, D. (2002). Telkom lights ADSL, but is the price right? Broadband Markets. Retrieved September 3, 2006 from http://www.iec.org/ events/2002/ dslwfeurope/newsletter/feature.html

Southwood, R. (2006). Satellite dominates cellular backhaul. Balancing-act Africa. Retrieved November 6, 2006 from http://www.mybroadband. co. za/nephp/?m=show&id=4799

Vapi, X. (2005). Telkom drops ADSL and data prices further. Retrieved June 2, 2005 from http:// www.telkom.co.za/common/aboutus/ mediacentre/pressrelease/articles/article_789b.html

Wikipedia (2009). Neotel. Retrieved November 5, 2009 from http://en.wikipedia.org/wiki/Neotel

World Economic Forum. (2007). The Africa Competitiveness Report 2007. Retrieved June 13, 2007 from http://www.weforum.org/en/ initiatives/gcp/Africa %20Competitiveness%20 Report/PastReports/index.htm

KEY TERMS AND DEFINITIONS

CDMA2000: Code Division Multiple Access is a mobile digital radio technology which operates in competition with traditional 3G/HSDPA services. It is widely regarded as being a hybid 2.5G/3G technology and is said to be a more efficient (and cheaper) technology than 3G/HSDPA.

3G/HSPDA: 3G is the abbreviation for third-generation cellular technology. HSDPA is an acronym for High Speed Downlink Packet Access and is a further iteration of 3G – widely considered to be 3.5G. A number of applications are associated with 3G/HSDPA including video conferencing, high-speed web access and even Internet telephony.

SEACOM: SEACOM is a privately funded cable which runs along the coastline of east Africa. South Africa, Madagascar, Mozambique, Tanzania, and Kenya are amongst the countries inter-connected by the cable. The strategic vision of the cable's operator is to lower bandwidth prices on the continent through granting volume discounts to customers.

ICASA: Independent Communication Authority of South Africa. ICASA is the body responsible for regulating the broadcasting and telecommunication sectors of South Africa. It has the power to levy fines are operators who abuse their market power and may also intervene in the market when pricing levels are deemed exploitive.

ISDN: Integrated Services Digital Network. This is set of protocols which transform a regular copper telephone line into a digital platform that is inherently faster and more reliable than its analogue counterpart.

LLU: Local Loop Unbundling. This is a regulatory process of allowing multiple locally- and national-based telecommunications operators to make use of connections from the telephone exchange's central office to the customer's premises. The physical wire connection between customer and company is often referred to as the "local loop" and was historically owned by the incumbent local exchange carrier.

PSTN: Public Switched Telephone Network. This is an engineering standard which defines the majority of the world's traditional telephone systems. The technology operates by automatically connecting customers according to the number dialled.

SAT-3/SAFE: South Atlantic 3. This is an existing submarine fibre optic cable linking Portugal and Spain to South Africa, with connections to several West African countries along the route. It forms part of the SAT-3/WASC/SAFE cable system, where the SAFE cable links South Africa to Asia.

SME: Small and medium enterprises or SMEs are companies whose headcount or turnover falls below certain limits. In South Africa, SMEs typically have fewer than 100 employees.

Teledensity: Telephone Density. This is the number of telephone connections within a particular area, typically a square kilometre. Manhattan Island would have an extremely high teledensity whereas the outback regions in Australia would have an extremely low teledensity.

Web 2.0: The creation of multimedia rich platforms through online sharing of content and interaction between individual users

Chapter 2
Strategic Interaction under Asymmetric Regulation:
The Case of New Zealand

Bronwyn Howell
New Zealand Institute for the Study of Competition and Regulation & Victoria University of Wellington, New Zealand

ABSTRACT

Regulation binds incumbent firms to a different set of obligations from their entrant-competitors, thereby creating an asymmetric set of options from which the firms may select the strategies under which they will interact. Whilst most regulatory obligations are specified in law, some take the form of contractual agreements. New Zealand's 'Kiwi Share' obligations bind the telecommunications market incumbent to a set of retail tariff structures and levels that have both restricted its choices and opened up a range of new strategic opportunities for its rivals that have had a significant effect upon the development of the New Zealand industry. This paper examines the specific consequences of the asymmetric tariff obligations and ensuing strategic interaction amongst sector participants on sector development – namely the effect of universal service retail prices and the allocation of the ensuing costs in determining the ongoing regulatory agenda; the role of a 'free local calling' obligation on the evolution of New Zealand's broadband market; and the consequent application of further asymmetric legislative obligations on the incumbent to address apparent 'problems' for which the asymmetric tariffs and rivals' strategic choices provide more credible explanations than the incumbent's exertion of its dominant position.

INTRODUCTION

All business sagacity reduces itself in the last analysis to a judicious use of sabotage.

– Thorstein Veblen

DOI: 10.4018/978-1-60960-011-2.ch002

In the application of competition law and industry-specific regulation, the obligations upon dominant firms typically differ from those imposed on firms which do not have significant market power. Asymmetric legislative and regulatory obligations undoubtedly influence the strategic actions of the firms concerned and other participants in the markets in which they trade (e.g. Hausman (2002), with respect to the broadband market developments in the United States and Korea). By extension, the contracts that firms enter into may also impose asymmetric obligations affecting subsequent strategic choices made by the firms concerned (e.g. Evans & Quigley (2000), with respect to the different litigation options competition law offers incumbents entering into contracts which may be subject to the exertion of market power).

In 1990, a contractual deed entered into by the New Zealand Government and the newly-privatised incumbent telecommunications provider Telecom New Zealand Limited (hereinafter 'Telecom'), known as the 'Kiwi Share', imposed additional obligations upon Telecom over and above those required under either the Commerce Act or the Telecommunications Act regulatory provisions[1] and to which no other firm was bound (Howell, 2008). The 'Kiwi Share' bound Telecom to three explicit obligations: that the price of residential telephone rentals would not rise faster than the Consumer Price Index (CPI) unless profits were unreasonably impaired (the 'price cap' obligation); that rural residential line rental prices would not exceed urban residential line rentals (the 'universal service' obligation); and that residential customers would continue to be offered a tariff with no charges for local calls (the 'free local calling' obligation) [2] (Boles de Boer & Evans, 1996).

As the 'Kiwi Share' obligations applied only to one firm in a market where the development of competition was an explicit *a priori* expectation, it would be expected that both Telecom and its competitors, customers, regulators and other stakeholders would take the asymmetric obligations into account when selecting the strategies upon which they would interact with each other. The contract thus formed part of a bundle of asymmetric 'regulatory obligations' deriving from all legislative and contractual origins that bind one firm differently from its competitors. The asymmetric obligations will therefore have affected the choice of firm strategies not just in respect of their interactions with each other, but also their interactions with markets (consumers), policy-makers and the technologies that they choose to adopt (as per Figure 1).

This chapter takes the form of a case study, examining the effect of the asymmetric 'Kiwi Share' obligations upon strategic interaction between all of Telecom, its competitors, customers and policy-makers in the New Zealand telecommunications sector during the period 1990 to 2008. Section 1 traces the origins of the 'Kiwi Share' provisions and places them in their institutional

Figure 1. Telecommunications Sector Interaction (Source Melody (2002:9))

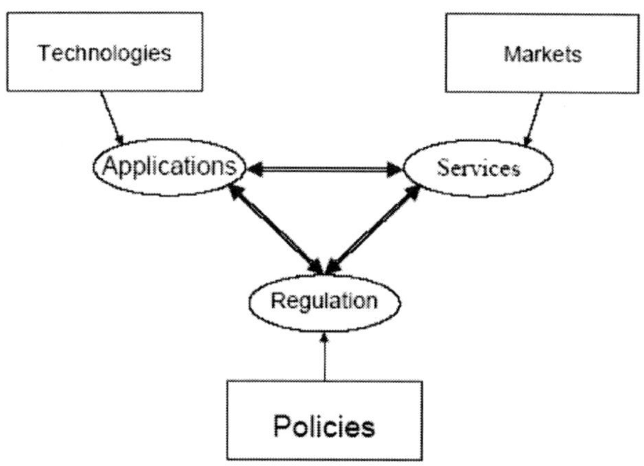

context in the sector. Section 2 discusses how, when competition emerged, strategic interaction arising from differing expectations about the allocation of 'Kiwi Share' costs between Telecom and its competitors influenced the subsequent direction of regulatory policy and legislation. In Section 3, the asymmetric effect of the 'free local calling' obligation upon Telecom in the face of both competition and dynamic technological change with the emergence of the internet is discussed. Section 4 then considers the ongoing strategic interaction arising between the firms and government specifically as a consequence of the historic asymmetric application of the obligations. Section 5 concludes with a discussion of the broader implications of asymmetric contractual obligations that emerge from the New Zealand case study.

THE 'KIWI SHARE' OBLIGATIONS

Although the three obligations comprising the 'Kiwi Share' were formalised in the deed between the New Zealand government and Telecom on September 12, 1990 (Howell, 2007:15), each of the individual elements had been long-established artefacts of the New Zealand telecommunications sector. The 'Kiwi Share' principles were effectively informal institutions arising historically from government ownership, control, regulation and provision of all New Zealand telecommunications services. The conflation of politics, state asset ownership and government service provision between 1880 and 1990 resulted in all technological and market decisions being subjugated to political activities (Howell & Sangekar, 2009). For example, Howell (2007) documents the political petitioning processes that rural residents had to undertake in order to become connected to the network[3]. Political priorities, rather than commercial imperatives, dominated the determination of budget sizes, technological choices, prices and the geographical location and order in which investments occurred.

The consequence was a sector inured to the exertion of both market and political power by a provider backed by the force of formal rule-making. The sole recourse for redress for any consequences of the inappropriate exertion of those powers open to sector participants was via political mediation (Howell & Sangekar, 2009). When Telecom was privatised, although the government retained the power to create the legislative framework under which the firm operated (level 2), most of the obligations and expectations historically undertaken under government ownership were transferred to Telecom, either formally via contracts, or informally via cultures, expectations and norms (Williamson, 2000). Indeed, a popular public expectation prevailed that the newly-privatised firm would continue to exert its monopoly powers in the same manner as had its former government owner, by actively preventing the development of competition.

However, it is also probable that in the public mind, and despite privatisation and market liberalisation, many of the informal expectations of government (for example, the power to exercise redistributive social agendas via telecommunications service pricing and provision) remained in force. As government had retained the legislative power to govern the legal and regulatory environment in which Telecom and its competitors would operate, then electoral pressure would inevitably be applied to ensure Telecom continued to deliver the social objectives to which consumers had become inured to under government ownership and operation. For example, the expectation existed that the government's obligation to safeguard social inclusion by ensuring telecommunications connectivity for rural residents did not become unaffordable (TCF, 2008:4) would not just continue unchanged under the new arrangements, but would be enforced via political processes.

The 'Universal Service' Obligation

The 'universal service' obligation had its origins in 1856, when the fledgling New Zealand colonial government, as one of its first independent actions, regulated postal tariffs to a single, uniform national rate. When the government assumed regulatory responsibility for telegraph services in 1864, universal 'postalised' telegraph tariffs were imposed. In addition to rationalising a plethora of individual, provincially-based tariffs, universal pricing satisfied a social policy objective to "establish a system commensurate with the rapid increase of the population nationwide … and to encourage and facilitate settlement of the interior" (Wilson, 1994:22). From the dissolution of provincial government in 1876, when all provincial telegraph assets passed to central government ownership, universal 'postalised' telegraph tariffs were imposed by the government as monopoly provider rather than via legislated price controls.

In 1880, the government assumed, via the Electric Telegraph Act, the monopoly for ownership and provision of all telephony services in New Zealand. Consistent with postal and telegraph pricing, universal telephony tariffs were imposed from the very outset. Courtesy of the 'Kiwi Share', universal tariffs remain today despite New Zealand's very different social, economic and demographic characteristics (for example, the current population is around 80% urban (OECD, 2007), indicating that the historic policy objective of "settling the interior" is no longer a primary justification for maintaining universal telephony tariffs). Any change to the traditional pricing pattern has been considered politically 'too difficult' to bring about[4], although a recent review by the Telecommunications Carriers' Forum (TCF) has recommended that the government revisit the possibility of funding the obligation from general taxation rather than as a charge on the industry, and making the obligation contestable, rather than binding Telecom alone (TCF, 2008).

The 'Free Local Calling' Obligation

The 'free local calling' obligation arose in 1880, when the incumbent Superintendent of Telegraphs, upon assuming responsibilities for telephone services, eschewed charging for calls between subscribers connected to the same exchange as he deemed the practice "too complex and onerous for exchange staff who would have to log the calls" (Wilson, 1994:66). When the issue was addressed four years later by the Postmaster General, any change in the pricing policy was deemed to be too "politically unpopular" to undertake (*ibid*). When technological change led to exchanges covering larger areas, the 'free local calling' zones expanded commensurately.

Following World War Two, 'free local calling' zones became a matter for political determination, independent of exchange or technological restrictions. Burgeoning use of long-distance telephony offered opportunities to politicians seeking political support to trade off politically-determined long-distance charges against 'free' local calling benefits, particularly in marginal electorates (Wilson, 1994:151-3). Consequently, New Zealand has historically exhibited some of the OECD's largest local calling zones (NZIER, 2005).

Once again, the embeddedness of 'free local calling' in New Zealand's institutional framework resulted in its inclusion in the 'Kiwi Share' obligations, as it was deemed too politically difficult to do otherwise. In the 2008 TCF review of the 'Kiwi Share' obligations, the economic and social implications of free local calling were excluded from the terms of reference for the review, as the government had directed that such services must continue to be provided (TCF, 2008:25).

The 'Price Cap' Obligation

The 'price cap' obligation draws on New Zealand's long history of political price-setting. From the inception of telephony services in 1880 until the passing of the Telecommunications Act in

1987, the legislative right to determine charges lay with the Postmaster General (politician), and was subject to Parliamentary scrutiny. Whilst commercial considerations could inform the decision, ultimately charges were determined as a matter of politics. When telecommunications service provision was separated from the other Post Office services as a State-Owned Enterprise (SOE) in 1987, the direct nexus to political price-setting was severed. However, as SOEs were still bound by a government agency agreement, broad political oversight of and accountability for price-setting remained.

When Telecom was privatised in 1990, the 'price cap' obligation was invoked as a pragmatic move to satisfy largely political concerns that the Commerce Act alone might be insufficient to constrain a private owner with market power from setting excessive prices in a newly-privatised and newly-competitive market where it was uncertain what information would be available to adequately assess market performance. In effect, the 'price cap' protected consumers from the risk of expropriation via prices to any greater extent than that which they might have been exposed under government ownership (at least in respect of the residential line rental charge).

That the Minister of Communications (political successor to the Postmaster General) *could* approve price rises above the cap in the event of sufficient compelling evidence being provided offered some protection to Telecom's owners from exposure to a legitimate risk of costs rising above 1990 levels (presuming the Minister would take into account the public benefit of Telecom continuing to be financially viable when making such a decision). That Ministerial approval *must be obtained* for tariff increases above the cap or any other change to the 'Kiwi Share' contractual obligations preserved the long tradition of political input into price-setting, at least in respect of determining the tariff structure and upper price bound. The requirement also ensured an ongoing need for ministerial monitoring of the sector (initially via the Ministry of Commerce (MoC), subsequently the Ministry of Economic Development (MED)). Ministerial approval to alter 'Kiwi Share' terms has been sought only once, in 1999 when Telecom applied for a variation to the 'free local calling' obligation in respect of dial-up internet calls in excess of ten hours per month. The Minister approved the charge.

The 'price cap' obligation also signalled important information about the margins the government believed were incorporated into Telecom's prices at the time of sale. As the government had provided all telephony services up to 1990, and had been operating the telephony segment as a stand-alone entity separate from other Post Office operations since 1987, it had access to all information available about the relationship between Telecom's prices and costs. This information could logically be assumed to have been incorporated into the decision-making at the time of sale.

If the government believed that the line rental prices in place on September 12 1990 incorporated profits in excess of those required to cover 'Kiwi Share' 'universal service' and 'free local calling' costs (which were embedded in existing product cost and price structures), then it begs the question of why the 'price cap' obligation established a cpi-x incentive with x=0. If it had been known that excessive profits were being garnered, then it would appear irresponsible for the government to set x at 0. Rather, the rational action would have been to set x at a small positive value, thereby 'ratcheting' allowable prices down over time and eliminating supra-normal profits (Milgrom & Roberts, 1992:232-6). That this was not done could be interpreted as a signal that the government believed the prices at the time did not incorporate any supra-normal returns, meaning a 'price cap' was required only to prevent opportunistic price-setting above this historically 'fair' level[5]. The price paid for the firm and accepted by the government would have reflected these assumptions.

THE 'KIWI SHARE' AND COMPETITIVE ENTRY

Competitive interaction is underpinned by the participants' actions, derived from strategic thinking – "the art of outdoing an adversary, knowing that the adversary is trying to do the same to you" (Dixit & Nalebuff, 1991:ix). The knowledge that another party is constrained in its choices by different obligations necessarily affects the range of strategies available to a party in selecting the one that will, in its assessment, leave it best off after the interaction has occurred. It would be expected that Telecom's competitors would utilise the knowledge that the 'universal service' obligation distorts competitive interaction in order to appropriate as large a proportion of the gains available from competitive entry as possible. Indeed, competitors' failure to take advantage of the asymmetry for their own purposes would be a deviation from expected behaviour warranting further investigation and explanation.

Strategic Responses to Asymmetrical Universal Service Obligations

The incongruity of the asymmetric allocation of universal service obligations on the prices of only one firm in a competitive market is highlighted by Farrell (1996). Requiring the incumbent to charge above-cost prices in urban markets and subsidise below-cost prices in rural markets induces inefficient selective infrastructure-based entry. As a consequence of the high margins on offer, competitors will enter only in the low cost urban markets (competitive entry will not occur in the high-cost rural areas as at the prevailing subsidised prices, losses will be made). Furthermore, competitors can enter in urban markets even though their cost structures are higher than those of the incumbent (i.e. it would have been more efficient if the incumbent continued to serve the urban market than the more costly competitor). As consumers switch to the entrant, the incumbent loses revenues from which the unprofitable (high-cost rural) services were subsidised. If the incumbent is also subsidising line rentals from calling revenues, even entry into calling markets (e.g. long distance) may impede the incumbent's ability to subsidise unprofitable access services and remain financially viable. Such distortions can be overcome only by levying a tax on the entrant, equivalent to the amount the incumbent used to extract from customers now served by the entrant, in order to continue subsidising the unprofitable services at the requisite level (Armstrong, 2001).

When an incumbent faces asymmetric universal service pricing obligations, an entrant seeking to maximise its own profits would logically seek to avoid any price or tax liability for universal service obligations, thereby forcing the incumbent to bear all of the costs uncompensated. If a tax can be avoided, the entrant can selectively enter only the most profitable (incumbent's capped prices above cost) areas, and undercut the incumbent's prices.

If the incumbent did not face a 'universal service' obligation, it would respond to selective entry by competing on price, down to the higher of its own or the entrant's costs. With a 'universal service' obligation in place, however, the incumbent's ability to respond is restricted. If the incumbent cannot lower its prices, a single entrant can charge approximately the same price as the incumbent, without fear of invoking price competition. As customers are induced to switch from the incumbent, profits previously earned by the incumbent and used to offset connections charged below cost are now extracted by the entrant as 'free profits' (although if there is more than one entrant, they may compete amongst themselves on price, thereby lowering prices to the competitive level – but as the fixed and sunk costs of infrastructure-based entry are substantial, only limited entry is likely to occur, increasing the likelihood that entrants may co-operate to appropriate the profits).

The incumbent must respond by raising its prices to all of its remaining customers in order to cover the entry-induced losses incurred in fulfilling the universal service obligation. If the entrant follows the incumbent and raises its prices too, it makes even higher 'free' profits'. If the entrant does not also raise its prices, the incumbent loses all low-cost consumers to the entrant, and must exit the low-cost market. All incumbent costs must then be recovered from the high-cost market, where it faces no competition. Prices in the high-cost market rise to at least their actual cost. A division arises between competitive entrants serving the low-cost market and the incumbent serving the high-cost market, each charging prices determined by their own costs, and removing any vestige of a 'universal service' price across the two markets. Moreover, if the entrants' technology is more costly than the incumbent's (e.g. they cannot capitalise upon economies of scale), then not only are universal service prices eliminated, but total welfare is lower, as the total cost of provision is higher than if the incumbent served all consumers.

From a public policy perspective, therefore, asymmetric levying of a 'universal service' obligation on the incumbent whilst simultaneously preventing the incumbent from sharing the costs of meeting the obligation with its entrant-competitors actually induces the exact geographically-based price differentials that the universal service obligation has been imposed to circumvent. It is therefore implausible to assume that the government, given its dual objectives of increasing competition in the market and retaining universal service prices to end consumers, could legitimately expect the 'Kiwi Share' to be borne by Telecom alone. If there was any question of prices exceeding costs, then the simplest solution was a ratchet on Telecom's fixed line rental prices in the 'price cap'. If there was an intention to selectively extract Telecom's accumulated profits for consumer and taxpayer purposes as a consequence of any other policy, then this should have been explicitly and separately levied. That neither of these was undertaken is consistent with an expectation that Telecom would incorporate 'Kiwi Share' costs in the prices charged to both consumers and its competitors*.

Contesting in Court: Clear v Telecom, 1991-94

When Telecom faced competitive entry in the local calling market from Clear Communications[6] (hereinafter Clear) in 1991, the firm used the Efficient Component Pricing Rule (ECPR) to set interconnection prices (Blanchard, 1995). The benefit of ECPR when the costs of social obligations are imposed asymmetrically is that the firm bearing the obligation alone can include the social obligation costs into the interconnection price charged to its rivals (Economides & White, 1995). Clear, however, contended that Telecom alone should bear the 'Kiwi Share' costs, and alleged that any charge levied by Telecom above its service delivery cost (i.e. excluding any component to recover 'Kiwi Share' costs) constituted an exertion of market power and was therefore illegal under Section 36 of the Commerce Act 1986.

From a strategic perspective, Telecom's use of ECPR in the presence of the 'Kiwi Share' universal service obligation is rational. Clear's assertion that entrants should not bear any of costs of the social obligations assumes a political distributive intention clearly inconsistent with the government's 'universal service' objectives. It is also inconsistent with the ambit of the Commerce Act, which concerns itself with the long-term welfare of consumers (a dynamic function of total welfare) and not the distribution of producer surplus amongst producers. Neither was it signalled anywhere in the agreement between Telecom and the government that there was any intention to tax Telecom any differently than any other market participant. These observations draw into question Clear's strategic logic in using litigation to contest the point, except perhaps to test the Court's interpretation, in the absence of any other New Zealand precedents, of the Commerce

Act Section 36 provisions regarding expectations about the behaviour of dominant firms trading in markets governed by competition law.

The New Zealand High Court ruled in 1992[7] that Telecom's use of ECPR prices did not constitute a breach of Section 36 of the Commerce Act. As Section 36 prevents monopoly pricing only when used with the intent of restricting, preventing or eliminating competition in a market, the Court deemed that setting interconnection prices above marginal cost to recover 'Kiwi Share' costs did not of itself constitute restriction, prevention or elimination of competition. However, when Clear contested the decision, the Court of Appeal[8] found in 1993 that Telecom could not lawfully charge an interconnection price that included a component of monopoly rent (i.e. the only legal interconnection price was marginal cost excluding the 'Kiwi Share' costs). The consistency of the Court of Appeal decision with the contention that Telecom alone must bear the 'Kiwi Share' costs, and the deleterious effect that the decision would have upon efficient entry incentives and the ultimate ability to deliver upon the universal service social objective, caused some dismay in both economic and policy circles (Blanchard, 1995). That the decision also appeared contrary to the signals of government intentions in the 'Kiwi Share' obligations themselves gave rise to comments from some quarters of the Court of Appeal's apparent judicial activism in coming to the conclusion that it did.

Telecom then appealed to the final arbiter, the Judicial Committee of the Privy Council of Great Britain[9]. The case was heard in 1994. The Privy Council decision argued that Clear's Section 36 case rested upon showing that Telecom had used its dominant position for the purpose of preventing, deterring or excluding competitive conduct. As Telecom had set its prices to Clear on the basis of opportunity cost, it was deemed to be behaving exactly as would any firm in a competitive market, so the appeal was upheld. Further, it was held that the application of the ECPR would put Clear in a position to compete out over time any monopoly profits obtained by Telecom, and that, if they were not competed out, the Government had the ability to introduce price controls under Part IV of the Commerce Act. Moreover, the Privy Council found that Clear had not demonstrated that Telecom's prices included monopoly rent, above that necessary to meet the social obligations of the 'Kiwi Share' (Blanchard, 1995).

Recourse to Political Strategising

Whilst from a legal and commercial perspective, the Privy Council resolution gave clarity and ultimately resulted in a significant number of new entrants participating in the market on the basis of ECPR-based interconnection contracts (Howell, 2007:26-7) considerable residual resentment remained, both amongst Telecom's competitors and some parts of the (voting) public. Having failed to outmanoeuvre Telecom either in the market or via the courts, a group of competitors, led by Clear and the Telecommunications Users' Association of New Zealand (TUANZ)[10], began lobbying politicians for the introduction of industry-specific regulation whereby interconnection prices would be set by a regulator rather than relying solely upon negotiation between the firms concerned. Telecom countered by lobbying for the retention of the existing competition law-based arrangements.

In the short-term, the contesters' lobbying efforts were not rewarded. In an inquiry in 1995, The Treasury and Ministry of Commerce found no reason arising from the 1991-94 court cases to change either the Commerce Act or the 'Kiwi Share' arrangements. The Minister endorsed the inquiry's recommendations (MoC/Treasury, 1995). In the long run, however, a sympathetic ear was found in the Parliamentary opposition (predominantly members of the social democratic Labour Party), many of whom had opposed both the privatisation of Telecom and its initial sale to a consortium of American[11] firms (although the consortium sold 724.5 million shares (a majority) into the market on 19 July 1991, as agreed in the terms of sale). As part of its manifesto for the 1999 election, the Labour Party promised a new inquiry into New Zealand's telecommunications regulatory framework. Support from the opposition naturally invoked political and regulatory uncertainty, due to the inability of a government to bind its successors to maintain any specific policies or laws (Howell, 2008).

A Labour-led coalition formed the government after the 1999 election and moved rapidly in 2000 to instigate a formal inquiry. Given the nature of its political origins, the 2000 Inquiry was instructed in its terms of reference to address (amongst other issues): "alternative means of establishing interconnection terms and conditions; pricing principles and other terms and conditions (such as service quality standards) for current and future forms of interconnection; processes applying to interconnection negotiations, including dispute resolution and enforcement mechanisms; local loop unbundling; resale of telecommunications services; information disclosure;"[12] and the 'Kiwi Share' obligations.

The Inquiry reported back on September 27 2000[13], with the finding that, as Telecom's interconnection prices still contained elements above cost, an industry-specific regulator with the power to regulate prices using TSLRIC (cost-based) methodology for a range of Telecom-provided non-internet services should be installed, and that Telecom alone should bear all of the costs of the 'Kiwi Share' until such time as it could satisfy the Minister that it should be allowed to increase its charges in order to maintain financial solvency. Telecom's internet-based retail services were spared from TSLRIC pricing, instead to be offered as wholesaled products to competitors at the retail price minus a regulator-determined discount (Howell, 2007:38)[14]. The inquiry also recommended that local loop unbundling should not proceed, as there was evidence of competition already emerging in the broadband market. However, a further inquiry into the feasibility of unbundling access was to be undertaken within two years.

Industry-Specific Regulation and the TSO Tax

The Labour-led government accepted most of the Inquiry recommendations. The Telecommunications Act 2001 established the office of the Telecommunications Commissioner. The Commissioner was granted powers to regulate prices for designated services and to advise on other services which should become subject to regulation. However, due to their clear incompatibility with universal service objectives, the Inquiry's 'Kiwi Share' recommendations were rejected. Whilst Telecom was still required to maintain its existing contractual obligations in respect of its retail pricing and service quality levels, the Act created a new obligation, the 'Telecommunications Service Obligation' (TSO). Under the TSO, the Commissioner would annually allocate the costs of Telecom meeting its social obligations across all firms participating in the market.

The price regulations removed any question of Telecom charging competitors prices above (TSLRIC-based) cost, but the TSO obligations introduced a 'tax' on entrants. The TSO 'tax' effectively appropriated the 'free profits' entrants made from arbitraging upon social costs that Telecom alone had to bear, and transferred them to Telecom, thereby restoring it to the financial position that would have prevailed to what it would have been had selective entry not occurred in the low-cost areas. As per Armstrong (2001), the incidence of the TSO tax thus substantially reduced the risk of inefficient entry occurring in low-cost markets by firms with higher costs than Telecom.

However, the TSO has been very unpopular with Telecom's competitors, who have continued to engage in political lobbying on the basis that they should not be taxed to compensate Telecom's losses, especially when they provide services in the same areas using alternative technologies (e.g.

mobile and wireless) and could service rural consumers more cost-effectively with these technologies[15]. The process of determining the level and allocation of the tax has also been problematic. The complexity of establishing the relevant facts have combined with other workload pressures at the Commission, to the extent that delays of up to three years in levying the liabilities have been experienced (Howell, 2007). Entrants routinely challenge the determinations, both via legal and public media channels. Consequently, the TSO remains a political issue.

Continued political pressure resulted in a further review by the TCF in 2008. The government's response to the recommendations that the government meet the costs of the social obligations from general taxation revenue and that firms be allowed to contest for the right to provide services in 'uneconomic' areas is yet to be delivered. However, as long as the government has social policy objectives in respect of telecommunications, the question of how to allocate the costs remains. As Farrell (1996) identifies, universal service prices are antithetic to competition and competition is antithetic to universal service principles. The New Zealand example illustrates that as long as the obligation is asymmetrically embodied in the prices charged by only one participant in the market, there will be continual strategic interaction between all participants in all of the market, the courts and the political arena.

THE 'KIWI SHARE' AND TECHNOLOGICAL CHANGE

Whilst the 'universal service' obligation has led to tensions regarding the allocation of costs, the 'free local calling' obligation has introduced further tensions into the New Zealand telecommunications market, which have arguably had a far bigger effect upon shaping firm behaviour and dynamic sector outcomes. The effects arise from the interaction of 'free local calling' and the consequences of technological change.

'Free Local Calling', Voice and Internet Usage

'Free local calling' is a special case of a two-part tariff, where a single price is charged to cover both connection to the network and usage of it. Whereas a classic optimal two-part tariff prices each component at cost, 'free local calling' requires connection be priced above cost (a call cannot be made unless a connection is purchased), and the surplus used to pay for all costs of usage, which is priced below cost, at zero. The consequence is that fewer connections are sold (as the connection price charged rises relative to an optimal two-part tariff, thereby excluding all users with a combined value of connection and usage who would have purchased at the cost-based prices but not at the flat-rate price), but usage is greater, as all those purchasing a connection now consume all calls or call minutes with a positive benefit, rather than only those where the benefit exceeds the marginal cost, priced as the usage charge (Laffont & Tirole, 2002).

The 'free local calling' obligation has had a profound effect upon patterns of telecommunications usage in New Zealand. NZIER (2005) finds that the volume of voice minutes consumed per account in New Zealand is one of the highest in the OECD. Howell & Sangekar (2009) find that the combined volumes of voice minutes and internet usage per account in New Zealand are five times those observed in Finland, where two-part tariffs and cost-based, region-specific (i.e. not universal) line rental prices prevail.

With the emergence of the internet, first accessed via dial-up telephony lines, 'free local calling' offered a consumer welfare boon in those markets where the tariff structure prevailed. Not only was a dial-up internet connection 'gifted' with an extant voice connection, but dial-up internet access could be consumed up to the point where the marginal benefit of the telephony component was zero, not the marginal price of the call (Howell, 2008a). The four countries with widespread 'unmetered' voice telephony pricing' (i.e. 'free local calling') – New Zealand, Australia, Canada and the United States – exhibited some of the earliest and highest rates of connection and usage of dial-up internet access in the OECD, leading to an endorsement of such pricing policies in order to increase internet usage (OECD, 2000). Such endorsement has likely been a significant factor in the commonly-observed practice of offering 'unmetered' or 'flat-rate' broadband tariffs.

'Free Local Calling' and Interconnection Contract Arbitrage

To compensate for the costs of carrying the call to its destination, classic interconnection contracts (ICAs) typically require the owner of the network on which a call originates to make a payment to the owner of the network on which the call terminates (Laffont & Tirole, 2002). Interconnection can be billed by call or by call duration. The most common billing is by call duration. The rate will vary depending upon the relative size of the networks. Such ICA agreements have proved reasonably acceptable to both parties as long as the calls made on each network are of the same length.

However, if the call origin or duration patterns are asymmetric, the network operators can engage in strategic selection of customers in order to arbitrage upon the gains and losses from interconnection revenues (classic 'adverse selection' or 'cream-skimming'). Networks have an incentive to sign on those customers who, on average, receive longer calls from customers of other networks than they make to customers of other networks (e.g. call centres), as these calling patterns generate net positive ICA cash flows to the receiving network. Indeed, it is strategically advantageous for a network operator receiving net positive ICA revenues to share some of the gains with the customers generating the revenues. The revenue-sharing acts as an inducement for the otherwise hard-to-identify relevant customers to 'self-select' by subscribing to the network in the first place. Such inducements are also likely associated with contract terms that discourage signing with a rival 'locking customers in' for an extended period of time[16].

In 1996, Telecom and Clear negotiated a five-year ICA. The agreement required Telecom to pay 2c per minute for all calls originating on the Telecom network and terminating on the Clear network. The Clear payment varied from 1c to 2c per minute over time, on the assumption that Clear's network size would increase. Discounts on both sides were offered for off-peak calling. The Telecom-Clear ICA became the prototype upon which Telstra, Saturn, Compass and others subsequently negotiated ICAs with Telecom (Karel, 2003). Whereas call origination and duration was likely approximately symmetric when the ICA was signed (as traffic was exclusively voice), the emergence of the internet dramatically changed calling patterns. Firstly, traffic generated on the PSTN grew exponentially (Figure 2), induced by 'free local calling' as new internet users dialed up to their internet service providers (ISPs). Secondly, as the vast majority of internet users were Telecom fixed line customers, the vast majority of this traffic originated on Telecom's network. Thirdly, internet calls were very much longer in duration than average voice calls.

Together, these factors created a classic ICA arbitrage opportunity for Telecom's competitors. As long as they could sign on the majority of ISPs as customers, there would be a net flow of ICA cash from Telecom to its competitors. Competitors moved rapidly to sign on ISPs, offering substantial cash payments as an inducement (Karel, 2003). These payments were further shared by ISPs with their customers, via reduced-cost or even free ISP accounts. The heavier the internet use of an ISP customer, the more valuable they were to Telecom's rivals, so there was intense competition amongst non-Telecom ISPs to sign on the heavi-

Figure 2. New Zealand Telephony Network Traffic 1996-2003 (Source Howell & Obren (2003:33))

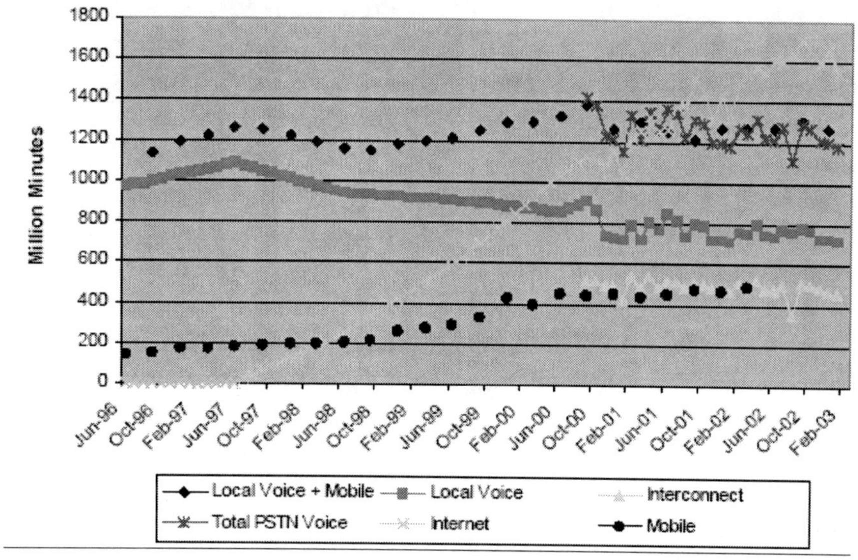

est internet users. Over the period 1996-99, whilst the market share of ISPs signed up to Telecom's rivals grew to slightly over 50% (Enright, 2000), undoubtedly more than 50% of the internet traffic was terminating on rival networks (Howell, 2008). The effect of reduced ISP fees is reflected in Boles de Boer, Enright & Evans' (2000) comparative analysis of the ISP market in Australia and New Zealand. Whilst New Zealand had fewer than one third of the number of ISPs per capita of Australia, the number of ISP accounts per capita was 10% higher and prices on average 30% lower.

The consequences of ICA arbitrage, however, were financially crippling for Telecom. Due to the 'free local calling' obligation, the firm was prevented from generating any additional revenue from residential dial-up internet usage. At the time, the average monthly residential line rental was \$30 per month. The consumption of an average dial-up internet connection customer at peak times was generating an ICA liability of \$36 per month per dial-up internet account. As ICA incentives likely resulted in a skew such that ISPs aligned with networks other than Telecom signed up a majority of the heavy users, the average monthly liability in excess of income per account for Telecom likely substantially exceeded the implied \$6 cash deficit (Howell, 2008a).

Telecom's response to the strategic threat posed by its rivals was two-pronged. The first response was commercial: technological bypass of the dial-up PSTN with a new, unregulated ADSL broadband service. The second response was political: renegotiating the terms of the 'Kiwi Share' to enable recovery of the costs imposed by burgeoning internet use and strategic arbitrage on the ICA by its rivals.

Telecom's Response #1: Strategic Broadband Investment

The first response was rapid investment in the successor broadband ADSL technology (New Zealand was the third country in the OECD to offer a commercial ADSL service when it went live in January 1999 – Howell, 2003). High speed (2Mbps was the basic offering) ADSL was made available widely (85% of customers by 2002 - *ibid*) and low-priced (one of the cheapest per megabit second in the OECD in 2000 – OECD, 2001) in order to induce rapid substitution by principally the heavy users (especially those of its rivals) from costly, unchargeable dial-up internet access to relatively less costly, chargeable and, importantly, 'Kiwi Share'-exempt ADSL. Furthermore, even though the firm faced different costs of installation in different parts of the country, universal nationwide tariffs prevailed (Howell, 2003).

Moreover, in order to induce substitution from dial-up as quickly as possible, Telecom offered ADSL under two-part tariffs. Howell (2008) shows that, as flat-rate tariffs for legacy technologies result in substitution to the frontier occurring at higher individual valuations of connection and usage volumes than under an optimal two-part tariff, earlier substitution could be induced by pricing ADSL connection below cost and subsidising it from either ADSL usage or prices above cost from another product in a bundle.

Under Telecom's offers, ADSL could be purchased only if the consumer also purchased a Telecom dial-up voice telephony account (contemporaneously, other infrastructure providers – notably Saturn – were offering broadband and voice telephony in a pure bundle), although ISP services could continue to be purchased from third parties. A high volume internet user substituting from dial-up to ADSL would be moving all internet usage to the new network (i.e. massively reducing consumption on the PSTN), but under 'free local calling', still paid exactly the same monthly fee for a very much lower volume of usage on the voice network. In a 'bundle' of voice telephony and ADSL, voice telephony payments above cost could subsidise ADSL charged below cost. If the PSTN internet usage was also incurred to an ISP aligned to a network other than Telecom's, then not only were PSTN internet usage costs avoided, but the costly ICA obligation imposed by the consumer's ISP choice was also eliminated. Telecom could thus afford to discount ADSL connection charges below cost to counter its rivals' strategic ICA arbitrage, simply because the counterfactual was a substantial loss in any case. Evidence that Telecom's ADSL connection prices charged were below cost comes from Covec (2004). International benchmarking of TSLRIC prices in similar markets reveals that Telecom's entry-level ADSL price for access with only a minimal usage bundle of 600 Mb/month was lower than prevailing regulated monthly bitstream access prices in most comparable regimes at that time.

However, due to the high value of the 'connection gift' from bundling dial-up internet with voice telephony and the comparatively low valuation of the internet usage of most consumers, rapid substitution to ASDL did not occur. Indeed, as heavy internet users faced positive connection and usage prices under the two-part ADSL tariffs compared to zero connection and usage under dial-up access using a free ISP account, and most of the usage was low-valued, dial-up usage per account continued to grow exponentially (Figures 2 and Figure 3). Rather, it was the high-valuing, low-using internet consumers who were most likely to substitute (Howell, 2008a). Whilst this reduced some pressure on the PSTN it did not address the fundamental problem of huge cash losses due to asymmetric allocation of the high dial-up internet users and the perverse incentives operating under the ICA and 'free local calling' obligations.

Telecom's Response #2: '0867'

Given the inability to substantially stem the cash flows via technological substitution, by September 1999, Telecom was forced to adopt its second response: petitioning the Minister to alter the prices charged under the "Kiwi Share' agreement. The proposal was to charge residential consumers 2c per minute (the ICA liability) for calls in excess of 10 hours per month per account made to ISPs not connected to a separate Telecom dial-up internet PSTN network (IPNET).

The response came to be known as '0867' after the calling prefix of IPNET accounts. As well as securing its financial viability, Telecom claimed

Figure 3. Source data Statistics New Zealand ISP Surveys and Telecom Management Commentaries

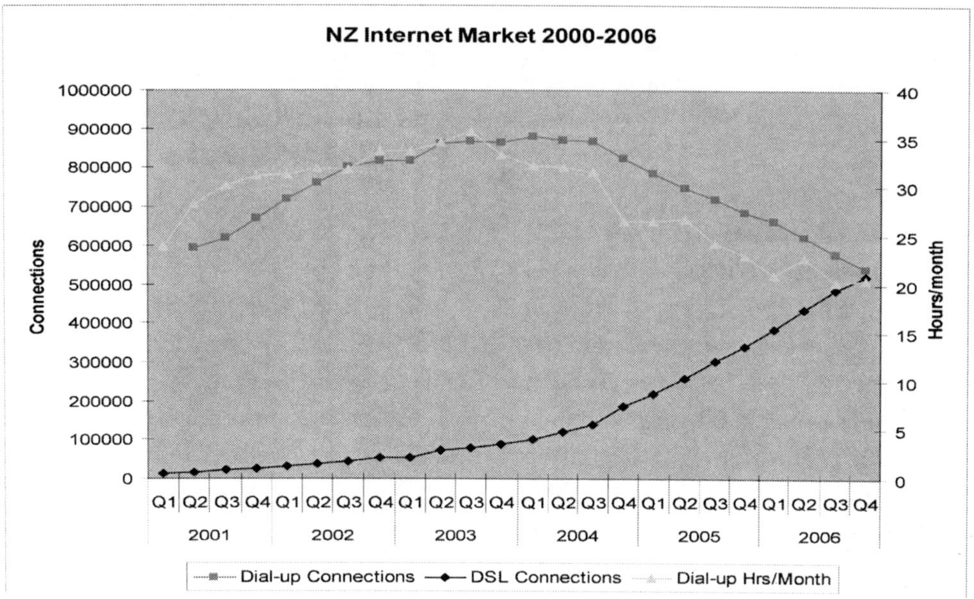

that '0867' enabled separation of internet traffic from voice traffic, thereby enabling the firm to maintain voice service quality levels as agreed under the 'Kiwi Share' terms. That consumers could still use 10 hours per month of internet connection free of charge was offered as a sign of good faith (credible commitment) that Telecom was not acting anti-competitively in order to shut down rival ISPs, as such usage would still incur an ICA liability to Telecom of up to $12 per month for each consumer dialling a non-IPNET ISP. The Minister, in one of his last acts before being removed from office in the 1999 election, approved the variation to the 'Kiwi Share' agreement[17].

Competitor and Administrative Strategic Responses to '0867'

Despite numerous complaints about Telecom's action over '0867' being a potential breach under Section 36 of the Commerce Act, no other firm (either ISPs or network operators) brought any litigation. All, however, petitioned politicians and requested the Commerce Commission to use its powers to bring a case – in effect, 'free-riding' off the Crown to cover the costs of bringing a suit against Telecom. In August 2000, one month before the 2000 Inquiry reported its findings, the Commission laid charges under the Commerce Act. The Commission alleged that "in introducing 0867 Telecom sought to prevent or deter competitive conduct by other telecommunications network operators and Internet service providers" (MED, 2001). However, both the High Court (2008) and the Court of Appeal (2009) found that the '0867' action did not breach the Act[18].

Despite the pending litigation, the original 1996 ICA that had stimulated the strategic exchanges ran its full course of five years. It was replaced first in 2000 by a 'bill and keep' agreement and in 2002 with regulated contracts specifying a termination rate of 1.13c/minute. The majority of ISPs acquiesced and acquired Telecom '0867' accounts, at least in respect of servicing their heaviest dial-up internet customers. Dial-up internet traffic continued increasing unabated, reaching a peak of 35 hours average consumption per month for each of the 850,000 dial-up internet access accounts in 2003 (Figure 3). Whilst low-priced ISP accounts remained the norm, in the absence of extensive interconnection cash flows, the 'free' ISP model induced by the contractual arbitrage had effectively disappeared by 2001, with minimal effect upon the number of internet accounts per capita (Karel, 2003).

Even though Telecom's ADSL prices were very low by international standards (New Zealand is still amongst the OECD top quartile in registering low ADSL prices (per month and per Mb/s), highest speed offered by an incumbent, and low price per additional MB over prescribed caps – OECD, 2008), substitution from dial-up to broadband continued to be sluggish until 2004. Figure 3 indicates that New Zealand broadband uptake is driven predominantly by substitution from dial-up usage rather than new internet users purchasing accounts. As prices did not change substantially in 2004, the most logical explanation for the increase lies in changes in demand arising from new applications conferring greater consumer benefit from broadband than dial-up. Howell (2008) identifies that increases in usage of New Zealand's most popular web property TradeMe and the emergence of social networking sites such as Facebook, YouTube Twitter and Bebo are nearly perfectly correlated with the upswing in New Zealand broadband connections.

THE 'KIWI SHARE', REGULATORY THREAT AND STRATEGIC GAMING

The preceding two sections have illustrated the role played by the 'Kiwi Share' in shaping strategic interaction in both the political and commercial institutions of the New Zealand telecommunications sector environment. The 'free local calling' obligation played a significant role in contributing towards both New Zealand's world-leading dial-up internet access and utilisation statistics, and the comparative absence of supply-side impediments to widespread broadband uptake, but depressed actual broadband uptake by distorting the price signals given to internet users of the real costs of their usage.

The LLU Inquiry and Broadband Uptake Politics

However, high levels of broadband uptake were a political priority for the Government as part of its Digital Strategy. The Labour-led government elected in 1999 had already shown its propensity to be influenced by political lobbying from Telecom's competitors when it instituted the 2000 inquiry (albeit rejecting its more radical TSO recommendations as a consequence of their inconsistency with universal service objectives). As well as introducing industry-specific regulation, the Telecommunications Act 2001 included an obligation (Section 64) for the Commissioner to revisit before the end of 2003 whether Local Loop Unbundling (LLU) should be imposed. A key consideration for the inquiry was the potential effect of LLU on infrastructure investment and broadband uptake levels.

The LLU inquiry, conducted over the course of 2003, provided a platform for Telecom's competitors to increase their lobbying efforts. A key theme underpinning the submissions made to the Commission by TUANZ and Telecom's rivals was to implicate Telecom's position as the dominant firm in the industry (i.e. an absence of actual competition, measured by the market share of entrants) in New Zealand's sluggish broadband uptake rate. Rivals further asserted that LLU, by increasing the market share of competitors, would redress the situation[19] - the implication being that Telecom, due to its dominance had strategically restricted investment in and supply of broadband and priced it highly via the use of two-part tariffs. By contrast, Telecom's submissions included theoretical and empirical evidence from the New Zealand and international markets indicating lack of substantiated evidence of a supply side problem in the New Zealand broadband market (i.e. prices were low, availability widespread, entry had occurred using alternative technologies). Furthermore, the 'Kiwi Share' obligations – principally 'free local calling' – could satisfactorily explain observed uptake variations, and that even if there was a broadband uptake 'problem', there was no substantive evidence that LLU would have any material effect in addressing it[20]. Moreover, LLU would impede Telecom's plans to invest in a Next Generation IP-based network (the NGN).

Conditional Agreements and Political Threats of Legislative Action

After considering all the submissions, and conducting its own investigations (including a cost-benefit analysis), the Commissioner recommended against full LLU: "the overall benefits from unbundling are not sufficiently persuasive to satisfy the Commission that a regulated solution is warranted"[21]. Rather, the Commissioner instead accepted Telecom's offer to make available a regulated ADSL bitstream product. This product would enable competitors to provide a wider range of services, but would avoid the risk of competitors sinking investment into exchange equipment that would become stranded when exchanges were closed as the NGN became operational. Furthermore, Telecom's incentives to persevere with its investment in the NGN would be greater under bitstream unbundling than under full LLU. A condition of the Commissioner's acceptance of the offer was that Telecom should be obligated to meet targets of 250,000 broadband connections sold by the end of 2005, of which 33.3% would be sold by Telecom's competitors.

To the considerable dismay of Telecom's competitors[22], the Minister accepted the Commissioner's recommendations. Nonetheless, political lobbying continued, with notable effect as Telecom's market dominance became a key issue in the 2005 election campaign. The Labour Party manifesto stated "this Labour-led government has ended the destructive period of ultra-light handed regulation that stifled competition, growth and consumer choice in ICT markets" and promised to "closely monitor and enforce commitments made by Telecom New Zealand under the local loop unbundling decisions and ensure targets for broadband uptake for the next three years as outlined in the Digital Strategy are met".[23] Upon re-election, the threat of political action if predetermined levels of competition and broadband uptake did not emerge was reiterated in the Governor General's November 9 speech from the throne outlining the government's legislative programme: "with respect to ICT, my government will be advancing policies to ensure that the telecommunications sector becomes more competitive and that we achieve faster broadband uptake in line with our competitors".[24]

Whilst the government's acceptance of the cause of low broadband uptake being a consequence of the competitive environment is questionable, and is addressed in other papers (see Howell, 2008; Howell, 2007), the implications for Telecom and its rivals were clear. Telecom was on notice that legislative action would ensue if the targets set by the Commissioner were not met.

Competitors' Strategic Responses

Whilst Telecom had every incentive to meet the targets, the *ex ante* signalling to competitors of both the nature of Telecom's obligations and the 'notice' that if they were not met legislative action would follow, opened up new strategic opportunities for Telecom's competitors. Competitors faced few incentives to work with Telecom to sell bitstream broadband accounts, as doing so would facilitate Telecom in meeting its obligations to the Commissioner, thereby making further political intervention less likely to occur. Rather, in a two-stage game incorporating likely future legislative actions, it was in the competitors' interests in the short term to thwart Telecom's ability to meet the targets (regardless of the consequences for consumers). If the government's threat was credible, the likely legislative response to telecom's failure to meet targets would be imposition of full LLU. As securing full LLU was the rivals' overriding objective, Telecom's success would prevent the objective from being realised (in the event of the government taking legislative action even though Telecom met the targets, government integrity and political credibility would suffer, so Telecom's success would likely rule out LLU in the short to medium term horizon). By sacrificing short term market share, longer-term the benefits of gaining regulated access to Telecom's infrastructure were more likely to be secured.

Moreover, even if the political legislation threat did not exist, simple economics militated against Telecom's rivals actively marketing bitstream broadband accounts to existing dial-up internet customers. As Figure 3 shows, New Zealand broadband purchase almost exclusively occurs as a consequence of existing dial-up internet users substituting to broadband once a threshold of usage is reached Telecom's rivals had slightly more than 50% of the dial-up internet market in 2003 (Howell, 2003). The most popular Telecom broadband product was the entry level connection and low data cap bundle retailing at $29.95 per month, resold by competitors under a regulated wholesale agreement based on retail price minus a discount. Howell (2003) indicates that average monthly megabyte consumption was low, even for those consumers buying large usage bundles. As Miravete (2003) suggests consumers are quite adept at picking the most cost-effective two-part tariff for their usage, it was not surprising that most New Zealand broadband consumers purchased the low-cost, entry level products with low data caps (Howell, 2008a).

The national regulated bitstream access price was set at $27.87 per month for the best available service on the line[25] - that is, of equivalent speed to the service Telecom was offering its own customers. Due to Telecom's 1999 decision to price ADSL under two-part tariffs and (likely) discount connection below cost to induce substitution, (and as average per-account consumption levels of substituting New Zealand users were low – Howell, 2003), the margins available to competitors under the bitstream offers were so small that they would likely be worse off by encouraging their existing dial-up customers to substitute to ADSL. Even with the 'retail-minus' resold products, the margins were likely very small compared to the margins available from continuing to sell dial-up accounts (these were routinely sold at around $10 per month, and still generated ICA revenues at least in respect of own-network consumers with usage less than 10 hours per month which under the prevailing regulated ICA rates, would generate cash flows from Telecom of up to $11.30 per month). Thus, they had no incentive to sell broadband accounts to customers other than the small proportion of high-valuing individuals who opted for the more expensive high-volume packages, which generated bigger margins under the resale agreements[26].

The consequences of both the economic and legislative incentives are evident in Figure 4. When broadband uptake was low (pre 2004, when only very high-valuing individuals were purchasing), Telecom's rivals sold around 37% of accounts under the resale arrangements. However, when demand started burgeoning in 2004 and 2005 (New Zealand exhibited the 4th-highest growth rate in connections per capita in the OECD in 2005 and 2006 – OECD, 2007), despite having access to both bitstream and resale products, and over 50% of the addressable market already as dial-up customers, competitors' combined market share plunged, to a low of 19% in Q4 2004, and then stabilised at around 25% across 2005-2007. That is, contrary to theoretical expectations in the regulation literature, as access possibilities increased as a consequence of bitstream regulation being imposed, competition, measured as entrant market share, actually *decreased*. This otherwise aberrant observation is most plausibly explained by Telecom's competitors responding to the dual incentives of low margins as a consequence of historic pricing strategies and strategising to invoke the political threat of stronger future legislative action against Telecom.

Political and Commercial Consequences

When the Commissioner reported to the Minister on February 2 2006 that, by the end of 2005, Telecom had exceeded the target of connections sold by 11.6%, but only 24.5% of the connections were sold by its competitors[27], the government acted upon its threat. On May 1, the Cabinet voted to proceed with both full LLU and accounting separation of Telecom (MED, 2006). On April 5 2007 – some ten months before the first unbundled circuit was handed over to a competitor – the Minister announced that Telecom would be required to proceed to full functional separation of its network, wholesale and retail divisions[28].

As a consequence of the 2006 LLU legislation, Telecom's ability to forcibly tie a Telecom-provided and billed fixed line voice calling connection to a Telecom-provided ADSL line sold by either Telecom or any other provider under wholesale, bitstream or LLU arrangements was broken. Competitors can now sell Telecom-provided fixed line voice services as well as ADSL, enabling them to offer similar pure bundles. By actively marketing to fixed line customers to induce switching of the voice account, competitors get access to the revenues derived from the 'flat rate' connection and calling bundle. If the customer already has a broadband account, then the rival gets access to the surplus of voice revenues over cost that Telecom previously used to (potentially) subsidise the (below-cost) broadband account. That is, the margins garnered over cost for fixed voice line services are now making it profitable for competitors to engage in the provision of broadband services, either by resale, bolting on equipment into Telecom's exchanges (LLU) or providing the service over their own infrastructures (e.g. cable, mobile and wireless). These high margins exist solely due to the historic obligation on Telecom to offer flat-rate voice telephony tariffs. That Telecom charged low ADSL prices in the first place, thereby necessitating competitor entry into the fixed line voice market to stimulate competition

Figure 4. Source data Telecom Management Commentaries

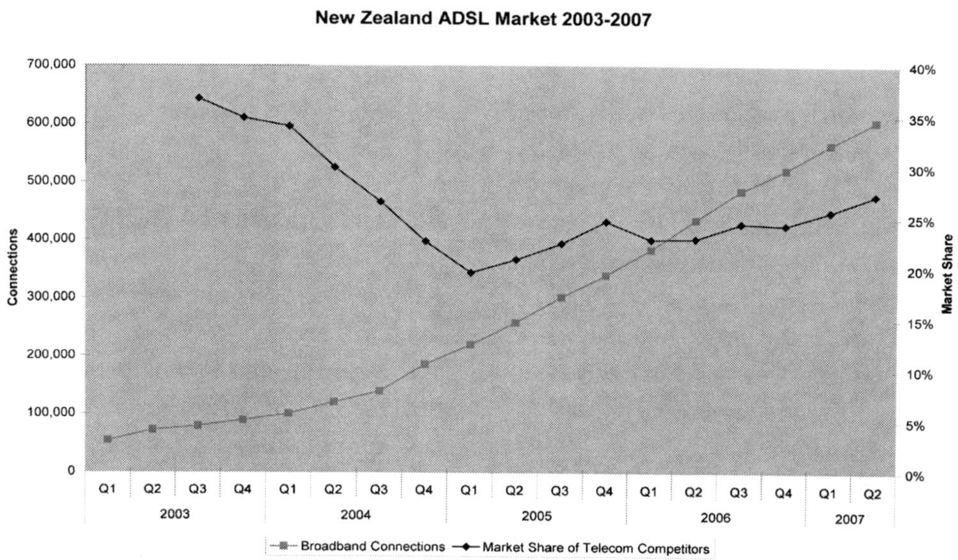

in the bundled voice-ADSL market, is also directly an artefact of the same mandatory tariff policy.

The pattern of strategic interaction evidenced in the New Zealand market leads to the conclusion that further distortions will occur as a consequence of the adverse selection possibilities that the historic tariff structure has enabled. Competitors will find it financially most profitable to target high-volume existing internet users to substitute their fixed line local voice calling away from Telecom. As competitor ISPs have historically served over 50% of dial-up internet customers (all major ISPs are now either structurally integrated or strategically aligned with a competing infrastructure provider), they have information regarding usage patterns of their existing customers. By targeting only those customers with high internet usage to switch to the new bundles, but encouraging the remainder of low-volume dial-up internet users to remain on Telecom fixed line services, both bundling benefits and ICA revenues of competitors are maximised. Telecom will be left with a disproportionately large share of the low-volume internet users and voice consumers with no internet demand at all. With more of the newly 'low-cost' broadband-purchasing voice subscribers leaving to join other providers, taking their 'notional subsidies and discretionary ADSL payments with them, Telecom has little choice but to raise its prices to the remaining high-cost voice-only and dial-up internet-using customers. Such adverse selection is exacerbated by selective competitor entry in low-cost urban areas, making it even more difficult for Telecom to meet its universal service obligations and for the regulator to ascertain how to allocate the annual entrant tax obligations via the TSO. After minimal changes to the residential fixed line charge 1990 and 2007, Telecom has raised its prices twice in the twelve months following LLU and separation becoming operational.

Perhaps unsurprisingly, given the pre-existing conditions in the New Zealand market, unlike other jurisdictions, competition in the more highly regulated market post 2006 has not been based around competition for broadband customers *per se*. Rather, competition for broadband customers has been couched in competition for fixed line voice customers, but predominantly those with high levels of existing internet usage (indeed, low volume dial-up ISP accounts are being advertised again by some providers). Practically all firms now market bundles of voice and broadband internet access, with broadband internet access alone being charged at a premium of $10 per month more if not purchased with a fixed line voice account and long-distance calling package. Reminiscent of past years, more extensive free local calling zones than Telecom's are being used by at least one competitor as an inducement for fixed line voice consumers to switch. Interestingly, no competitor is offering a two-part tariff for fixed line voice services.

Ironically, the substantial changes in market share engendered by real competition for fixed line voice customers reignites the debate over the 'universal service' aspects of the 'Kiwi Share'. If margins previously garnered by Telecom in the voice market to offset investment in its infrastructures and to underwrite its own universal service pricing of ADSL services across the country irrespective of the costs of providing those services are now transferred to competitors, Telecom will be unable to continue offering a universal broadband price and still continue to provide equivalent quality services in high cost areas.

FUTURE IMPLICATIONS

Asymmetric assignment of the 'Kiwi Share' obligations, and the strategic responses to them by both Telecom and its competitors, have clearly had a significant effect in shaping the New Zealand telecommunications sector as it exists today. Regardless of the nature of the legislative infrastructure prevailing – either competition law-based or industry-specific regulation – the

'Kiwi Share' provisions have been the underlying causes of strategic tensions, the outcomes of which have determined both the nature of competitive interaction undertaken and the extent of political intervention observed. The asymmetric nature of the obligations has undoubtedly created many more potential opportunities for strategic interaction, arbitrage and conflict than if they did not exist. The consequence of their presence has been a very different pattern of strategic interaction in New Zealand than has been observed in other countries where more conventional industry-specific regulation, arbitrated by a regulator, has prevailed.

The 'Kiwi Share' obligations have been unique to the New Zealand market, and the conflicts that have emerged have also been unique to the circumstances in which they emerged. Nonetheless, when conflicts or enigmas in sector performance have emerged, the response of both the industry participants and policy-makers has been to look 'downward' in the institutional hierarchy to find 'causes' and hence 'solutions' for the enigmas in the firms, markets and access regulations, rather than 'upwards' to institutions embodied in the 'Kiwi Share'. Whilst there has been a complete reversal in New Zealand's industry regulations, from the OECD's most liberal competition law-predominant governance in the 1990s, to one of the most radical industry-specific regulatory regimes with mandatory functional separation in the late 2000s, the 'Kiwi Share' provisions have remained essentially unchanged.

Despite theoretical and empirical economic evidence of the incompatibility of universal service with competitive outcomes, and a body of literature detailing the effects of tariff structure on technology diffusion, the embeddedness of the 'Kiwi Share' provisions in the New Zealand institutional framework appears to have rendered them beyond the scope of consideration as possible explanators of observed strategic interactions and outcome. Yet as this paper shows, they have been pivotal. Although Williamson (2000) observes that such embedded institutions themselves may be long-lived, the asymmetric contracts that embody them as obligations on Telecom alone need not be immutable. Indeed, the extent to which the 'universal service' obligation has been made a charge on the industry as a whole via the TSO reflects the ability to bring about changes.

A question remains, however, about the sanctity of the 'free local calling' obligation. Its economic implications for the sector have arguably been greater than those of 'universal service', yet it has been specifically excluded from consideration in the latest review of services. Its sanctity has likely contributed to the proclivity to seek institutional solutions to the problems it has engendered, rather than addressing its features as a potential cause of the problems evidenced. That the institutional solutions adopted so far have failed to adequately address the 'problems', and have led to the emergence of even more complex problems that are even less likely to be addressed by the solutions adopted, is unsurprising.

Whilst it might be hoped that the 'free local calling' obligation will be eventually resolved by the replacement of traditional PSTN services with internet-based voice telephony services, its spectre remains in the form of mandatory 'flat rate' broadband tariffs. Aside from the difficulties of mandating flat-rate tariffs for one application on a multi-application platform, the New Zealand experience suggests that extensive mandatory and voluntary adoption of flat-rate tariffs poses challenges for the strategies of both firms and regulators as traffic volumes become more asymmetric between networks (e.g. with extensive use of video streaming as a substitute for broadcast television). The New Zealand lesson for the rest of the world is that embedding 'flat rate' broadband tariffs into broadband market cultures may be both distorting and costly, and potentially even more dangerously, overlooked as an underlying cause of 'problems' that emerge, simply because of common acceptance of their sanctity. In the New Zealand case, the extent of the distortions was exacerbated by the regulatory imposition of tariff structures that prevented market-based adjustments. The hope is that, in more competitive broadband markets, the temptations to impose asymmetric retail tariff obligations are less pressing.

In sum, therefore, the New Zealand case study serves to further reinforce that asymmetric obligations on an incumbent, be they embodied in legislation, regulation or contract, do not create a 'level playing field' on which competition can play out. Whilst they may exclude the incumbent from adopting some strategic options that might otherwise be available, they open up the possibility of other strategic options to its entrant-competitors that would not otherwise exist. The ensuing interaction will not necessarily conform to textbook patterns of behaviour or outcomes, and will be contingent upon the emergence of other exogenous factors (e.g. technologies) that affect the balance of powers in the industry. Dynamic competition is an inherently unbalanced game of strategic interaction by parties each striving to gain the upper hand. If asymmetric obligations provide the opportunity, either party will utilise the asymmetry to gain competitive advantage, as has been evidenced in the case of New Zealand's 'Kiwi Share'. If legislation, regulation and contract design fail to take account of the strategic possibilities invoked, the outcomes may be very different from those expected.

REFERENCES

Armstrong, M. (2001). Access pricing, bypass and universal service. *The American Economic Review, 91*(2), 297–301. doi:10.1257/aer.91.2.297

Blanchard, C. (1995). Telecommunications regulation in New Zealand: light-handed regulation and the Privy Council's judgement. *Telecommunications Policy, 19*(6), 456–475. doi:10.1016/0308-5961(95)00024-Z

Boles de Boer, D., Enright, C., & Evans, L. (2000). The performance of internet service provider (ISP) markets of Australia and New Zealand. *Info, 2*(5), 487–495. doi:10.1108/14636690010801672

Boles de Boer, D., & Evans, L. (1996). The economic efficiency of telecommunications in a deregulated market: The case of New Zealand. *The Economic Record, 72*(216), 24–39. doi:10.1111/j.1475-4932.1996.tb02606.x

Commerce Commission. (2003). Telecommunications Act 2001 Section 64 review and schedule 3 investigation into unbundling the local loop network and the fixed public data network Final Report. Wellington, New Zealand. Retrieved from http://www.comcom.govt.nz/Industry Regulation/Telecommunications/I nvestigations/ LocalLoopUnbundling/ContentFiles/Documents/ finalreport.PDF

Covec (2004). Benchmarking Telecom's UPC service. Retrieved from http://www.comcom.govt.nz//IndustryRegu lation/Telecommunications/Investigations/Unbundling-PartialCircuits/ ContentFiles/Documents/Final%20Draft%20 Report%20(August%204%202004)%20-%20 UPC%20Benchmarking.pdf).PDF

Dixit, A., & Nalebuff, B. (1991). *Thinking strategically: the competitive edge in business, politics and everyday life*. New York, London: W. W. Norton & Company.

Economides, N., & White, L. (1995). Access and interconnection pricing? How efficient is the "efficient component pricing rule"? *Antitrust Bulletin, 40*(3), 557–579.

Enright, C. (2000). *Strategic behaviour of internet service providers in New Zealand and the performance of this market.* Unpublished paper based on Masters thesis, New Zealand Institute for the Study of Competition and Regulation, Wellington, New Zealand. Retrieved from http://www.iscr.org.nz/f240,4833 /4833_enright_210600.pdf

Evans, L., Grimes, A., Wilkinson, B., & Teece, D. (1996, December). Economic reforms in New Zealand 1984-1995: the pursuit of efficiency. *Journal of Economic Literature*, *34*, 1856–1902.

Evans, L., & Quigley, N. (2000). Contracting, incentives for breach and the impact of competition. *Journal of World Competition*, *23*(2).

Farrell, J. (1996). Creating local competition. Retrieved from http://law.indiana.edu/fclj/pubs/v49/no1/farrell.html

Hausman, J. (2002). Asymmetric regulation of related services. In Crandall, R., & Alleman, J. (Eds.), *Broadband: should we regulate high-speed internet access?* (pp. 129–156). Washington, DC: AEI-Brookings Joint Center for Regulatory Studies.

Howell, B. (2003). *Building best practice broadband: bringing infrastructure supply and demand together.* Wellington, New Zealand: New Zealand Institute for the Study of Competition and Regulation. Retrieved from http://www.iscr.org.nz/f213,4379/4379_ building_best_practice_060603.pdf

Howell, B. (2007). *A Pendulous Progression: New Zealand's Telecommunications Regulation 1987-2007*. Wellington, New Zealand: New Zealand Institute for the Study of Competition and Regulation. Retrieved from http://www.iscr.org.nz/research

Howell, B. (2008). *From competition to regulation: New Zealand's telecommunications sector performance 1987-2007.* Paper presented at the International Telecommunications Society European Regional Conference, Rome, Italy. Retrieved from http://www.iscr.org.nz/f439,12724/12724_From_Competition_to_Regulationbh230708.pdf

Howell, B. (2008a). *The role of price structure in telecommunications technology diffusion.* Paper presented at the International Telecommunications Society European Regional Conference, Rome, Italy. Retrieved from http://www.iscr.org.nz/research

Howell, B., & Obren, M. (2003). *Telecommunications usage in New Zealand 1993-2003. Broadband diffusion*. Wellington, New Zealand: New Zealand Institute for the Study of Competition and Regulation. Retrieved from http://www.iscr.org.nz/f212,4362/4362_nz_telecommunications_usag_090703.pdf

Howell, B., & Sangekar, M. (2009). Beyond Surface Similarities: Telecommunications Industry Structure Evolution in Finland and New Zealand. *Prometheus*, *27*(2), 99–115. doi:10.1080/08109020902895243

Karel, A. (2003). *The development and implications of free ISPs in New Zealand.* Wellington, New Zealand: New Zealand Institute for the Study of Competition and Regulation. Retrieved from http://www.iscr.org.nz/f209,4313 /4313_Free_ISPs_220903.pdf

Laffont, J.-J., & Tirole, J. (2002). *Competition in telecommunications*. Cambridge, Massachusetts: MIT Press.

Melody, W. (2002). *Building the regulatory foundations for growth in network economies*. World dialogue on regulation for network economies discussion paper #0201.

Milgrom, P., & Roberts, J. (1992). *Economics, organization and management*. Englewood Cliffs, New Jersey: Prentice-Hall.

Ministry of Commerce & The Treasury. (1995). *Regulation of access to vertically-integrated natural monopolies*. Wellington, New Zealand: Ministry of Commerce; The Treasury. Retrieved from http://www.med.govt.nz/templates /Multi-pageDocumentPage____4560.aspx

Ministry of Economic Development. (2006). *Promoting competition in the market for broadband services*. Wellington, New Zealand: Ministry of Economic Development. Retrieved from http://www.med.govt.nz/uplo ad/36537/promoting-competition.pdf.

Miravete, E. (2003). Choosing the wrong calling plan? Ignorance and learning. *The American Economic Review*, *93*(1), 287–310. doi:10.1257/000282803321455304

New Zealand Institute of Economic Research. (2005). *Telecommunications pricing in New Zealand: a comparison with OECD countries*. Wellington, New Zealand: NZIER.

Organization for Economic Co-Operation and Development. (2000). *Local access pricing and e-commerce*. Paris: OECD.

Organization for Economic Co-Operation and Development. (2001). *The development of broadband access in OECD countries*. Paper DSTI/ICCP/TISP(2001)2/FINAL. Paris: OECD. Retrieved from http://www.oecd.org

Organization for Economic Co-Operation and Development. (2007). *Communications Outlook 2007*. Paris: OECD. Retrieved from http://www.oecd.org

Organization for Economic Co-Operation and Development. (2008). *Broadband growth and policies in OECD countries*. Paris: OECD. Retrieved from http://www.oecd.org

Spiller, P., & Cardilli, C. (1997). The frontier of telecommunications deregulation: small countries leading the pack. *The Journal of Economic Perspectives*, *11*(4), 127–138.

Telecommunications Carriers' Forum. (2008). *Report on the Telecommunications Service Obligation for local service*. 23 July 2008. Retrieved from http://www.tcf.org.nz/library/df101bd9-0668-452d-9227-e9e3b319d95d.cmr

Williamson, O. (2000). New institutional economics: taking stock, looking ahead. *Journal of Economic Literature*, *38*, 595–613.

Wilson, A. (1994). *Wire and wireless: a history of telecommunications in New Zealand, 1860-1987*. Palmerston North, New Zealand: Dunmore Press.

ADDITIONAL READING

Especially relevant theoretical pieces are:Armstrong, M. (2001). Access pricing, bypass and universal service. *The American Economic Review*, *91*(2), 297–301. doi:10.1257/aer.91.2.297

Dixit, A., & Nalebuff, B. (1991). *Thinking strategically: the competitive edge in business, politics and everyday life*. New York, London: W. W. Norton & Company.

Economides, N., & White, L. (1995). Access and interconnection pricing? How efficient is the "efficient component pricing rule"? *Antitrust Bulletin*, *40*(3), 557–579.

Howell, B. (2003). *Building best practice broadband: bringing infrastructure supply and demand together.* Wellington, New Zealand: New Zealand Institute for the Study of Competition and Regulation. Retrieved from http://www.iscr.org.nz/f213,4379/437 9_building_best_practice_060603.pdf

New Zealand-specific factors are discussed extensively in: Howell, B. (2007). *A Pendulous Progression: New Zealand's Telecommunications Regulation 1987-2007*. Wellington, New Zealand: New Zealand Institute for the Study of Competition and Regulation. Retrieved from http://www.iscr.org.nz/research

Laffont, J.-J., & Tirole, J. (2002). *Competition in telecommunications*. Cambridge, MA: MIT Press.

Melody, W. (2002). *Building the regulatory foundations for growth in network economies*. World dialogue on regulation for network economies discussion paper #0201.

Milgrom, P., & Roberts, J. (1992). *Economics, organization and management*. Englewood Cliffs, New Jersey: Prentice-Hall.

Williamson, O. (2000). New institutional economics: taking stock, looking ahead. *Journal of Economic Literature, 38*, 595–613.

Wilson, A. (1994). *Wire and wireless: a history of telecommunications in New Zealand, 1860-1987*. Palmerston North, New Zealand: Dunmore Press.

KEY TERMS AND DEFINITIONS

Competition Law: A form of over-arching industry governance whereby the legitimate means by which firms may interact in competitive processes are specified. Breaches are usually pursued in court-based anti-trust actions: where the plaintiff is required to demonstrate that the defendant with market power has acted unilaterally to exert its market power in a manner other than that exercised by a competitive firm in a competitive market.

Industry-Specific Regulation: A form of over-arching industry governance whereby the market power of a dominant firm is constrained by a specific set of prescriptive prohibitions or conditions determined by an industry regulator. Provisions and prohibitions are typically imposed following an extensive politically-based lobbying process: so may or may not be consistent with longer-term objectives of increased welfare or competitive market outcomes.

Universal Service: The requirement that all consumers are charged the same price for a good or service, regardless of any differences in the costs of providing the good or service to the different customer groups.

Price Cap: A maximum retail price set for the provision of a good or service. Typically set by an industry-specific regulator.

Two-Part Tariff: A tariff set such that there is a fixed component, usually for access to an infrastructure service, and a variable component proportional to the amount of usage of the infrastructure services.

Flat-Rate Tariff: A special case of a two-part tariff whereby the usage component is set to zero.:

Interconnection: The agreement between two network operators enabling customers of one network to enter into exchanges with customers of the other network.

Local Loop Unbundling: A regulatory requirement whereby the owner of the bottleneck local loop of telecommunications network is required to lease access to competitors in order to enable them to compete in the provision of services to end consumers. May also include arrangements whereby competitors are allowed to attach some of their own components to the regulated firm's network in order to provide differentiated services.

Functional Separation: A regulatory obligation imposed upon a dominant telecommunications firm whereby it is required to separate its network, wholesale and retail operations into functionally separate firms which are unable to interact with each other in any manner that is different from its competitors and customers (i.e. the firm is unable to use shared ownership as a means of coordinating activities).

ENDNOTES

* It cannot be discounted that the Government was so poorly informed that it had no conception of these ramifications. However, accepting this premise requires the additional assumption that the government's policy advisers were largely ignorant of the economic fundamentals of both regulation

and competitive market operation. Given that the government's overriding policy objective since 1984 had been the economic reform of the New Zealand economy (Evans, Grimes, Wilkinson & Teece, 1996), there is little apparent justification for accepting this hypothesis.

1 Although the obligations for funding the 'Kiwi Share' obligations were subsequently incorporated into the Telecommunications Act, where they were renamed the Telecommunications Service Obligation ('TSO'), the contractual obligations with respect to Telecom's pricing obligations remain in place.

2 Changes to prices (outside these guidelines) or any of the other terms, required the approval of the Minister of Communications.

3 At their own expense, even though the Post Office appropriated the assets paid for by residents under the petition system.

4 Personal communication with an opposition Member of Parliament with responsibility for telecommunications policy.

5 It is noted that these assumptions rely upon a static technological environment, and that there were undoubtedly productive inefficiencies within the firm that remained to be competed away. However, given that the firm was sold in a fully deregulated environment, and privatization would undoubtedly eliminate some of these inefficiencies, the price at which the firm changed ownership would have reflected some middle ground between each party's assessment of the value of potential savings. Any margins remaining for Telecom's shareholders once the productive efficiency gains were realized would be expected to be competed away when entry occurred.

6 Clear was subsequently merged into Telstra-Saturn, a company formed by the purchase by Telstra of Saturn Communications, to form TelstraClear. (Howell, 2007:17).

7 Clear Communications v Telecom Corporation (1993) 5 TCLR 166 (HC) 25, 27, 35, 103

8 Clear Communications v Telecom Corporation (1993) 5 TCLR 413 (CA) 25

9 Telecom Corporation v Clear Communications [1994] 5 NZBLC 103, 552 (PC); [1995] 1 NZLR 385 (PC) passim

10 Despite the implications of its title, TUANZ is an association comprised principally of Telecom's wholesale (i.e. competitor) customers and Internet Service Providers (ISPs). It has no mandate to represent Telecom's retail customers.

11 The Opposition at the time was comprised largely of members of the previous government, which had severed the ANZUS strategic military alliance with the United States on the basis of its inconsistency with New Zealand's anti-nuclear policy.

12 http://www.med.govt.nz/templates/Page____16432.aspx#tor

13 http://www.med.govt.nz/templates/MultipageDocumentTOC____16484.aspx

14 These recommendations were made despite an absence of any compelling evidence that the New Zealand industry had performed any worse than the regimes imposing the types of regulation recommended, and in particular in the presence of clear evidence that by dint of the 'price cap', real residential telephone prices had fallen by a greater margin in New Zealand than the OECD average (Howell, 2008).

15 http://www.scoop.co.nz/stories/BU0804/S00241.htm

16 It is no coincidence that TUANZ has a separate membership division catering solely for operators of call centres, as such customers would be most beneficial to small, new competitors to a large, incumbent network such as Telecom.

17 http://www.med.govt.nz/templates/MultipageDocumentPage____4850.aspx

18 The Commission has indicated its intention to appeal the Court of Appeal decision in the Supreme Court. http://www.nbr.co.nz/article/com-com-seeks-supreme-court-appeal-against-telecom-0867-ruling-109703

19 http://www.comcom.govt.nz/IndustryRegulation/Telecommunications/Investigations/LocalLoopUnbundling/submissions.aspx

20 The author acknowledges that her early work in this area was extensively cited by Telecom in submissions presented.

21 Executive Summary, (v). http://www.comcom.govt.nz//IndustryRegulation/Telecommunications/Investigations/LocalLoopUnbundling/ContentFiles/Documents/Finalreportexecutivesummary.PDF

22 See, for example, iHug's submission to the Minister on the matter http://www.med.govt.nz/upload/5898/tcl-rsp-to-comcom-llu-rpt-submission040209.pdf

23 http://www.labour.org.nz/policy/jobs_and_economy/2005policy/Pol05-Comms/index.html

24 http://www.scoop.co.nz/stories/PA0511/S00104.htm (Despite successive attempts in June 2007 to retrieve the official version from http://www.dia.govt.nz/Pubforms.nsf/NZGZT/Speech187Nov05.pdf/$file/Speech187Nov05.pdf, it could not be retrieved.) This source appears to have reproduced the text complete, but this fact cannot be verified.

25 http://www.comcom.govt.nz//IndustryRegulation/Telecommunications/Wholesale/WholesaleDeterminatons/ContentFiles/Documents/Bitstream%20Determination%20Decision%20568.pdf

26 http://computerworld.co.nz/news.nsf/fry/C047FEC66E9B47B3CC25723900423E2A

27 http://www.comcom.govt.nz/IndustryRegulation/Telecommunications/telecommunicationsactcompletionofb.aspx

28 http://www.med.govt.nz/upload/45925/11.pdf

Chapter 3
Examining the Influence of Government Policy on Broadband Internet Access: The Case of China

Qiuyan Fan
University of Western Sydney, Australia

ABSTRACT

Governments worldwide are driving initiatives to deploy broadband infrastructure and services. China has topped the world broadband market with the total broadband user base exceeding 320 million (CNNIC, 2009). However, a closer analysis indicates that broadband in China is less advanced in terms of penetration rate, access speed and price. Why China is lagging? Despite the fact that broadband performance and adoption are determined by both non-policy factors and policy factors, the OECD experience shows that smart national broadband policies are the key to broadband success. This chapter reviews and analyzes relevant regulations and government policies pertaining to broadband development in China using a multiple perspective model. The primary goal of this chapter is to examine how various policy factors interplay to affect broadband deployment and performance in the case of China. This research shows that the Chinese approach to broadband development has not produced desirable results. Several factors affect the growth of broadband. The most significant is lack of a detailed national broadband plan with clearly stated goals, performance targets and action plans for implementation. China's regulatory arrangement and approaches to regulating broadband industry have failed to establish a well functioning and competitive market and to drive high levels of broadband performance, investment and innovation.

INTRODUCTION

Broadband has been considered as a way of boosting national competitiveness and bringing new opportunities to individuals and businesses in China over the past decade. China's broadband sector has experienced an exceptional annual growth rate of 80 percent in the last three years. As of 30 June 2009, the broadband users reached 320 million while the Internet users reached 338 million

DOI: 10.4018/978-1-60960-011-2.ch003

(CNNIC, 2009). While the number of broadband users is an important metric, a more complete measure of a country's position in broadband would consider penetration rate, access speed and cost. The China Internet Network Information Centre (CNNIC) reported that the broadband penetration rate in China was less than 25% as of 2009 (CNNIC,2009) while the Information Technology and Innovation Foundation (ITIF) data showed that more than 50% of households subscribed to broadband among the 30 OECD countries as of 2007 (Atkinson, Correa & Hedlund, 2008). Eighty percent of Chinese broadband users use ADSL access technology at a speed of less than 2Mbps, compared with the average download speed of 9.2Mbps in the OECD member countries (Benkler, 2009). The fastest broadband connection widely available in Japan is 100 mbps fiber optic service at a price of about $27 per month (Correa, 2007). In contrast, on average, a Chinese broadband user pays $25 per month for 1.5mbps ADSL service. With poor broadband coverage and performance, China today still lags behind in broadband internationally despite the fact that it has topped global broadband market with the total broadband user base exceeding 320 million.

Access to high-speed and always-on of the broadband network is determined by a combination of the widely available telecommunication infrastructure and the accessibility and affordability of broadband services, which are closely related to government policies (IT Group, 1999). Other literature suggests that the underlying reason for broadband Internet access can be traced back to the policy factors in each country, because government policies governing the telecommunications service market and promoting information infrastructure have a significant impact on the performance of broadband Internet access.

In common with other national governments, the Chinese government recognizes that rapid roll-out and adoption of broadband access across China is critical to long term economic development. Many national approaches to promoting ICT development have been taken since the 1990s. How have those strategic and regulatory developments impacted on broadband Internet access? To answer this question, this research investigates the relationship between policy factors, market environment and broadband access by using a multiple perspective model.

The minimum necessary condition for participation in this information age is access to adequate network infrastructure. Without access to global communication networks, no country can participate in the networked world. Although much of the press coverage on the information economy has focused on applications such as e-health, e-learning, e-government and e-commerce, these content rich and data intensive applications will not realise its full potential unless quality broadband is widely accessible and more and more household adopt broadband. Correa (2007) states that 'in the digital economy it is the speed and capacity of the network that matters' (p. 1). There are many studies focusing directly on regulatory impacts on broadband supply such as Fan (2008), Howell (2008) and Prieger and Lee (2008). Fan (2008) looks at the influence of Australia's government policy on broadband Internet access while Howell (2008) examines the effect of some regulatory factors on broadband diffusion rates in the case of New Zealand. Using U.S. data covering all areas served by major broadband providers, Prieger and Lee (2008) investigate the impact of telecommunications regulatory policy on broadband availability and the role of incentive regulation and local telecommunications competition policy in stimulating broadband deployment. This study is primarily looking at how broadband infrastructure and services have been provided in China and how the Chinese government policies can help to develop Internet infrastructure and promote affordable and equitable broadband Internet access in China. The objectives of this chapter are to examine how policy factors influence broadband access and identify what role the government policies have played in promoting broadband deployment and adoption in the case of China.

The following section will briefly discuss a framework for analysis followed by an in-depth examination of the impact of the government policies, including national strategies, regulatory frameworks and competition, on broadband Internet access in the case of China.

A FRAMEWORK FOR ANALYSIS

Governments around the world have taken a multi-faceted approach to promoting broadband development, including reliance on market forces, regulations and grants program as well as national broadband policies. Figure 1. shows how the various policy elements interplay to affect broadband service deployment and performance. For instances, the elements of regulations, market forces and national strategies have a direct impact on broadband supply and access while regulations, industry structure and national policies affect market forces directly. Regulations, national broadband strategies have some influence on telecom industry structure.

This study attempts to use the above multiple-perspective model (Figure 1) to examine the influence of the Chinese government policies on broadband Internet access. The following sections discuss how various government efforts and actions China has taken affect broadband deployment and performance.

Regulations of Telecommunications and Broadband Internet Access

China's telecommunications and Internet sector is largely governed by Telecommunications Regulation of the Peoples' Republic of China, which was passed at the Thirty-first Session of the Standing Committee of the State Council held on September 20, 2000. According to the regulation, telecommunications service is categorised into basic telecommunications service and value-added telecommunications service. Basic telecommunications service means the provision of public network infrastructure, the transmission of public data and basic voice communication. Value-added telecoms service means the provision of telecommunications and information services through public network. The provision of the transmission network infrastructure and the Interconnecting networks fall into the category of basic telecommunications service. The service licences are to be approved by State Council and restricted to State controlled enterprises. Specifi-

Figure 1. Model of the policy factors and broadband service deployment

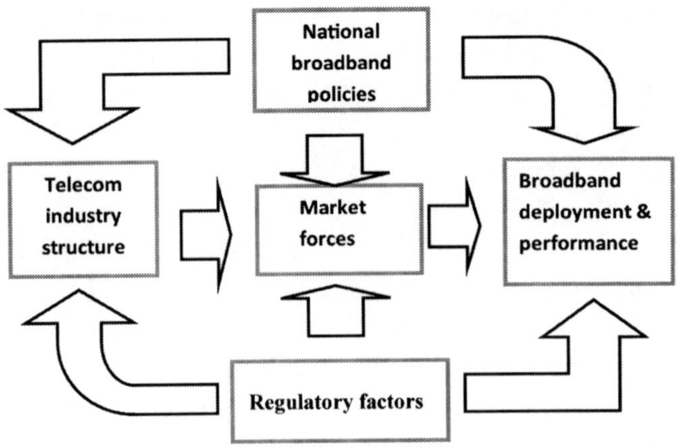

cally, the regulation prescribes that China owns no less than 51% of the stakes.

Access networks are categorised as value-added telecommunications services. Any Chinese corporation can be licensed as an ISP by the Ministry of Industry and Information Technology (MIIT) as long as it meets legal, technological and financial requirements stipulated in the Telecommunications Regulation of the People's Republic of China. Importantly, interconnection rules are laid down in the telecommunications regulation, which requires that telecommunications networks shall be interconnected on the basic principles of technological feasibility, economical reasonableness, fairness and justice, and reciprocal cooperation. A major telecommunications carrier should not refuse the request of other telecommunications carriers and dedicated network operators for interconnection. In the case where a network interconnection agreement is not arrived at through negotiation, MIIT is responsible for coordination.

The regulation requires that all ISPs have to gain global Internet access through one of the Interconnecting Networks and those network providers must lease their international links from China Telecom, China Unicom or China Mobile. These regulatory arrangements have the consequence of limiting competition at broadband backbone networks. China Telecom and China Unicom together have over 95% of China's international outlet bandwidth and enjoy their huge market power with around over 90% market share. As a result, the cost of an ISP access to the backbone networks is relatively expensive in China. The connectivity cost is a major component of the costs that ISPs pass along to subscribers. Higher prices from the carriers directly translate into higher price for users, which in turn retards broadband adoption. Jew and Nicholls (2000) argued that the level of pricing and product innovation that is available to consumers would be impeded without competition at backbone layer.

In addition to fair regulation, an independent regulator is considered by the international telecom community to be one of the three fundamental requirements for the introduction of a liberalised and competitive telecommunications environment (Lovelock, 1999). According to the telecommunications regulation, MIIT is entitled to oversee and control the telecommunications industry of China. As a government department, MIIT is responsible for strategy planning, policy-making and overall regulation of the information industry. This status makes it difficult for MIIT to be an independent regulator without any political obligation and political intervention. Therefore China does not have an independent regulatory agency in a real sense.

As Kennard (1999) points out, establishing an independent regulatory authority is a crucial factor in the success of any country's effort to introduce competition and to liberalize the telecommunications sector. OECD experience has shown that more effective regulation can result where there is a certain degree of structural independence allowing the regulator to implement its regulatory mandate without any political intervention. In the case of China, the government is essentially the only real basic telecommunications service provider as all telecom carriers are state owned. 'It is for this reason that WTO argues that an independent regulatory agency is particularly important where the government retains any ownership and/or control of domestic telecommunications carriers' (Lovelock, 1999). This less competitive telecommunication environment also discourages private investment in the broadband industry. Although foreign investment in telecommunications service enterprises has become legally possible since 2001, very few foreign owned broadband service providers are operating in China. There is definitely a need to establish an independent regulator in order to ensure a fair and transparent competitive environment in China.

Regulatory Initiatives and Market Forces in Broadband Industry

The Telecommunications sector, in particular the public networks, had been a monopoly of the Ministry of Post and Telecommunications (MPT) until in the early 1990s the Chinese government introduced competition policy to the telecommunication infrastructure. In an attempt to promote competition in China's broadband Internet access and telecommunications market, the government has implemented regulatory initiatives and undertaken four rounds of industry restructure since 1994.

In 1993, the State Council approved the setting up of the JiTong Communications Corporation, which was responsible for the 'Golden Bridge' VSAT–based ISDN backbone. The Golden Bridge network would be interconnected with the MPT data network and private networks and be connected to the Internet via satellite. In 1994, the state council approved the setting up of Unicom, the second state owned player. It was run by a consortium of rival ministries and operates all telecommunications services except the international service. The formation of two new enterprises outside MPT to provide advanced data communication services and long-distance voice service represents a milestone in the development of China's telecommunications sector.

The State Council decided to restructure China Telecom and Unicom in 1999. As a result, China Telecom was split into China Mobile, China Telecom (for fixed line services) and China Satellite. In 2000, the China Communications Network Corporation (CCNC) was founded by the Chinese Academy of Science (CAS), the Shanghai government, the Ministry of Railways, and the State Administration of Radio, Film, and Television (SARFT). By the end of 2000, a new competitive structure of networks has emerged in telecommunications infrastructure market. China Mobile, China Telecom (for fixed line services), China Satellite, China Netcom, China Jitong and China Unicom were licensed by the Ministry of Information Industry (former MIIT) to supply telecommunication and Internet service. However, China Unicom was the only operator who was allowed to offer a full range of telecom services such as cellular, paging, long distance, satellite, data and Internet services.

As the country joined the World Trade Organization, the Government made a decision on further reforms in China's telecommunication industry. In December 2001, the State Council approved China Telecom's restructuring plan, under which China Telecom was split into the north sector and the south sector. The south sector has the 21 provincial (municipal and autonomous regional) corporations, holding 70% of the national trunk-line transmission network assets owned by the former China Telecom in the south and northwest of the country and operates under the name China Telecom. The north sector covers former China Telecom's operations in the 10 provinces in the north and the northeast of the country and regroups with the former China Netcom and Jitong Telecom to form a new China Netcom Group. The two new groups, China Telecom Group (China Telecom) and China Network Communications Group, (China Netcom) were officially established in May 2002. They retain the existing business scope of the former China Telecom and are allowed to build local networks, operate in areas under each other's regime.

In May 2008, the MIIT regrouped the telecommunication service providers. China Telecom acquired China Satellite and the CDMA mobile network of China Unicom. China Unicom retained its GSM mobile operations and merged with China Netcom to form China United Network Communications Corporation. China Mobile Group took over the China Railway Communications Group. As a result, the number of carrier is decreased from 6 in 2002 to 3 in 2009. All of these three firms were granted 3G licenses with using different technology and allowed to offer a full range of telecommunication services including fixed line,

3G wireless and data and broadband service from January 2009. A decline in the number of licensed carriers indicates that the government regulatory initiatives and competition policy have failed to address market concentration, which will have an impact on competitive entry, investment and innovation in broadband.

Industry Structure and Broadband Market Environment

The limited nature of competition in China has often been attributed to industry structure issues. As the result of the industry restructuring, China Telecom and China Unicom currently have control over key existing facilities and the local access networks. The two companies dominate the Chinese broadband services market, with a combined market share of over 90%. Both China Telecom and China Unicom are not only the major backbone services operators at the wholesale level but also the Internet service providers at the retail level. As Budde pointed out, the vertically integrated model protects the backbone providers from competition, and it allows them to drive up Internet access costs for other independent ISPs (Budde, 2002).Consequently smaller ISPs are not able to compete in the price and quality of broadband services due to market dominance and market power, and Chinese users have to access to broadband services at lower connection speed and higher prices. 86.6% of Internet users in China are those whose monthly income is under $500 and they pay $25 per month for 1.5mbps broadband services. Apparently, the costs are too high as a percentage of an average Chinese income. In comparison, a monthly subscription fee for broadband is about 2% of the monthly average income of Australians.

China has a very low level of intermodal competition. The government has focused on promoting competition in the broadband market within the existing telecom networks but not between telecom and cable networks (e.g. Digital subscriber line services and cable modem services). Cable TV operators, who are regulated by the State Administration of Radio, Film and Television (SARFT), were not allowed to enter telecommunications markets until a few years ago. Despite widespread cable coverage and over150 million cable TV subscribers in China, the cable TV companies who operate locally instead of nationally have difficulties to enter broadband access markets due to lack of funding and expertise to upgrade their networks to provide broadband services. Cable TV companies provide little competition to China Telecom and Unicom in broadband services.

It is apparent that a well functioning and competitive market in China's telecommunications industry has not really developed. Indeed, the Chinese broadband environment is characterized by:

- all basic service providers are state owned and controlled by the government.
- the broadband market is a largely duopoly of China Telecom and China Unicom in both the wholesale and retail market.

National Strategies, Funding Programs and Broadband Internet Access

The Chinese government has developed some national strategies for ICT development since the early 1990s. However, none of them specifically address the challenge of promoting broadband infrastructure development. Despite the fact that the 2006-2020 National Informationization Development Strategy set the priority areas for broadband development (MIIT,2006), there were no specific goals and action plans supporting the rollout of broadband infrastructure and the range of more affordable broadband services under the strategy. The Chinese government's strategies relying on the telecom operators to deploy broadband infrastructure and deliver high capacity bandwidth have failed to ensure that the benefits

Table 1. Internet users and penetrations rate in urban and rural areas from 2005-2009

		2005	2006	2007	2008	2009
Urban	Internet users	91.69 million	113.89 million	157.38 million	213.40 million	224.35 million
	penetration rate	16.9%	20.2%	27.3%	35.2%	37%
Rural	Internet users	19.31 million	23.11 million	52.62 million	84.60 million	95.65 million
	penetration rate	2.6%	3.1%	7.1%	11.7%	13.2%

(Source: CNNIC 2005 -2009)

of affordable broadband access flow through to 250 million rural households. Table 1 shows pronounced divisions between urban and rural areas in terms of Internet adoption.

The governments in leading broadband nations have used financial incentives in improving broadband services. For instance, the Swedish government initiated various funding programs to roll out broadband infrastructure in rural areas and treat broadband as a universal service. However, China has no funding programs that are truly aimed at improving broadband services and developing competitive broadband infrastructure in areas underserved. Although over 70% of populations live in rural areas, the rural broadband Internet users only account for 28% of the total broadband Internet users as of June 2009.

Many countries have developed the public policy necessary both to invest in and deploy broadband networks or upgrade their existing networks. One of the most important factors accounting for the successful rollout of broadband in the countries with the highest levels of household broadband penetration such as Japan, Korea and Sweden is government strategic commitments and explicit broadband policies. For example, the Japanese government clearly set a goal to provide broadband services to every household by 2010 and provides low-interest loans to any carrier with a fiber optic network installation plan for rural areas and tax deductions for broadband investment under its 2006 IT New Reform Strategy (IT Strategic Headquarter,2006). In contrast, in China, there is lack of national broadband policies that focus on universal broadband access and encouraging competition across broadband infrastructure and services. The Chinese government primarily uses regulatory mandates and official documents to pressure the state-owned telecom providers to deploy broadband networks. The 2002-2006 National Informationization Development Strategy not only provides no financial support for broadband but also does not include specific goals, timelines, action plans for deploying broadband.

To stimulate economy in the times of global recession, in February 2009 China's State Council approved a stimulus plan of the ICT industry which involves the investment of RMB 600 billion to promote 3G mobile communication services, next generation Internet and digital TV networks. Although China has made the public commitments to next generation broadband, due to lack of detailed action plans, how the government will implement the planned investment remains unclear at the time of finalizing this chapter.

CONCLUSION

Based on this research and other studies, clearly national broadband policies and competitive broadband environment can have a significant influence on broadband development. The state of broadband Internet access is closely related to the regulatory framework. Several policy factors affect the growth of broadband Internet access in

China. The most significant is that China has failed to establish a robust and effective national broadband policy that aim at getting as many Chinese households as possible using broadband services at a more affordable price and faster speed. For example, there is virtually no broadband policy that focuses on deploying broadband services in rural areas where the great majority of Chinese live. As a result, the rural broadband adoption is three times lower than the urban adoption.

Another significant factor is the level of competition permitted in the broadband sector. Competition between the two state-owned carriers is insufficient to develop a truly national broadband network and to improve the delivery of broadband services across China. The lack of robust competition among broadband service providers may explain why the broadband performance and deployment are relatively behind many countries. Political intervention and political decisions over interconnection and pricing have reduced competition. The major regulations of telecommunications and broadband Internet access are a reflection of the Chinese government's desire to keep its hands on the reins through central planning of national backbone networks and careful structuring of the telecommunications industry in China. Competition with government control has been a path adopted by China. The constraints of this approach indicate that the benefit of competition can not be realised fully. It has become fairly clear that proactive policy initiatives are needed in China to establish a more competitive environment for broadband Internet access such as deregulatory mechanisms and an independent regulator.

The Chinese government could do more to improve its broadband performance and adoption. OECD experiences suggest that nations with more effective and well funded broadband policies and competitive broadband environment appear to have better adoption, lower price, and higher level of broadband services.

REFERENCES

Atkinson, R. D., Correa, D. K., & Hedlund, J. A. (2008). Explaining international broadband leadership. Retrieved October 13, 2009, from http://www.itif.org/files/Exp lainingBBLeadership.pdf

Benkler, Y. (2009). Next Generation Connectivity: A review of broadband Internet transitions and policy from around the world. Retrieved November 2, 2009, from http://www.fcc.gov/stage/pdf/Berkman_ Center_Broadband_Study_13Oct09.pdf

Budde, P. (2002). Australia- Internet wholesale market. Retrieved November 10, 2009, from http://budde.com.au

CNNIC (The China Internet Network Information Centre). Semiannual survey report on Internet development in China, 1997-2009. Retrieved November 5, 2009, from http://www.cnni.net.cn

Correa, D. K. (2007). Assessing broadband in America: OECD and ITIF Broadband Rankings. Retrieved November 18, 2009, from http://www.itif.org/files/Broa dbandRankings.pdf

Fan, Q. (2008). Analysis of the Influence of Australia's Government Policy on Broadband Internet Access. In Dwivedi, Y. K., Papazafeiropoulou, A., & Choudrie, J. (Eds.), *Handbook of Research on Global Diffusion of Broadband Data Transmission* (pp. 109–125). Hershey, PA: IGI Global.

Group, I. T. (Information Technology Group Center for Information Development at Harvard University) (1999). Readiness for the networked world – A guide for developing countries. Retrieved November 12, 2009 from http://www.harvard.edu

Howell, B. (2008). Competition, Regulation and Broadband Diffusion: the Case of New Zealand. In Dwivedi, Y. K., Papazafeiropoulou, A., & Choudrie, J. (Eds.), *Handbook of Research on Global Diffusion of Broadband Data Transmission* (pp. 139–154). Hershey, PA: IGI Global.

Jew, B., & Nicholls, R. (2000). Competition and regulatory issues. In N. Roehrich (Ed.), *Internet carriage, peering and chaos.* Sydney, NSW: Network Insight Group, RMIT.

Kennard, W. (1999). Connecting The Globe: A Regulator's Guide to Building A Global Information Community (pp. IX-3 &4). Washington, DC: Federal Communications Commission.

Lovelock, P. (1999). *The Evolution of China's NII Initiative: A policy making analysis.* University of Hong Kong: HongKong

MIIT (Ministry of Industry and Information Technology). (2006). The 2006-2020 National Informationization Development Strategy. Retrieved October 8, 2008, from http://www.miit.gov.cn/n112 93472/index.html

Prieger, J. E., & Lee, S. (2008). Regulation and the Deployment of Broadband. In Dwivedi, Y. K., Papazafeiropoulou, A., & Choudrie, J. (Eds.), *Handbook of Research on Global Diffusion of Broadband Data Transmission* (pp. 278–297). Hershey, PA: IGI Global.

Strategic Headquarter, I. T. (2006). New IT reform strategy: realizing ubiquitous and universal network society where everyone can enjoy the benefits of IT. Retrieved October 7, 2009, from http://www.kantei.go.jp/foreign/ policy/it/ITstrategy2006.pdf

Chapter 4
The Federal Communication Commission's National Broadband Plan

John B. Meisel
Southern Illinois University Edwardsville, USA

John C. Navin
Southern Illinois University Edwardsville, USA

Timothy S. Sullivan
Southern Illinois University Edwardsville, USA

ABSTRACT

The American Recovery and Reinvestment Act of 2009 charged the US Federal Communications Commission (FCC) to develop and deliver to Congress a national broadband plan by February 17, 2010. The FCC formally commenced the process of developing the plan by issuing a Notice of Inquiry (NOI) on April 8, 2009. The NOI identified broadband issues and critical questions and asked stakeholders to respond to these issues and questions with data and analysis. The purpose of this chapter is to analyze the written documents generated by stakeholders' responses concerning the specific issues of open networks and competition and to make recommendations to the FCC in its formulation of federal policy as to the position that makes the most economic sense on these issues. We find that many of the arguments and concerns of stakeholders are dependent upon predictions regarding the competitiveness of ISP markets. We predict with confidence that technological innovations are likely to make many legal arguments (on all sides) obsolete in the near future.

INTRODUCTION

The American Recovery and Reinvestment Act of 2009 (Public Law No. 111-5, 123) charged the Federal Communications Commission (FCC) to develop and deliver to Congress a national broadband plan by February 17, 2010. The FCC formally commenced the process of developing the plan by issuing a Notice of Inquiry (NOI) on April 8, 2009 (FCC, 2009). The NOI identified broadband issues and critical questions and asked stakeholders to respond to these issues and

DOI: 10.4018/978-1-60960-011-2.ch004

questions with data and analysis. As of the end of September 2009, over 1,000 comments and replies have been submitted to the FCC.

The purpose of this chapter is to examine the written documents generated by stakeholders' responses concerning the specific issues of open networks and competition (sections C. 5. and 6. in the NOI) and indicate how these issues can aid in the achievement of the nation's broadband goals. Also, the chapter makes recommendations to the FCC in its formulation of federal policy as to the position that makes the most economic sense on these issues.

While the focus of this chapter is on how public policy in the Unites States can promote adoption, diffusion, and utilization of broadband, it should be recognized that there is a significant body of earlier research that examines demand and supply factors that influence broadband adoption and diffusion. For instance, Papazafeiropoulou and Dwivedi (2009) review 49 studies examining broadband adoption and diffusion in developed and developing economies and construct a framework that consists of a listing of and number of studies associated with factors that influence adoption and diffusion decisions at the national level, the consumer level, and for small and medium size enterprises. In particular, with respect to consumer adoption and usage, the authors identify 34 key micro factors that include demographics/user characteristics (14 papers), cost/price/subscription fee (11 papers), and content/applications (6 papers). Another line of empirical research examines households who have chosen not to subscribe to broadband and compares them to consumers who have adopted broadband. For example, Dwivedi and Irani (2009) find empirical support to suggest that there are factors other than subscription fee and speed of access that matter to non-adopters. In particular, to increase consumer adoption of broadband, non-adopters must perceive that broadband is better than narrowband and that broadband will enhance the effectiveness of household activities.

Residential consumers' purchase of broadband service has grown remarkably this century. In 2000, only 3% of American residential homes subscribed to broadband service, whereas by August 2009 the percentage of broadband Internet residential subscribers was about 67% (Horrigan, 2009). However, the provision of broadband Internet *access* appears to be heavily concentrated. According to FCC and network provider data, incumbent cable broadband providers (providing cable modem service) have a 57% share of the fixed residential broadband *access* market and incumbent phone companies (providing digital subscriber line service (DSL) or fiber-optic broadband service) have 39%, with the remainder spread equally among cable overbuilders, competitive local exchange carriers, and others (satellite, wireless, and powerline providers) (Free Press, 2009). Generally, Congress hopes that the creation and implementation of a broadband plan will achieve the following national goals:

a. Increased deployment of advanced broadband networks to unserved and underserved regions of the nation, such as the states of Mississippi, West Virginia, Alabama, New Mexico, and Montana. (We refer to this as a supply side goal.) The FCC estimates that between 3 and 6 million homes do not have access to a broadband connection, defined as speed in one direction of 768kbps or more (Schatz, 2009b).
b. Increased adoption of advanced broadband service by end users. Specifically, what needs to be done to convince the remaining 33% of the residential home population to perceive sufficient value to justify a subscription for broadband service?
c. Increased utilization of advanced broadband networks by end users. (We refer to goals b. and c. as demand side goals.) As end users increase their utilization of broadband service by demanding higher bandwidth services, this goal will require network

providers to increase the capacity of their existing advanced networks.

To what extent can the implementation of the concepts of openness and competition help the FCC achieve its broadband goals in an effective and efficient manner? For instance, one major thrust of this chapter focuses on the business relationships between broadband Internet *access* service providers (ISPs), such as Comcast, Time Warner, Cox, AT&T, Verizon, and Qwest and applications and content provider, such as Google, Amazon, and Vonage. Specifically, one critical issue involves the meaning of the concept of openness to advanced broadband networks. No stakeholder seems to dispute the value of an open Internet. Nevertheless, there is disagreement about the exact meaning of the term and whether exceptions to the term should exist.

In one sense, the concept of openness can be disaggregated into two components: (1) mandated open access to the physical layer of an ISP's network involving openness of broadband transmission conduits (cable modem, DSL, fiber-optic line) to unaffiliated Internet access service providers, and (2) a version of network neutrality which includes mandating openness by incumbent ISPs to unaffiliated upper layer applications and content providers of services such as streaming video, Voice over Internet Protocol (VoIP), and interactive e-commerce.

A related issue to openness involves the meaning of non-discriminatory behavior by ISPs. Recent Supreme Court and FCC orders have freed ISPs from Title 2, common carriers obligations that have traditionally applied to telephone companies' provision of dial-up and DSL service. These recent agency and judicial actions involving ISPs' provision of broadband Internet access service will be chronicled. These obligations were intended to prevent telephone companies from leveraging their market power in Internet *access* to upper layer applications and content markets. Incumbent ISPs such as AT&T, Verizon, and Time Warner argue in their comments in the NOI that no such regulatory protection is necessary given the competitiveness (as a result of inter-modal platform competition) of the advanced broadband *access* service market. Moreover, they argue that such regulations are affirmatively harmful for they reduce the incentive of network providers to invest and innovate in next generation broadband access networks. There is a clear disagreement among stakeholders regarding the competitiveness of the advanced broadband *access* service market that will be explored in this chapter.

At present, ISPs' conduct involving the ways that they can manage their advanced broadband networks is only constrained by the Internet Policy Statement that identifies four principles the FCC issued in 2005 (FCC, 2005b). In 2008, the FCC used the Internet Policy Statement to sanction Comcast for engaging in unreasonable network management in its use of its broadband access network to block/delay a peer-to-peer video applications protocol (FCC, 2008). This policy statement is not a set of rules but rather a set of four principles. Whether these principles should be enacted in a rulemaking process by the FCC is an outstanding issue to be explored in this chapter. In addition, there is controversy to be explored over whether there should be a fifth principle of non-discrimination added to the FCC's enforcement arsenal for broadband access services.

Recent trends have elevated into the political and public spotlight the issue of the degree of freedom that broadband Internet *access* service providers should have to interfere with the transmission of data to manage their networks. First, there has been and continues to be an explosion in the growth of global Internet traffic, led by bandwidth-hungry applications such as online video and file-sharing. Second, there is a desire by network providers to vertically integrate into applications and content markets (e.g., Internet Protocol television service) raising concerns that ISPs have the ability and incentive to favor their own offerings over unaffiliated competitive service providers (The Rights of Bits, 2009).

Amplifying these concerns, the chairman of the FCC, Julius Genachowski, delivered a significant broadband policy speech on September 21, 2009 in which he outlined a plan to maintain the openness of the Internet by proposing to initiate a Notice of Proposed Rulemaking to codify the four principles identified in the Internet Policy Statement and to add two more to that list (Genachowski, 2009). The chairman expressed three reasons to be concerned about whether the Internet of the future will continue to be an open platform:

(a) Limited competition among broadband Internet *access* service providers.
(b) Economic incentives of ISPs.
(c) Explosion of traffic on the Internet (Genachowski, 2009).

The stage is now set to investigate the role that openness and competition can play in achieving the nation's broadband goals. This chapter is organized as follows: The first subsection provides a brief history of broadband deregulation in the USA. The following subsection describes the current economic, political, and regulatory environment surrounding broadband issues. The final subsection presents the opposing views of major stakeholders on critical openness and competition issues. The final subsection provides a conclusion that outlines several general economic points that should be incorporated into the FCC's National Broadband Plan.

THE FCC ANDTHE NATIONAL BROADBAND PLAN

Brief History of Broadband Deregulation

A fundamental economic concern that the FCC has addressed since the incorporation of computer technology into the telecommunications network is what regulations, if any, are necessary to protect data processing firms from the threat of monopoly leveraging by network operators that may have market power in the provision of last-mile transmission services. For over 30 years, the FCC had generally been consistent in fashioning a federal policy that actively pursued a goal of an open telecommunications platform for the users of the platform. The debate has manifested itself in several ways through regulatory policies to promote competition among ISPs, to trying to increase competition in the last mile by mandating network elements to be unbundled, to proposals to protect directly end users and applications and content providers against overt discriminatory behavior. Until 2005, one critical element of the openness policy included a regulatory obligation that facilities-based wireline telecommunications providers provide nondiscriminatory transmission *access* service to independent providers of enhanced/information services, such as e-mail service, World Wide Web searches, Internet access providers, and voice mail service that necessarily include a telecommunications service. For instance, this regulatory mandate facilitated competition in the provision of access to the Internet.

This *access* transmission obligation developed as an outcome of the Computer Inquiry proceedings that spanned the years 1966 to 1986. These proceedings tried to modernize federal regulatory policy in an environment of rapid advances in computer and data processing technology. In particular, in the Computer Inquiry II proceedings, the FCC radically revised its regulatory framework by classifying communications services into two mutually exclusive categories: basic services that were subject to Title 2 regulation as outlined in the Telecommunications Act and enhanced services which were deemed competitive, thus, unregulated, and subject to FCC jurisdiction under Title 1 of the Act. The FCC was concerned that wireline telecommunications providers of basic service possess the ability and incentive to engage in anticompetitive behavior in providing basic services for independent providers of enhanced/information services that rely on the underlying communications platform to provision their services. Thus, the openness access policy was intended to benefit directly competitive enhanced/information service providers (Cannon, 2003) and to facilitate innovation at the edge of a network, innovation that did not require the permission of facilities-based wireline network providers. It is often argued that the spectacular growth of the Internet benefited from a federal policy of 'unregulation' (Oxman, 1999) but such a policy clearly did not apply to conditions governing *access* to the Internet.

In the mid 1990s, end users wishing to access the Internet initially relied on dial-up (narrowband) wireline connections over the telephone companies legacy circuit-switched networks, networks centrally designed for the single purpose to provide high quality voice services. The Computer Inquiry decisions in combination with the decision to exempt ISPs from having to pay interstate access charges to local telephone providers stimulated the growth of an independent ISP industry. According to a government report in 2001 (Kruger and Gilroy, 2001), there were about 6,000 Internet Service Providers, (classified as providing an unregulated enhanced/information service), such as AOL and Earthlink, in the United States and this competition resulted in benefits to end users for Internet access service such as price reductions, quality enhancements, and a choice among a wide variety of differentiated access services. Independent ISPs generally relied on the telephone company to provide the facilities to provision the underlying transmission component upon which they added upper layer Internet services for the end user.

As the supply of and the demand for high bandwidth (broadband) connections grew at the end of the 20th century, changes in technology and federal policy caused the market structure of the ISP market to change dramatically. First, in response to the emergence of satellite television competition, cable television service providers upgraded their one-way video distribution networks to provide more channels and to provide two-way services such as high bandwidth connections, called cable modem service. Secondly, telephone companies upgraded the copper local loop in order to be able to provision high bandwidth digital subscriber service (DSL) over the same local loop as they provision voice service. As these technological developments occurred, independent ISPs called for the wireline transmission component of Internet access, supplied by incumbent cable and telephone companies, to remain a mandated wholesale transmission service (referred to as an open access policy) which, of course, had historically been true for telecommunications companies but not for cable companies, other than for several cable mergers in the late 1990s and early 2000s where the merged cable company agreed to provide open access to the cable transmission component in order for the federal agency to endorse the merger.

In 2002, the FCC took its first concrete step in constructing a deregulatory policy for wireline broadband Internet access services by determining that cable modem service (which the FCC portrayed as a service that inextricably intertwined transmission with information processing capabilities) was an information service and, thus, free of Title 2, common carrier obligations (FCC, 2002). The Supreme Court in its Brand X decision in 2005 upheld this judgment (National Cable and Telecommunications Association, 2005). In 2003, the FCC eliminated open access to the DSL component of the local loop (called line sharing) by deciding that the high frequency portion of the local loop was no longer a mandatory unbundled network element (FCC, 2003). Prior to this reversal of policy, it was possible for a competitive local exchange company to share the local loop with the incumbent telephone company in provision of one of the services (voice and DSL service) provided over the local loop. Also, in a highly controversial order released in 2005 (FCC, 2005), citing changed technological and market conditions, the FCC continued the deregulatory trend by eliminating the obligation of facilities-based wireline telecommunications companies to provide, as a regulated service, the broadband transmission component for independent ISPs. This effectively ended the Computer Inquiry II rules that had governed the industry for 25 years. To a large extent, these regulatory decisions and technological changes have devastated the independent ISP market. Today, end users purchasing wireline broadband service generally purchase an integrated transmission and Internet access service from a facilities-based telecommunications or cable company.

In the controversial 2005 order, the FCC provided several reasons for its policy of deregulation of the facilities-based wireline broadband Internet access market (FCC, 2005a). In general, the FCC determined that federal regulations, as outline in Computer II, were no longer necessary to protect against anticompetitive behavior and, moreover, were believed to be affirmatively harmful for the development of next generation wireline networks.

First, the FCC recognized that, in general, the telecommunications platform is no longer monopolized, given the emergence of competition between telephone and cable platforms as a result of the convergence of different platform technologies adopting the Internet Protocol (IP). As communications platforms originally built to provide one service were upgraded with the inclusion of digital technologies, packet switches, and fiber optic technology, the next generation version of these networks were capable of providing multiple services by use of a common standard employing the IP. Adherence to a common standard enables various publicly and privately owned networks to communicate with each other and creates an open platform for applications innovation. Telephone companies now provide video services as well as voice, and cable companies provide voice services as well as video. All of these services become the equivalent of data services. More importantly, these two platforms compete in the provision of broadband Internet access services. Modern communication platforms are now composed of an underlying physical transmission layer (such as fiber optic lines, copper lines, coaxial cable lines) upon which upper layer applications (such as instant messaging, Internet browser, Voice over Internet Protocol, Video over Internet Protocol) and content services (such as movies, short film clips and television shows) are provisioned. The lower physical layer and the upper layer applications and content are built to work with each other using the IP.

Second, the FCC predicted that emerging technologies would provide new facilities-based platforms to compete with the wireline telecommunications and cable platforms. For instance, publicly and privately funded wireless networks are developing Internet access capabilities, satellite companies offer Internet access, and power companies were predicted to upgrade their networks to provide Internet access service. Thus, the FCC viewed the Internet access market as one with two existing, strong competitors with their own facilities and potential competitors on the precipice of becoming significant competitors, all to the benefit of demanders for broadband Internet access service. This reflects a tendency on the part of the FCC to merely count the number of competing platforms rather than conduct a real economic market analysis to determine what forms of Internet access are close substitutes in the eyes of end users in actual geographic markets.

Third, the FCC, citing Section 706 of the 1996 Telecommunications Act, placed a very high priority on stimulating investments in next generation broadband networks and wished to eliminate what it considered to be unnecessary hindrances to their deployment. To order to accomplish this infrastructure goal, the FCC eliminated the sharing of facilities regulations that the incumbent telecommunications companies had argued reduced their incentive and that of non-facilities-based ISPs to invest in next-generation networks. In a version of the belief that there is such a thing as a free lunch, the FCC predicted that the incumbents would still possess incentives to make their facilities available to independent ISPs and, thus, there would exist a healthy wholesale market for the critical transmission access service. In short, the FCC imagined a world in which there would generally be competition in the provision of broadband Internet access between several facilities-based platforms (telephone, cable and potential entrants) and that these platforms would remain open and reasonably priced for non-facilities-based competitors. (Non-facilities-based competitors provide what is termed intramodal competition since they rely on a specific platform utilizing a particular technology.)

The FCC focus on increasing the incentives to construct next generation broadband networks also was revealed in its 2003 order in which the FCC eliminated all network unbundling obligations for incumbent telephone companies when they construct fiber local loop facilities-to-the-premises (FCC, 2003). This deregulatory treatment of fiber investments was predicted by the FCC to jumpstart a race to build next-generation broadband networks by both incumbents and potential entrants. What the FCC calls predictive judgments were thus critical in providing the underlying rationale for such a significant deregulatory broadband policy.

Current Economic, Political, and Regulatory Environment

In light of the virtual extinction of a role for independent ISPs to provide broadband Internet access service competition to facilities-based incumbents, the FCC deregulatory broadband policies unnerved two critical stakeholders, end users and independent applications and content companies. Each group relies on broadband Internet access providers in order to conduct voluntary exchanges with the other. The two-sided nature of the broadband industry is illustrated in Figure 1. Traditionally, Internet charging arrangements involve each side (points A and D) of the Internet platform paying for its broadband connection to the Internet and then being free to utilize that connection in any way it sees fit (Stifel Nicolaus, 2006).

Each side of the Internet platform is concerned with the continuing openness of the platform. These concerns, lumped together under the general term "net neutrality," have been explored, to some extent, by recent FCC actions or inactions. First, recognizing the controversial nature of its decision to deregulate wireline broadband facilities-based companies, the FCC created a set of four principles to protect end users of Internet services since ISP competition that resulted from regulatory policies and the industry norm of treating packets in a neutral manner combined to constrain the ability of network operators to engage in unreasonable transmission discrimination. The

Figure 1. Two-sided nature of broadband markets

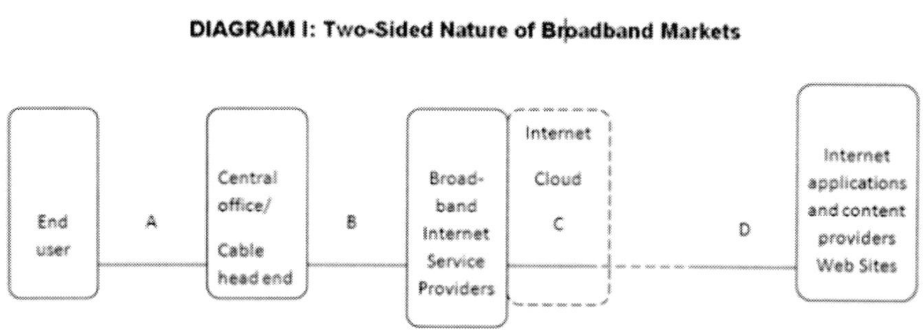

Internet Policy Statement that contains these principles (FCC, 2005b) was released at the same time that the FCC eliminated the Computer Inquiry rules. The Internet Policy Statement (FCC, 2005b) contains the following protections for end users:

(1) Consumers are entitled to access the lawful Internet content of their choice.
(2) Consumers are entitled to run applications and use services of their choice subject to the needs of law enforcement.
(3) Consumers are entitled to connect their choice of legal devices that do not harm the network.
(4) Consumers are entitled to competition among network providers, applications and service providers, and content providers.

These principles can be viewed as protections for all consumers of Internet services. Each of the entitlement principles is subject to reasonable network management.

Major players in the applications and content industry, such as Google, Amazon, Vonage, and Skype were concerned and continue to be concerned with statements made by top management in telecommunication companies when they characterized these Internet companies as free riders on Internet infrastructure, a criticism that continues in the leading business media today (Jenkins, 2009 and Review & Outlook, 2009). Such allegations overlook the substantial payments that Internet companies make to IPSs (point D) to access the entire Internet. In turn, these ISPs have voluntary commercial relationships with backbone providers to ensure delivery of data to end users from and to the Internet companies.

It appears that telecommunications companies wish to import to the Internet the pricing arrangements of the legacy circuit-switched telephone network where a local telephone company received payment for terminating telephone calls from other telecommunications companies. In fact, incumbent telecommunications providers of local telephone service have consistently argued for a requirement that ISPs pay termination charges suggesting that ISPs use the local network in the same manner as do long distance companies. This traditional pricing arrangement and the exertion of centralized control of the network work in the opposite direction to the decentralized culture of the Internet. In the circuit-switched network the intelligence of the network lies within the network allowing the network provider to dictate the range of permissible applications and to ensure the quality of each application. It appears that telephone and cable companies wish to impose a similar structure on their advanced broadband networks. Then, a new application provider would need to adapt the application to the particular specifications of a network provider in order to reach successfully the end users of that ISP.

An attempt to shift value from Internet companies to ISPs is reflected in the desire of ISPs to sell to Internet companies a premium service (referred to as better-than-best-efforts where best-efforts is the standard for the delivery of packets over the public Internet) for the delivery of data to end users. ISPs argue that as congestion on the public Internet increases and the demand for real time, bandwidth-hungry applications grows (e.g., Internet video), they need to have flexibility to manage their networks and to have incentives to invest and innovate within the networks that connect users to the Internet. ISPs argue that to improve network performance they should be able to offer different levels of service at different prices just as firms in other industries offering other kinds of service are able to do. Internet companies are concerned about unconstrained network management practices given the lack of effective competition among ISPs in local markets. Specifically, Internet companies worry that the creation of a two-tier public Internet could provide the mechanism for ISPs to engage in anticompetitive conduct if broadband network operators favor particular applications and content companies. This has led to calls for a fifth principle of nondiscrimination to protect independent applications and content companies in their commercial relationships with ISPs.

The FCC delivered a landmark broadband decision in August 2008 (FCC, 2008) regarding the reasonableness of a non-standard network management practice utilized by the leading cable operator, Comcast, a decision that has been called "an extraordinary change of direction in FCC policy" (Sieradzki and Maxwell, 2008). This practice implicated the concept of openness for it resulted in degradation in performance of a popular file-sharing applications protocol. This regulatory decision began the difficult process of delineating the boundaries for reasonable network management practices. Comcast was found to have utilized a network practice that singled out and secretly interfered with the performance of a peer-to-peer protocol used commonly to share files among end users. In particular, the peer-to-peer applications protocol is widely used to distribute in a decentralized manner video files, competing to some extent with Comcast's video-on-demand service. Comcast was able to identify the contents of a packet (similar to the Postal Service opening a letter to determine its content) by using a technology called deep packet inspection. Comcast first argued that the protocol interference only occurred when there was congestion on the network, an assertion that they later retracted. During the course of the FCC's investigation of the practice Comcast and BitTorrent (one Internet service provider that used the protocol) reached a settlement on the practice which some assert today means that the market, without government intervention, corrected the problem (The FCC's Heavy Hand, 2009). It should be noted that competition among ISPs did not solve the problem of the degradation of performance of this protocol for end users. Comcast promised to replace by the end of 2008 the practice with a protocol-agnostic network practice to address congestion problems on the network. This practice focuses on the bandwidth use of end users and only intervenes in times of a congested network by restricting the activity of heavy users of the scarce bandwidth resource. The FCC found the new practice to be reasonable for it furthered a critically important interest (network congestion) and was narrowly tailored (focused specifically on heavy users) to further that interest. This decision, in effect, creates a safe harbor for this protocol-agnostic network practice. Such a standard preserves discretion for ISPs to select among reasonable practices to manage their networks. It should be noted that the Comcast decision did not address the legality of a management practice that gives higher priority to some packets, such as real-time communications packets (e.g., VoIP), than to other packets as negotiated in a commercial agreement (FCC, 2008, footnote 202).

The decision (narrowly passed by the FCC with a 3-2 vote) has been challenged by Comcast in court arguing that the FCC did not possess the legal authority to take such action for a service classified as an information service and basing its decision on principles not rules. It has been argued that this assertion of regulatory power by the FCC is "unprecedented, sweeping in its breadth, and seemingly unbounded by conventional rules of interpretation and procedure" (Esbin and Marcus, 2009). Recognizing the somewhat murky nature of the FCC's authority to regulate the behavior of broadband ISPs, there are outstanding bills now in Congress to provide the agency clear jurisdiction over these services.

Another FCC decision provides a model for the FCC to implement targeted regulation of broadband ISPs. The FCC approved the merger of AT&T and BellSouth in 2007. The approval was subject to several conditions implicating net neutrality concerns. First, the merged company agreed to conduct business in a manner consistent with the Internet Policy Statement for 30 months. Second, AT&T agreed to not provide a prioritized service that would favor packet transmissions based on source, ownership, or destination for two years. In short, this condition banned the offering of a two-tier public Internet. Third, AT&T retained the authority to provided managed IP-based services (such as IP television service) over a private Internet utilizing its transmission lines. A possible precedent set in this merger agreement forbids a two-tier public Internet while allowing the network provider to reserve a portion of its infrastructure to operate a private Internet.

It is commonly argued that net neutrality is a solution in search of a problem given the lack of widespread net neutrality violations. Several reasons can account for the scarcity of net neutrality violations. First, previous merger agreements, AT&T-SBC (2005), Verizon-MCI (2005), and AT&T-BellSouth (2006), contained provisions in which the merged companies agreed to abide by the four principles identified in the Internet Policy Statement for, generally, a two-year period. Second, only since 2005 have deep packet inspection (DPI) technologies gained widespread adoption within broadband networks. DPI enables an ISP to identify the source of a packet, the destination of a packet, the content of a packet, and the application used by the packet (Gilroy, 2009). This information puts the ISPs in a position where they can, if they so desire, discriminate among packets based on the preference of the network operator. Third, Internet video is becoming a growing threat to the television delivery business of cable and telephone companies and promises to erode the revenues of this significant source of future profits. Fourth, there no longer are significant competitors providing broadband Internet access service as there had been in the period before the elimination of Title 2 requirements on facilities-based telecommunication companies. In light of the renewed Congressional and FCC interest in the appropriate regulatory framework for broadband access service, the time is ripe for reconsideration of the deregulatory path that the FCC has been on the last seven years for broadband services.

Opposing Views on Openness and Competition Issues

Much of the debate surrounding these issues can be summarized in terms of five questions:

1. Are the four principles identified in the Internet Policy Statement sufficient to protect the openness of the Internet, or do these principles need to become formal rules.
2. Do ISPs have the incentive to discriminate against independent applications and content providers?
3. Would net neutrality rules significantly diminish the network providers' incentive to invest in the next generation of networks?
4. Should potential net neutrality rules only pertain to the infrastructure layer of the Internet, or should they apply at any layer of the Internet stack where a bottleneck exists?
5. What role, if any, should managed IP services play?

In this section, we address these questions in turn and summarize the opposing economic arguments.

1. Are the four principles identified in the Internet Policy Statement sufficient to protect the openness of the Internet, or do these principles need to become formal rules?

Those opposed to formalizing the principles point to the costs and feasibility of formalization and argue that the principles are sufficient. In their view, fast-changing technology mixed with slow-moving regulators is a recipe for stifled innovation. Improvements are likely to be slowed when technical experts are distracted from innovating by being forced to constantly evaluate the legitimacy of the innovation. Furthermore, the international nature of the Internet makes regulation problematic. How, they ask, can the FCC regulate content and transmission that works its way across numerous national borders? Finally, opponents to formalization question the appropriateness of prohibiting activities that aren't taking place (Time Warner, 2009b).

Proponents of formalization argue that ISPs do have a history of violating the principles, and that only explicit regulations can meet the FCC's goals. They point to the 2008 Comcast case as evidence that ISPs are making attempts to circumvent the principles. In their view a strict enforcement regime is needed to combat ISPs' incentives to defy the FCC.

2. Do ISPs have the incentive to discriminate against independent applications and content providers?

There are a number of profit-driven rationales for ISPs to discriminate against particular applications and content providers. First, discrimination against applications with high bandwidth requirements enables ISPs to lower operating costs. Discrimination can take the form of blocking the content/application, reallocating it to the public Internet or reallocating it temporally by delaying its delivery. In all cases, the ability to discriminate allows the possibility of lowering the delivery cost. Second, discrimination opens the door for charging content providers for access to the ISPs delivery system, opening a new revenue source. Neutrality advocates fear that dominant ISPs may have market power in negotiating the fee, potentially making this a lucrative source of revenue – one that ISPs are unlikely to forgo voluntarily. Finally, ISPs that vertically integrate with content providers can increase profits by bundling and placing un-partnered content providers and (in the case of exclusive arrangements) other ISPs at a competitive disadvantage.

Neutrality skeptics doubt that ISPs will find it profitable to discriminate in the delivery of content and applications. First, an ISP that routinely discriminates against popular content and applications will be at a competitive disadvantage. Most of the effective differentiation among ISPs focuses on speed (and to a lesser extent reliability). Customers who discover that a desired site or application is blocked or slowed will have an incentive to switch their ISPs. As the duopoly that exists in most markets (where a cable company competes with a phone company) moves towards multiple providers (perhaps including wireless), ISP's will have even less ability to discriminate. Further, it's not entirely clear that ISPs will be in the dominant negotiating position *visa vis* content and application providers. As seen in a recent episode involving ESPN 360, content providers may be quite willing to discriminate against ISPs that are unwilling to pay a fee (Bode, 2009).

3. Would net neutrality rules significantly diminish the network providers' incentives to invest in next generation networks?

Neutrality opponents argue that the inability to discriminate against certain content would reduce the incentive to upgrade networks (Time Warner, 2009b). The fear among ISPs is that an inability to discriminate will commoditize Internet service. An ISP in this case will only be able to differentiate its service on the basis of speed and reliability. Microeconomic theory predicts that (in this case) an ISP will continue to upgrade its service only to the point that the cost of such upgrades can be passed along to customers as a higher price. The ability to discriminate in favor of particular content increases the likelihood of upgrading on two fronts.

First, the ability to discriminate against high-bandwidth content will provide an incentive for customers who access such content to either upgrade their service levels with their current ISP or switch to a provider that doesn't discriminate – improving the ability of ISPs to price discriminate. Under a neutrality system, users of high-bandwidth content already have an incentive to upgrade service to a faster speed. In fact providers often advertise their services as being appropriate for email and basic web browsing at the low end, with higher speeds recommended for customers accessing high-bandwidth content such as video. The ability to discriminate against (all) high-bandwidth content (regardless of the content's provider) can (if designed properly) be used to further induce customers accessing this content to upgrade their service. This increased willingness to pay for higher-speed access increases the ability of providers to pass the cost of infrastructure upgrades on to customers.

The ability to discriminate would also allow ISPs to charge content providers – which in some circumstances would increase the incentive for infrastructure upgrades. Any ability to charge content providers for transmission allows a new revenue source. However, economic theory would only predict that this would translate into a desire on the part of the ISP to upgrade the network if the upgrade would sufficiently increase the revenue that could be extracted from content providers. This is most likely to be the case if the most desirable content to consumers (and thus the most profitable to content providers) is the highest-bandwidth content. While this seems plausible on the surface, it's not clear whether this will ultimately be the case or not. In general, consumers seem fairly unwilling to pay for *any* content, whether low bandwidth (such as news, reference and search) or high (such as video or voice), and few providers have been able to use advertising to translate free content into a profitable business model.

Neutrality proponents discount these arguments. They (correctly) point out that ISPs are not predicted to spend any extra revenue on upgrades just because the money is available. As described above, economic theory would argue that the ability to earn extra revenue from discriminating would only encourage upgrades under particular circumstances. Absent this ability (where Internet access is largely commoditized) ISPs still have the primary incentive to upgrade because customers will still shop for speed and reliability in order to access the content they desire.

4. Should potential net neutrality rules only pertain to the infrastructure layer of the Internet, or should they apply at any layer of the Internet stack where a bottleneck exists?

Stakeholders on both sides of the issue have demonstrated a convenient ability to compartmentalize their arguments. If content providers object to ISPs having the ability to discriminate against particular content, it seems convenient (in the least), if not entirely disingenuous, for the content providers to not face the same requirement (Szoka and Thierer, 2009). Google Voice, for example, has recently become the target of complaints from AT&T, after acknowledging that the service blocks particular numbers (Schatz, 2009a). Many content sites discriminate – Google and other search engines, for example, do not present results based upon a blind algorithm, but place "sponsored" links near the top.

While clearly embarrassed over the apparent inconsistency, advocates for neutrality point out that the barriers to entry are several orders of magnitude higher at the ISP level than at the content level. It is far easier for a user to switch search engines than for a user to change ISPs. Once again, the argument centers on how competitive Internet service provision is today and how the level of concentration might change in the near future.

5. What role, if any, should managed IP services play?

Some suggest that a two-tiered Internet holds the solution to the neutrality debate. Under such a system, access providers, such as AT&T, would allow neutral access to a portion of its network, while reserving a portion of the infrastructure to preferred content. This would increase the incentive for the providers to invest in next-generation networks, as some providers and users would be willing to pay a premium to access the reserved portion. Economically, offering a menu of services at different prices is often good for both the producer and the consumer (as seen in industries as diverse as automobiles and financial services). Ironically, numerous Internet companies provide thriving examples of such a system. Many content providers offer free (advertising-driven) access (to news, games and entertainment), while withholding the better-quality content for paying members. Amazon "discriminates" by offering faster shipping speeds to those willing to pay for it on a one-time basis or through its Amazon Prime membership.

Neutrality advocates fear that the once providers are able to discriminate and charge a premium for access to the private tier, the public tier will be starved. Unless a constraint is in place, access providers will only maintain the public Internet to the extent that is profitable. Providers may, in fact, find it profitable to allow the public Internet to become congested – all the better to migrate users and content providers to the higher-priced reserved Internet.

CONCLUSION

The profound changes in the technology, content and applications now used by the US public have challenged the FCC's ability to regulate Internet service provision. In less than a decade the FCC has re-visited the industry several times. The effort to devise a National Broadband Policy has given the FCC the opportunity to reconsider several of the policies that it has established in recent years. These include policies that deregulated the facilities-based wireline broadband Internet access market.

Among the controversies surrounding the National Broadband Policy, many surround the obligations of ISPs to be neutral in their provision of Internet access for content and applications. On one side are proponents of re-regulation and net neutrality. Stakeholders in this group, led by Google and Free Press, would like to see formal regulations that explicitly prohibit the ISPs from discriminating against particular content or applications through blocking or slowing its delivery. On the other side are stakeholders, led by telephone and cable Internet providers, that see no need for this type of regulation and voice concerns that such policies would stifle technological innovation and jeopardize investments in the next generation of networks.

As discussed in this chapter, the economic arguments largely hinge on the competitiveness of the market, now and in the future. Competition will discourage ISPs from blocking or slowing particular content or applications (especially if the ISPs are forced to be transparent regarding their actions). Consumers will leave an ISP that makes particular content and applications unavailable or less functional. ISPs that operative in a competitive market will be largely unable to apply economic pressure to content and application providers. In fact, it's completely feasible that the content and application providers will be applying the pressure to ISPs in much the same way that desirable cable networks are able to charge cable and satellite television service providers.

Regarding the current level of competition, most ISPs in the US currently operate as duopolists; in most markets a cable television provider competes with an incumbent telephone service provider. It's not clear whether the lack of complete market power in such a market will act to discourage an ISP from discriminating against particular content or applications in a way that is detrimental to consumers. Traditional economic models are not in agreement regarding the behavior of firms in a duopoly market (Perloff, 2003).

While the future is always uncertain, technological innovations are likely to increase consumers' choices regarding Internet access. Improvements in wireless networks have already increased the number of Americans accessing the Internet through mobile phones and mobile cards plugged into laptops. While many technical and compatibility obstacles remain, it's not far-fetched to predict that the current wired infrastructure may become obsolete in consumers' minds. In addition to increasing the level of competition in the ISP market, it would significantly shift the ground under the current debate. We can't help but recall how the development of the PC in the 1970s superseded antitrust concerns regarding mainframe computers; similarly, the proliferation of mobile phones in the past 15 years has recast the debates in the 1980s regarding the appropriate level of regulation of telephone service. We do predict with confidence that stakeholders basing arguments (and regulators basing decisions) upon the current state of technology will find their concerns equally outdated in less than a decade.

REFERENCES

Bode, K. (2009, June 12). Small ISPs Want FCC To Ban ESPN 360 Model. *Broadband DSL Reports*. Retrieved from www.dslreprots.com.

Canon, R. (2003). The Legacy of the Federal Communications Commission's Computer Inquiries. *Federal Communications Law Journal*, *55*(2), 167–206.

Dwivedi, Y. K., & Irani, Z. (2009). Understanding the Adopters and Non-adopters of Broadband. *Communications of the ACM*, *52*(1), 122–125. doi:10.1145/1435417.1435445

Esbin, B., & Marcus, A. (2009). "The Law is Whatever the Nobles Do": Undue Process at the FCC, 17. *CommLaw Conspectus*, *1*, Retrieved from http://commlaw.cua.edu//artic les/v17/17.2/Esbin-Marcus-Revised.pdf.

Federal Communication Commission (2002). *Inquiry Concerning High-Speed Access to the Internet Over Cable and Other Facilities; Internet Over Cable Declaratory Ruling; Appropriate Regulatory Treatment for Broadband Access to the Internet Over Cable Facilities*. Declaratory Ruling, 17 FCC Rcd 4798.

Federal Communication Commission (2003). Review of the Section 251 Unbundling Obligations of Incumbent Local Exchange Carriers, Implementation of the Location Competition Provisions of the Telecommunications Act of 1996, Deployment of Wireline Services Offering Advanced Telecommunications Capability, CC Docket Nos. 01-338, 96-98, 98-147, Report and Order and Order on Remand and Further Notice of Proposed Rulemaking, 18 FCC Record 16978.

Federal Communications Commission. (2005a). *Appropriate Framework for Broadband Access to the Internet over Wireline Facilities*. Report and Order and Notice of Proposed Rulemaking.

Federal Communications Commission (2005b, September 23). *In the Matter of Appropriate Framework for Broadband Access to the Internet Over Wireline Facilities.* Policy Statement, 20 FCC Rcd 14986.

Federal Communications Commission (2008, August 20). *Formal Complaint of Free Press and Public Knowledge Against Comcast Corp. for Secretly Degrading Peer-to-Peer Applications; Broadband Industry Practices; Petition of Free Press et al. for Declaratory Ruling that Degrading an Internet Application Violates the FCC's Internet Policy Statement and Does Not Meet an Exception for "Reasonable Network Management*." Memorandum Opinion and Order, 23 FCC Rcd 13028.

Federal Communications Commission (2009, April 8). *In the Matter of A National Broadband Plan for Our Future*. GN Docket No. 09-51, Notice of Inquiry.

Free Press(2009, September4). In the Matter of Inquiry Concerning the Deployment of Advanced Telecommunications Capability to All Americans in a Reasonable and Timely Fashion, and Possible Steps to Accelerate Such Deployment Pursuant to Section 706 of the Telecommunications Act of 1996, as Amended by the Broadband Data Improvement Act, A National Broadband Plan for Our Future. *Comments of Free Press (p. 48).*

Genachowski, J. (2009, September 21). Preserving a Free and Open Internet: A Platform for Innovation, Opportunity, and Prosperity. The Brookings Institution, Washington, D.C.

Gilroy, A. (2009, June 1). Access to Broadband Networks: The Net Neutrality Debate. Congressional Research Service.

Horrigan, J. (2009, August 25). Broadband Adoption Barriers. Blogband, The Official Blog of the National Broadband Plan.

Jenkins, H., Jr. (2009, September 23). Neutering the 'Net. *Wall Street Journal*.

Kruger, L.G., & Gilroy, A.A. (2001, May 18). IB10045: Broadband Internet Access: Background and Issues. CRS Issue Brief for Congress.

National Cable & Telecommunications Ass'n v. Brand X Internet Services, 545 U.S. 967 (2005).

Oxman, J. (1999, July). The FCC and the Unregulation of the Internet. OPP Working Paper No. 31.

Papazafeiropoulou, A., & Dwivedi, Y. K. (2009). A Framework of Broadband Adoption and Diffusion (BA&D): Understanding Factors Relevant to Residential Consumers and Small & Medium Enterprises. In Dwivedi, Y. K. (Eds.), *Handbook of Research in Global Diffusion of Broadband Data Transmission*. Hershey, PA: IGI Global.

Perloff, J. M. (2003). *Microeconomics* (3rd ed.). Pearson-Addison Wesley.

Review & Outlook. (2009, September 22). Bad News for Broadband. *Wall Street Journal* (p. A24).

Schatz, A. (2009a, September 26). AT&T Asks for Curbs on Google. *The Wall Street Journal* (p. A5).

Schatz, A. (2009b, September 30). Full Web Access is Pegged at $20 Billion. *Wall Street Journal* (p. A2).

Sieradzki, D. L., & Maxwell, W. J. (2008). The FCC's Network Neutrality Ruling in the Comcast Case: Toward a Consensus with Europe? *Communications & Strategies*, *72*(4), 73.

Stifel Nicolaus (2006, March). Net Neutrality: Value Chain Tug of War. Equity Research, Industry Update, Telecom & Media Regulatory.

Szoka, B., & Theirer, A. (2009, October). Net Neutrality, Slippery Slopes and High-Tech Mutually Assured Destruction. *Progress Snapshot, The Progress and Freedom Foundation, 5*(11).

The, F. C. C's Heavy Hand (2009, September 28). *WashingtonPost.com*. Retrieved from http://www.washingtonpost.com/wp-dyn/content/article/2009/09/27/AR2009092703026.html.

The Rights of Bits. (2009, September 26). *The Economist* (pp. 78-79).

Time Warner Cable, Inc. (2009, July 21). Reply Comments of Time Warner Cable, Inc. GN Docket No. 09-51.

Verizon Communications, Inc. v. Law Offices of Curtis Trinko, LLP, 124 S. Ct. 872 (2004).

KEY TERMS AND DEFINITIONS

Internet Policy Statement: A set of four principles issued by the Federal Communications Commission in 2005 that provides end users specific entitlements with respect to accessing lawful content, running applications, connecting devices, and expecting competition when using the Internet. Each entitlement is subject to reasonable network management by an Internet service provider.

Two-tier Internet: When an Internet service provider offers independent content and applications developers two types of broadband access to its end users: (1) a normal best efforts path to end users and (2) a better than best efforts premium path to end users.

Comcast Decision: The FCC sanctioned Comcast Corporation in 2008 for allowing its Internet service provider to degrade the performance of a popular file-sharing application protocol used by its end users.

Network Neutrality: A network engineering design principle in which an Internet service provider treats all data it carries on its network in a neutral, non-discriminating manner regardless of the type of data.

Broadband Internet Access: The provision of a high speed Internet access line to the Internet for end users by Internet service providers (e.g. telephone DSL or fiber line, cable modem, or wireless connection).

Internet Service Provider: A company that provides to end users a high speed transmission connection and enhanced services enabling an end user to access the Internet.

Applications and Content Providers: Internet companies that provide functionalities (such as email, Internet browsers, search engines) and content (such as data, music, video) for end users to consume or as inputs into further production activities.

Next Generation Networks: Upgrades to existing Internet broadband access networks that frequently include fiber technology extending closer to and if not ending at an end user's premises.

Discrimination: When an Internet service provider degrades or blocks the use of an application or access to content or favors the use of application or access to content.

Internet Stack: A modular, layered conception of the Internet. One common specification includes four layers: an infrastructure layer (transmission lines, servers, routes) a logical layer (Internet protocol), applications layer (email program, Web browser, search engine, voice over Internet protocol) and content layer (Web page, music, text).

National Broadband Plan: An effort by the FCC to promote the adoption, diffusion, and utilization of the Internet by the United States.

Chapter 5
Broadband and Structural Separation from the Perspective of Transaction Cost Economics

Hidenori Fuke
Komazawa University, Japan

ABSTRACT

Conduct regulation and structural separation are often discussed in industrial organisation studies as options to prevent the abuse of market power by vertically integrated firms toward the downstream market. Both the structural separation of NTT and conduct regulation have been discussed in the Japanese telecommunications industry since the introduction of competition in 1985 and the issue is still being discussed, although the industry is going through a transition from POTS (Plain Old Telephone Service) to the broadband internet. Past discussions have been inclined toward elimination of the harmful effects of vertical integration. However, there is a benefit of vertical integration in the sense that it will promote the efficient management of the firm concerned. I will present a new contention that it is important to conclude a balanced analysis of costs and benefits of vertical integration based on transaction cost theory. Structural separation in the broadband market entails significant transaction costs between a carrier with access facilities and firms offering broadband services by renting these facilities as input. These kinds of transaction costs are comparatively negligible in POTS. I will make it clear that the balance analysis of costs and benefits of structural separation has become more important in broadband than in POTS based on the actual differences in network structure.

INTRODUCTION

Vertical Integration is one of the topics that have been much discussed in industrial organisation studies. The focus of the discussion here is the possible abuse of market power toward the downstream market by a vertically integrated firm that is operating in both upstream and downstream markets and is offering goods or services to downstream competitors as a monopolistic supplier. This is discussed not only in the case of a merger of a firm that has market power in the upstream

DOI: 10.4018/978-1-60960-011-2.ch005

market with a downstream firm but also in the case of the regulation of conduct of a vertically integrated firm. It has long been regarded as one of the important competition policies to prevent the abuse of market power in the upstream market. The measures adopted are largely classified into conduct regulation and structural separation.

Control over vertically integrated incumbent carriers has been an important regulatory consideration since the introduction of competition in the telecommunications industry. The rationale for the regulations is the fact that vertically integrated incumbent carriers are offering both local and long-distance services and newly entered long-distance carriers are offering their services by interconnecting with local telecommunications facilities of the incumbents. The measures considered are conduct regulation including the obligation to interconnect and structural separation of incumbents into local and long-distance businesses. Structural separation has not been introduced except in the U.S.A. where the courts ordered the old AT&T to separate into a new AT&T and several RBOCS (Regional Bell Operating Companies).

Structural separation of the incumbent NTT has been discussed since the privatisation of NTT Public Corporation and the introduction of competition in 1985 in Japan. It once seemed that the discussion had been settled by the reorganisation of NTT into two local companies, one long-distance company and a cellular company under a holding company structure in 1999. However, the discussion has been revived with the 'Koizumi-Reform' in 2006 (MIC (2006a)) and it was decided, through political compromise, to resume the discussion in 2010 ((MIC (2006b)).

The structure of the telecommunications industry has changed significantly with the diffusion of broadband internet. This chapter tries to re-evaluate the discussions on structural separation from the view point of transaction costs by taking into account the changes in industry structure from POTS to broadband.

BACKGROUND

The reason why vertical integration has been discussed so much is explained by the probability that a vertically integrated firm might abuse market power it has built in the upstream market to the detriment of the downstream market. The abuse of market power derived from vertical integration typically includes 'refusal to deal' and 'raising rivals' costs'. If a firm having market power in the upstream market refuses to supply goods or services to rivals in the downstream market, they are obliged to exit from the market when they cannot find other sources of supply.

Price discrimination is also frequently discussed in this connection. A vertically integrated firm can raise wholesale prices for competitors in the downstream market. This kind of price discrimination drives competitors into disadvantageous situations. This is called 'raising rivals' costs'. Margin squeeze refers to the same kind of situation.

Two measures have been discussed to prevent these kinds of abuse of market power. One is conduct regulation to control the supply conditions of input goods or services, and the other is structural separation to divide a vertically integrated firm into a monopolistic input supplier and a final goods or services provider in the downstream market.

Although anti-competitive problems have been widely discussed, we must take into account the fact that vertical integration promotes efficiency[1]. The merits of vertical integration include the elimination of inefficiency between firms in addition to the usual economies of scale and scope.

A typical example is the elimination of the 'hold-up' problem. Investment into assets that have a value only in relation to a specific firm is called 'asset specific investment'. Once a firm is committed to an asset specific investment, the counterpart might demand ex post unfavourable changes in the terms of the business or the termination of the business relationship. If a firm fears that it might fall into this kind of situation, it holds back from useful investment. Vertical integration will solve this kind of hold-up problem.

The second merit relates to the coordination problem and transaction costs. When separate firms are operating in the upstream and downstream markets, it is essential to adjust between them specifications and qualities of the input goods or services in producing the final goods or services. As analysed in detail in the following sections, the coordination between incumbent access facilities providers and their competitors in the service market is more complicated. Various new technologies are developing in the broadband market and huge investment in unstable technologies is required on the part of incumbents. The coordination between incumbents and competitors is complex and accompanies considerable transactions costs. Vertical integration will eliminate the coordination between firms and reduce transaction costs. Therefore, structural separation should be re-evaluated by taking transaction costs into account.

It seemed that the FCC understood this point as it performed balancing analyses between efficiency and public interests in the investigation of merger cases in the U.S.A. (Kamino (2009)). In the case of the EU, ADSL offered on POTS facilities dominates the broadband market. Therefore, structural separation of incumbents is still being discussed. In contrast, new technologies such as FTTx are widely deployed in the Japanese broadband market. However, structural separation of the incumbent NTT is still on the discussion table without paying due attention to the considerable transaction costs involved. I will analyse the differences in the broadband market structure in the U.S.A, the EU and Japan and evaluate structural separation policy from the view point of transaction costs in the following sections.

THE TELECOMMUNICATIONS MARKET AND STRUCTURAL SEPARATION

In the case of the telecommunications industry, former monopolistic government institutions were privatised with the introduction of competition in many countries[2]. New entry began from the long-distance market because of the cost and tariff structure of the industry. New entrants in the long-distance market could not offer their services without interconnecting with the local facilities of the incumbents. At the same time, a dominant carrier in the local market that is integrated into the long-distance market competes with new entrants that utilise its local services as input (Armstrong (1998)).

Vertical foreclosure by former state monopolies was a focus of regulation. There was a temptation for former monopolies to engage in such anti-competitive behaviour as the refusal to interconnect or setting higher interconnection charges to exclude new entrants. Therefore it was necessary to prevent vertical foreclosure by the former monopolies in order to secure fair competition.

The measures adopted to avoid these problems are largely classified into 'conduct regulation' and 'structural separation'. Conduct regulation includes the imposition of the obligation to interconnect and the regulation of interconnection charges. Structural separation means the separation of a monopolistic division (local businesses) and competitive division (long-distance businesses) as independent legal entities.

The actual interpretation of structural separation in the telecommunications industry ranges from narrowly defined separation with ownership unbundling to mere accounting separation as classified by Cave (Table 1). As the European Commission (EC) defines separation with ownership unbundling as 'structural separation' and separation without ownership unbundling as 'functional separation' in the EC 2007 telecommunications regulatory reform (EC (2007a)), I follow this EC

Table 1. Various structural separation

Classification		Examples
Ownership separation (in whole or part)		Divestiture of AT&T (1984)
Legal separation (separate legal entities under the same ownership)		Reorganisation of NTT (1999)
Business separation	With separate governance arrangements	BT's Openreach (2006)
	With localised incentives	
	Simple	NTT before reorganisation in 1999
Virtual separation		
Creation of a wholesale division		
Accounting separation		NTT local companies

Source: Examples added to Cave (2006)

definition hereafter in this chapter. The main measure adopted at the early stage of the introduction of competition was conduct regulation. Only the U.S.A. among major countries adopted structural separation with ownership unbundling.

However, discussions on structural separation have gained momentum in recent years based on the assertion that conduct regulation was not sufficient in preventing strategic anti-competitive behaviour of incumbents.

The classic model for structural separation was the divestiture of AT&T in the U.S.A. in 1984[3]. AT&T was divided into the new AT&T and 23 RBOCs. The line of business control was introduced on the new AT&T and RBOCs. The new AT&T was limited in its business areas to long-distance (inter LATA) businesses and the 23 newly born RBOCs offered local (intra LATA) services. Here I would stress the fact that telecommunications services at the time were basically POTS. It was comparatively easy to separate local and long-distance networks, as the POTS technology was mature and stable. As AT&T had several operating companies and a long-line division under a holding company, the implementation costs of structural separation were not significant. In the case of the divestiture of AT&T, it is not unreasonable to think that the implementation cost was not crippling due to these conditions.

BT's 'Openreach' in the U.K. that is often cited in the discussions on structural separation is classified as business separation, because BT set up Openreach as an autonomous business unit that is in charge of access facilities defined as bottleneck within the BT Corporation[4]. Functional separation was adopted in the U.K. Although conduct regulation has been implemented since the privatisation of BT in 1984, the progress of competition in the fixed communications market was slow to develop. Ofcom performed a series of 'Strategic Reviews of Telecommunications' in view of this situation and urged BT to secure 'equality of access' between competitors and its own retail divisions by suggesting ownership unbundling (Ofcom (2004a), (2004b)). As a result, BT voluntarily proposed the functional separation of its access division in February 2005 and set up Openreach in January 2006 (BT (2006)). Ofcom expected that the competitive problems resulting from BT's vertical integration of its access division with its retail division that offers services based on access services would be eliminated by functional separation (Ofcom (2005)).

Competition was introduced in 1985 with the privatisation of NTT Public Corporation in Japan[5]. However, conduct regulation including interconnection conditions was not well provided for. Although the divestiture of NTT was proposed in the study of the actual structure of privatisation for NTT, NTT was privatised as a single entity following a political compromise. Therefore discussions on structural separation lingered on after 1985. Once it seemed that the discussions were settled by the reorganisation of NTT in 1999 under a holding company structure. NTT was reorganised into NTT-East, NTT-West, NTT Communications, and NTTdocomo under this holding company structure. The reorganisation is classified as a separated subsidiary form of structural separation as it does not accompany ownership unbundling.

It became relatively easier to separate between local and long-distance businesses because NTT completed the digitalisation of the telecommunications network. However, this distinction was applied only to POTS and the diffusion of the Internet has blurred the distinction between local and long-distance services and the framework has become inconsistent with the development of the Internet.

In contrast to the POTS market where mature technologies are deployed, various access technologies have started to burgeon in the broadband market where the implementation of structural separation is not without its problems. We should objectively estimate the costs and benefits of structural separation in the broadband market. Before discussing the costs and benefits of structural separation in broadband market, I will analyse the differences in the structure of the broadband market between the U.S.A and the EU, and Japan. Remedies should be different depending on the difference in market structure.

As stated above, although functional separation is discussed in many countries, structural separation with ownership unbundling is not implemented except in the U.S.A. There are two reasons behind this. One is the existence of a trade off between the fair competition expected to be realised by structural separation and the efficiency accompanying vertical integration. The other is the indispensability of conduct regulation such as regulation on interconnection charges even after the implementation of structural separation.

More important are the changes in the structure of the telecommunications industry. In the POTS era, telecommunications services were vertically integrated from physical to content layers based on the established circuit switched network technology that was composed of metallic cables and exchanges. In the case of broadband that utilises IP and various access technologies such as ADSL, FTTx and wireless, the physical network, service and content layers are horizontally separated. It has become difficult to decide where to structurally separate. It is also feared that structural separation might prevent the development of new technologies. As a result, the discussions against vertical integration have retreated behind the scenes and mergers involving vertical integration such as between SBC and AT&T or between Verizon and MCI have been approved (Kamino (2009)) in the U.S.A.

However, the EU is continuing discussions on structural separation of former state monopolies and the discussions on structural separation with ownership unbundling of NTT are still being pursued in Japan. I submit that structural separation in the telecommunications industry should be re-evaluated by taking into account the changes in industry structure and technological innovation.

THE DIFFUSION OF BROADBAND AND THE CHANGES IN MARKET STRUCTURE

The American Broadband Market

One of the characteristics of the broadband market in the U.S.A. is that facilities based competition has developed between LECs (Local Exchange Carriers)' ADSL services and cable modem services offered by cable operators. According to the FCC's statistics (Table 2) that give the market shares by technology, ADSL is 39.6% and cable

Table 2. Broadband shares by technology in the U.S.A (Source: Calculated from FCC (2010)).

	June, 2007	June, 2009
Cable modem	52.4	53.1
ADSL	42.8	39.6
Fibre	2.0	4.5
SDSL etc.	0.9	0.9
Others	1.9	1.9

modem is 53.1% if we exclude cellular that is not regarded as broadband in many other countries.

The development of facilities based competition between cable modem and ADSL is explained by the asymmetric regulation between cable modem and ADSL. Although cable tariffs were regulated by the Cable Competition Act of 1992, new services offered over cable networks such as cable modem or VOIP are not regulated. Cable operators are not only free from an unbundling obligation but also from an 'open access' obligation. They can bundle their own ISP services with cable modem. Further, MSOs have long been authorised in the cable industry in the U.S.A. Big MSOs have been able to invest in cable modem facilities due to abundant financial resources.

In the case of ADSL, the FCC rule based on the Telecommunications Act of 1996 (FCC (1996)) imposed an unbundling obligation on LECs. LECs sued several times against the FCC rules and the FCC was obliged to review the rule, whenever it was ordered to suspend the rules. The litigations were settled finally by the new rules of the FCC[6]. Although the FCC maintained the unbundling obligation on copper loops, other unbundling obligations were greatly relaxed.

a) Although the Line-sharing obligation for existing CLECs (Competitive Local Exchange Carriers)' customers was maintained, the obligation for new customers were to be gradually relaxed and will be abolished after 3years.
b) The Unbundling obligation of FTTx loops was abolished except for 64 kbs voice grade passes.

Further, the FCC ruled that ADSL was included as an 'information service' in order to maintain consistency with its rule that cable modem is 'information service' (FCC (2005)). Owing to the revision, ADSL can compete under the same conditions as cable modem and LECs have started to invest in ADSL facilities. Facilities based competition between cable modem and ADSL has started to develop. As a result, the discussions on structural separation of LECs have ended. On the contrary, it seems that the FCC has given up its structural separation policy as mentioned in the former section. Vertical integration has progressed through its authorisation of the mergers between SBC and AT&T and between Verizon and MCI.

The main objective of structural separation is to prevent the abuse of market power of a dominant firm in the upstream market toward the downstream market. The progress of facilities based competition in the broadband market means that the bottleneck nature of the access facilities of LECs is phased out. LECs no more hold market power in the upstream market and the possibility of the abuse of market power disappears as shown in Figure 1. It is not necessary any more to introduce structural separation. As the competition between LECs and cable operators in broadband access services has progressed in the U.S.A., the discussions on structural separation have become low profile and the mergers between RBOCS and long-distance carriers were approved.

Here it is important to notice that this is the result of the FCC regulation that allowed the formation of big MSOs (Multiple System Operators) with strong financial resources and induced the investment in cable modem facilities by the classification of cable modem as an 'information service', while it kept its unbundling obligation on LECs. The regulation of the FCC has contributed to the formation of the specific broadband market structure in the U.S.A. The development of the facilities based competition in the broadband market has made it unnecessary to implement structural separation. The vertically integrated structure of LECs and cable operators brought about efficiency by eliminating the transaction costs of structural separation.

The Broadband Market in the EU

The introduction of competition implemented by the Full Competition Directive (EC (1996)) was rather late in many of the EU member states except for the U.K. The progress of competition in the POTS market lagged behind the U.S.A. or Japan. When regulatory policy on broadband was discussed in both countries, the discussion in the EU was still focused on POTS. For example, although the usage of LLU (Local Loop Unbundling) was expanded in Germany at the beginning of 2000, it was mainly used for voice services. It was around 2004 when LLU began to be utilised for broadband. Although the EU imposed a LLU obligation in 2000 (EC (2000)), it moved to a market analysis approach. It will decide whether or not to regulate based on market analysis. Therefore the regulation on LLU differs depending on the market structure of member states.

The penetration rate of cable was comparatively high in some of the member states and there existed the possibility of facilities based competition developing between LECs and cable operators. However, in the case of Germany where most of the cable network was owned by DT, cable modem was not widely diffused because DT was reluctant in the expansion of cable modem for fear that it might cannibalise its own ADSL.

As the diffusion of broadband in the EU lagged behind Japan or the U.S.A, and the ADSL offered by competitors that utilise LEC's LLU was not developed (Table 3), discussions on structural separation were more vigorous in the EU.

It was the U.K. that first started discussions on structural separation as stated in the previous section. The EU stance on structural separation has not been settled[7]. The EC proposed in the process of its '2006 Review of the Regulatory Framework (EC(2006))' to introduce functional separation as an exceptional measure when it is considered that conduct regulation by itself is not sufficient in securing fair competition (EC (2007b)). The ERG expressed the view in October 2007 that functional separation may be introduced after conducting a full cost and benefit analysis

Figure 1. Structural separation and broadband market structure in the U.S.A.

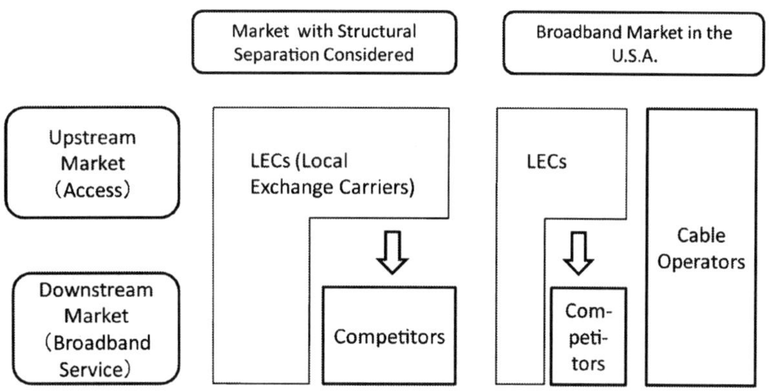

Table 3. Utilisation of LLU (As of September, 2008)

		Dry Copper		Line-Sharing		Total		Number of Metallic lines	Ratio among Total Lines	
		June, 2004	Sept, 2008	June, 2004	Sept, 2008	June, 2004	Sept, 2008		2004	2008
U.K.	Number of lines	7,580	1,448,407	6,270	3,635,323	13,850	5,083,730	21,035,169	0.1	24.2
	Distribution(%)	54.7	28.5	45.3	71.5					
Ger-many	Number of lines	650,000	7,900,000	0	100,000	650,000	8,000,000	29,200,000	2.2	27.4
	Distribution(%)	100.0	98.8	0.0	1.3					
France	Number of lines	13,066	4,574,000	711,754	1,434,000	724,820	6,008,000	21,000,000	3.5	28.6
	Distribution(%)	1.8	76.1	98.2	23.9					

Source Calculated from ECTA (2009)

only when conduct regulation is deemed insufficient in securing fair competition (ERG (2007)). It clearly opposed the introduction of structural separation with ownership unbundling.

Some countries including France and Spain rejected the introduction of functional separation in their own countries and the German Ministry of Economics and Technologies is against the introduction of functional separation. However, Sweden and Italy have introduced functional separation.

Thus the attitude toward structural separation varies from country to country. This is a reflection of the fact that the costs and benefits analysis of structural separation has remained abstract and it has not been proved that benefits exceed costs (OECD (2003)). Although OPTA, the Dutch regulatory authority, suggested steps to evaluate structural separation based on a survey of academic studies (OPTA (2006)), it has not yet made a proposal for actual evaluation. Here, I should like to point out the fact that although the utilisation level of LLU in the U.K. that adopted functional separation is catching up to the level of France and Germany (Table3), it is not demonstrated that the introduction of structural separation on top of conduct regulation is effective in promoting competition. The recent discussions on structural separation in the EU are a reflection of the slow development of broadband and the reliance on ADSL that is offered on POTS facilities.

The Japanese Broadband Market

The distinction between local and long-distance services disappeared with the diffusion of broadband based on new technologies, so that now we can only distinguish between access and the network. Thus the discussions on structural separation have now moved on to the treatment of access division of incumbents in Japan.

The Japanese broadband market started to grow in 2000 when the line-sharing and co-location obligation was introduced[8]. As the line-sharing charges were set very low (Table 4), service-based new entry was encouraged. Softbank Corp., the largest new entry in the broadband market, deployed aggressive marketing strategies based on low price. As a result, Japan has been leading the world broadband market in terms of price and speed.

FTTx in Japan started to grow replacing ADSL. Two NTT local companies and subsidiaries of electric power companies are competing on the facilities base in the FTTx market. They decreased the retail price of FTTx to compete with Softbank's cheap ADSL. As a result the number of users of FTTx exceeded that of ADSL in the first quarter of 2008 (Figure 2) and Japan is the first in the

Table 4. Comparison of LLU charges (October, 2004)

	Full unbundling	Shared access (Line Sharing)
Japan	NTT-East: 1,256 yen NTT-West: 1,318yen	NTT-East: 158 yen NTT-West: 165 yen
The UK	8.76 pound (1,752 yen)	2.26 pound (452 yen)
Germany	11.80 euro (1,652 yen)	2.43 euro (340 yen)
France	10.50 euro (1,470 yen)	2.90 euro (406 yen)

(Note) Calculated on the basis of the exchange rate at the time (1 pound = 200 yen, 1 euro = 140 yen)
Source: Kamino & Fuke (2010)

world in terms of the diffusion of FTTx. NTT local companies further started to offer commercial NGN (Next Generation Network) services from March 2008.

However, the rapid growth of FTTx brings regulatory distortion to light. Both an obligation to offer dark fibres and a margin squeeze rule are imposed on NTT local companies. If NTT local companies want to lower their retail prices, it is necessary for them to decrease wholesale prices. The wholesale prices are set on the level at which NTT local companies assert that they are losing money in the wholesale market. As a result, many competitors gave up facilities based competition except one subsidiary of an electric power company. Therefore, the share ofNTT local companies has been going up steadily as shown in Table 5 and their share of the retail market is also going up as shown in Table 6. The share of facilities based competitors such as subsidiaries of electric power companies and service based competitors like USEN and KDDI also decreased.

Here, I will refer to cable modem that is important in comparison with the broadband market

Figure 2. Number of broadband users in Japan (Source based on MIC data)

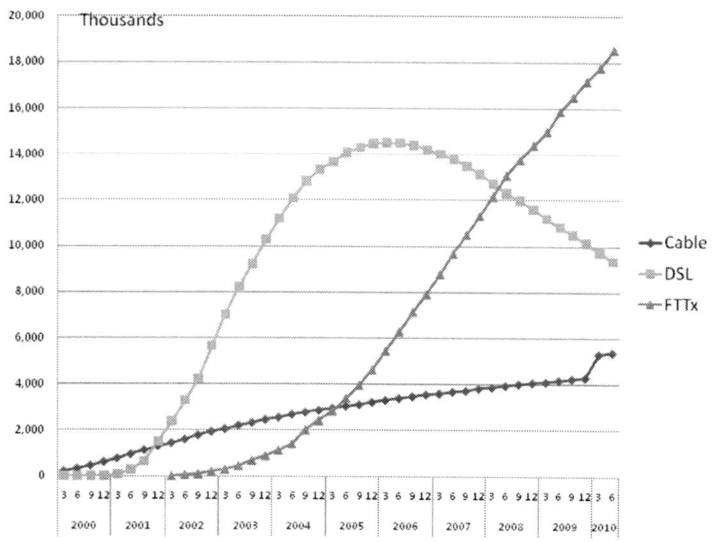

Table 5. NTT local companies' FTTH share (Wholesale) (%)

	2005	2006	2007	2008	2009
Metallic	94.8	93.9	92.8	92.2	90.6
FTTH	78.6	78.9	78.9	78.8	77.3
Total	93.8	92.5	91.0	90.0	87.9

Source: Calculated based on MIC data

in the U.S.A. Although cable modem users are increasing steadily, the percentages of cable modem users among broadband users in Japan is still 16.1% as shown in Figure 2. This reflects the late approval of MSOs in Japan. As the MIC (Ministry of Internal Affairs and Communications) limited the business areas of cable operators within each city, town or village and did not approve MSOs until 1993, the market was long dominated by a number of very small operators. Since the liberalisation of MSOs in 1993, MSOs like J: Com (Jupiter Telecommunications Co., Ltd.) was formed by Liberty Global, Inc. and Sumitomo Corporation. The number of users of J: Com is only 2.6 million as of March 2010compared to 24 million users of Comcast as of the end of 2007. Its revenue is about a tenth of NTT local companies (Table 7). It is hard to expect that such a financially weak company can compete effectively with the huge NTT local companies. We can conclude that cable regulation has retarded the development of inter modal competition between FTTx and cable modem[9].

However, this understanding is not necessarily shared by many and the discussion on the structural separation of NTT has been revitalised. The report of the 'Ministerial Panel on the Framework of Communications and Broadcasting' issued in June 2006 (MIC (2006a)) stated as follows.

Accounting separation, reinforcement of the system to secure the observance of the interconnection rules, realisation of equality of access to NTT local companies' bottleneck facilities are to be considered and measures to secure fair competition regarding video distribution through the IP network are also to be considered. Further it is necessary to functionally separate the bottleneck facilities divisions within NTT local companies.[10]

Further, it was decided to investigate whether it is appropriate to introduce ownership unbundling of NTT group companies in 2010 (MIC (2006b)). However, the measures taken in the U.K. where the main access technology to broadband is metallic cables and FTTx is little deployed and the market power of BT is still strong do not necessarily fit to Japan where the broadband market structure is different. The reason why ownership unbundling is

Table 6. FTTx share (Retail market) (%)

	Business & Independent Houses20042009		Flats20042009		Total20042009	
NTT Local Companies	74.6	77.5	35.0	70.1	57.5	74.4
Subsidiaries of Electric Power Companies	22.3	11.5	8.8	5.8	16.2	9.2
USEN	1.6		20.3	6.7	9.7	3.0
KDDI		8.0	7.3	8.0		8.0
Others	1.4	3.0	28.6	9.4	16.7	5.3

Source: MIC

Table 7. Business scale of NTT local companies and J:Com (Fiscal 2009) (Unit: Million Yen)

	NTT-East	NTT-West	J:Com
Revenue	1,928,652	1,780,821	333,724
Operating Profit	47,634	18,475	61,159

Source: Financial data published by each company

to be imposed on top of functional separation was not well explained. Here, it is important to note that costs and benefits analysis of structural separation have not been performed in the broadband market where new technologies such as FTTx are developing and transactions costs are significant.

THE BROADBAND MARKET AND STRUCTURAL SEPARATION

I have analysed the structure of the broadband market and policies on structural separation in the U.S.A., the main EU countries and Japan. The unique cable policy and asymmetric regulation between cable modem and ADSL promoted facilities based competition between cable operators and telecommunications carriers in the U.S.A. As a result, structural separation is not required any more in the U.S.A. This means that a particular institutional factor contributed to the formation of a particular market structure and the policy implications for other countries are limited. Therefore, I will concentrate on the comparison between Japan and the EU.

Although the diffusion of broadband is rather slow in the EU, the fact that the penetration of broadband is higher in France or Germany where there are negative attitudes towards structural separation than in the U.K. which implemented the functional separation of BT casts doubts on the effectiveness of structural separation in promoting broadband.

The analysis described in the above sections will lead to the following propositions. The imposition of line-sharing and co-location obligations on NTT local companies together with aggressive marketing by Softbank Corp. contributed to the rapid growth of ADSL in Japan. NTT local companies set the price of FTTx services very low to be competitive with the cheap ADSL services of Softbank Corp. and the FTTx services offered by subsidiaries of electric power companies. This has contributed to the rapid growth of FTTx services in Japan. The main factor behind the rapid growth of broadband is not the structural separation of NTT but conduct regulation including LLU and a dark fibre obligation[11].

I will elaborate on the differences in market structure between POTS and broadband in the following sections.

When we study structural separation in the broadband market, we should notice the differences in market structure between POTS and broadband. The distinction between local and long-distance has become obscure with the diffusion of broadband, as the Internet is characterised by the layered structure of physical, service and content.

The basic elements of the POTS network are composed of metallic subscriber cables and local exchanges (Figure 3). They deploy mature technologies and we cannot expect further significant innovations in these areas. When we consider structural separation, we can stipulate the dividing lines at the interface between subscriber cables and the entrance of local exchanges or at the exit of local exchanges. Further, access lines, i.e. metallic subscriber lines and local exchanges have been already installed. The incumbents acting in both the upstream and downstream markets need not invest in upstream facilities. This means that such problems as hold-up, coordination, or transaction costs are not significant in the POTS market.

As the Internet is a self-governed and dispersed network and voice is handled as one of the various applications over the Internet, the distinction between local and long-distance has disappeared. The only remaining distinction is between access and the network. Therefore, discussions on structural separation in the broadband market will be focused on access division.

Various access technologies in the broadband market are being introduced. Technologies such as ADSL, cable modem, Wi-Fi, WiMAX, FTTx and cellular have been appearing in the market one after another and it is difficult to predict which technology will dominate the market. Even in the case of FTTx, actual implementation varies. In the basic case, an individual subscriber uses exclusively a fibre core. In the case of VDSL, fibre cores are shared by some users by interconnecting with the inside metallic cables. NTT local companies are offering services for individual users based on the sharing of a fibre core by several users. In other cases a fibre core is highly utilised based on wave-length multiplexing technology. Therefore, the definition of bottleneck facilities depends on technological innovation. NTT local companies are offering shared fibre services. In this case, up to 8 users share a fibre core. Competitors are demanding the unbundling of each shared element of a fibre and NTT local companies are insisting that they should offer dark fibre only on a core basis. The designation of bottleneck facilities depends on the definition of the market. Unless we define the market, we cannot decide whether market power exists or not. Even if market power is admitted to exist, it is hard to decide where to separate the vertically integrated firm.

The Internet technologies are making steady progress. Both IPV4 and IPV6 are utilised with the introduction of NGN by NTT local companies and the coordination between access technologies and the network has become more complicated as shown in Figure 3. These facilities are in the process of being built up. Huge investment is required and it is necessary to coordinate on specifications and service qualities with downstream firms. Thus it is important to consider such problems as hold-up coordination and transaction costs in a broadband market.

Figure 3. Structure of POTS and broadband

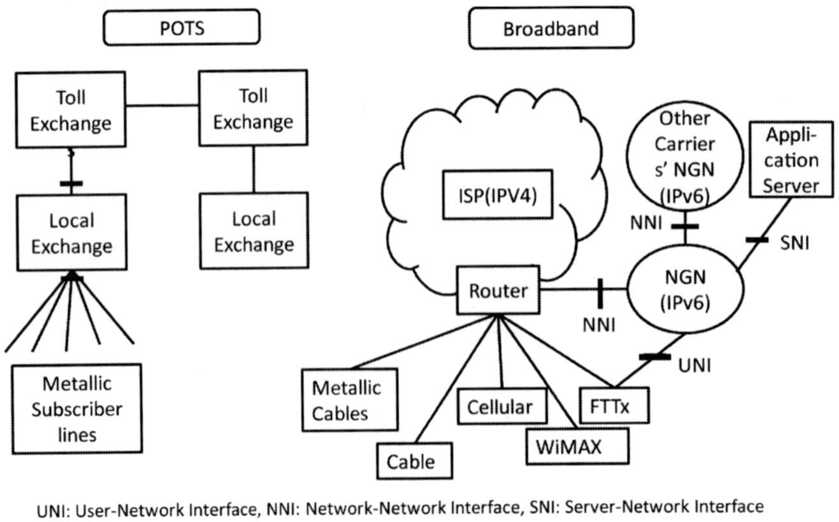

STRUCTURAL SEPARATION IN THE BROADBAND MARKET AND TRANSACTION COSTS

Structural separation has been discussed by many scholars from the view point of industrial organisation or competition law[12]. One of the characteristics of recent developments is the balancing analysis between costs and benefits of vertical integration. As stated in the above sections, it is important to take into account coordination problems and transaction costs in the broadband market where new technologies such as FTTx are newly introduced. There are many papers that have discussed the balancing analysis. I would elaborate on the topic by combining Cave's discussions (Cave (2008)) that evaluated structural separation in relation to NGN with transaction economics[13].

Cave's discussion is summarised under the following five points.

a) While regulation properly responds to industry structure it also shapes it.
b) Even the bottleneck parts of the network are characterised by periodic substantial re-investments.
c) Incentives should be put in place to ensure investment and innovation.
d) Separation creates challenges in the coordination of both operation and investment decisions.
e) Non-price discrimination can often be solved by behavioural remedies.

The first point means, as discussed earlier in this chapter, that regulation defines the market structure. Early studies by the Harvard scholars on industrial organisation relied on the structure, conduct and performance paradigm (Bain (1959)). The rationales for regulation in the telecommunications industry have been discussed based on a market structure in which incumbents had market power in the upstream market. However, it is important to recognise that the market structure itself might be shaped by regulation. Although both conduct regulation and structural separation were imposed on incumbents because they own bottleneck facilities, the facilities stay regulated as bottleneck even when the facilities that were considered bottleneck have become duplicable. Differently from other industries in which vertical integration was discussed, the telecommunications industry is characterised by dynamic technological changes. POTS is built on comparatively mature and stable technologies and the discussions on structural separation used to be comparatively simple[14]. Contrary to POTS, broadband is going through rapid technological changes and various access technologies including ADSL, cable modem, FTTx, Wi-Fi, WiMAX, and LTE (Long Term Evolution) of cellular services are utilised. If we add structural separation to conduct regulation on FTTx like Japan, the development of a burgeoning facilities based competition in the broadband market will be retarded and FTTx will continue to stay as a bottleneck facility. Cave, further indicates that functional separation by having set up Openreach in the U.K. is delaying the take up of FTTx.

Second, long-term large scale investment is required for bottleneck facilities in broadband. The quality of service of ADSL depends on the quality of copper cables and it has some limitations as follows.

a) Unstable transmission quality
b) Limited geographical availability
c) Slow up-link speed

FTTx is superior to ADSL as a fixed access technology to broadband[15]. However, FTTx is built on new facilities that require huge investment. This relates to Cave's third point about investment incentives.

Huge investment and investment incentives are closely linked to the fourth point about coordination problems between wholesale and retail businesses. The coordination problems arise when there is structural separation between the wholesale business managing bottleneck facilities and the retail business utilising bottleneck facilities. It is absolutely necessary to co-ordinate, in the case of highly enhanced services like broadband, between the access business that manages bottleneck facilities and the retail business that offers network services to end-users. Structural separation makes the coordination between wholesale and retail businesses complex and such problems as hold-up become real. These kinds of coordination problems will accompany transaction costs.

Williamson has proposed bounded rationality, complexity and uncertainty, opportunism, a small number condition and information imperfectness as causes of transaction costs and pointed out that vertical integration might contribute to the elimination of transaction costs (Williamson (1985) pp.15-102). He further stated that vertical integration was more favourable when the investment involved asset specificity. Here asset specificity is defined as durable investments that are undertaken in support of particular transactions (Williamson (1985) p.55). Although Cave's discussions are not necessarily developed on Williamson's study, I think his discussions will become more convincing if we look at them through Williamson's works.

The investment in FTTx is a typical asset specific investment. Williamson defined an asset specific investment in terms of site, physical, human and dedicated (Williamson (1985) p.55). FTTx belongs to three of these categories. Only the human factor is absent. The investment in wholesale facilities of FTTx is targeted towards retailers of FTTx. FTTx facilities are not transferrable for other uses and the investment becomes sunk once FTTx facilities have been constructed. When the obligation to offer dark fibres to retail carriers is imposed on incumbent carriers and there is no obligation to use it for a certain period, as in Japan, the regulation is likely to induce opportunistic behaviours on the side of retail carriers. When retail customers are geographically scattered, they offer services based on wholesale products. However, once the density of end users exceeds a certain level, it is possible for them to switch to their own facilities. As a result, wholesale carriers are obliged to keep not fully depreciated facilities without users. This kind of problem will become serious if the regulation to oblige wholesale carriers to offer on a shared branch basis of a fibre core is imposed as new competitors are demanding in Japan[16].

The level of the investment by wholesale carriers is likely to stay below a socially optimal level. If these kinds of problems are to be avoided, it is necessary to negotiate detailed dealing conditions within the contract. This will result in big transaction costs. If we want to avoid big transaction costs, it is preferable to keep vertical integration.

The fourth point regarding structural separation is that it becomes necessary to coordinate regarding interfaces between wholesale and retail facilities, investments and day to day operations. As mentioned in the previous section, interfaces between wholesale and retail facilities have become very complicated with the development of broadband and the introduction of NGN. NTT indicated three typical interfaces with retail carriers, i.e. UNI (User-Network Interface), NNI (Network-Network Interface) and SNI (Application-Server Interface). NNI is the most complicated, as NGN is expected to interconnect with other carriers' networks including, IP networks, Ethernets, ISP networks and traditional fixed and mobile networks. If NGN is regarded as a bottleneck facility as in Japan, it becomes necessary to settle interface conditions that vary according to the detailed specifications of networks that NGN interconnects with. The technological specifications of NGN have not been fully settled yet. It took a very long time to fix even temporary conditions. Prolonged discussions are still going on over the actual implementation of the interconnections between NGN (IPv.6) and other IPv.4 networks.

As NTT is not structurally separated at present, it has been able to manage the complicated negotiations based on the information obtained through discussions with its own prospective users. If it is structurally separated, it is obliged to rely on information supplied by competitors that is likely to be biased toward the interests of the competitors. Interface conditions reached might not be optimal. If we want to avoid this kind of unfavourable situation through negotiations between the parties concerned, the transaction costs involved could be huge.

As for the specifications, quality and scale of the investment and geographical areas for investment, these must be decided according to the users' demand in the market. Wholesale carriers cannot decide on the optimal investment because they have only inadequate information about the retail market. As wholesale carriers have no interface with end users, they are obliged to accept the size of the demand forecast given by retail carriers. The size of the demand forecast given by retail carriers tends to be too big and the quality of services is likely to be excessive as retail carriers bear no investment risks.

The coordination of day to day operations such as maintenance and the handling of failure reports will also incur huge transaction costs, because the interfaces between networks are complicated.

In summary, if we try to avoid these kinds of problems through contractual arrangements, huge transaction costs become inevitable.

The fifth point relates to Cave's claim that price discrimination by incumbent carriers can be dealt with by conduct regulation and that most forms of non-price discrimination can also be solved by conduct regulation (Cave (2008)). Conduct regulation such as unbundling of local loops, co-location and sharing of ducts contributed to the rapid diffusion of broadband in Japan. The diffusion of ADSL based on LLU in France and Germany where functional separation is denied is faster than the U.K. where functional separation was implemented. Thus, it is clear that conduct regulation plays an important role in promoting competition and the role of the structural separation is limited.

CONCLUSION

There are two important implications derivable from the above analysis of broadband markets in the U.S.A., the EU and Japan. Firstly, we must be cautious in adopting structural separation for the broadband market, as the market is going through rapid technological changes differently from the static POTS market. The uncertain prospects for technological changes are likely to bring serious hold-up and coordination problems and the application of structural separation to the broadband market is problematic. It is not appropriate to simply apply the traditional Harvard school's "structure, conduct and performance" paradigm to broadband markets. At least, we should reconsider the simple analysis that asserts that it is necessary to structurally separate the NTT group on the grounds that NTT's shares in the wholesale and retail markets are very high. At the same time it is not possible to simply conclude that vertical integration contributes to efficiency based on the Chicago theory. It is necessary to conduct full and specific cost and benefit analyses and to reach logical conclusions.

Second, we should appreciate that the present structure of the broadband market is shaped by regulation. It is not appropriate to justify structural separation based on the present market structure.

The broadband technologies including NGN and the broadband market on which my discussions have been based have not matured yet. My arguments are therefore tentative and need to be borne out by the actual development of the market. However, it is at least clear that structural separation discussions from the POTS era do not necessarily fit the broadband market. Once structural separation is implemented, it is hard to reverse the process, even if we find it unfavourable after its implementation.

REFERENCES

Armstrong, M. (1998). Network Interconnection in Telecommunications. *The Economic Journal*, *108*, 545–564. doi:10.1111/1468-0297.00304

Bain, J. S. (1959). *Industrial Organization*. Hoboken, NJ: John Wiley & Sons.

BT.(2006). *Openreach-Open for business*. News release (DC06-010, January 10, 2006).

Cave, M. (2006). Six Degrees of Separation: Operational Separation as a Remedy in European Telecommunications Regulation. *Communications & Strategies*, *64*, 89–103.

Cave, M. (2008). *Vertical Integration and the Construction of NGA Networks*. Paper commissioned by Telstra

EC. (1996). *Directive 96/19amending Directive 90/388 with regard to the implementation of full competition in telecommunications markets*. Full Competition Directive.

EC. (2000). Regulation (EC) No 2887/2000 of the European Parliament And Of The Council of 18 December 2000 on unbundled access to the local loop.

EC. (2006). Communication from the commission to the council, the European parliament, the European economic and social committee and the committee of the regions on the Review of the EU Regulatory Framework for electronic communications networks and services, *COM(2006) 334 final, 29.6.2006*

EC. (2007a). Proposal for a directive of the European parliament and of the council of … 2007 amending Directives 2002/21/EC on a common regulatory framework for electronic communications networks and services, 2002/19/EC on access to, and interconnection of, electronic communications networks and services, and 2002/20/EC on the authorisation of electronic communications networks and services.

EC. (2007b). *Commission staff working document, impact assessment*. Accompanying document to the Commission proposal for a Directive of the European Parliament and the Council, {COM (2007) 697, COM (2007)698, COM (2007)699, SEC (2007)1473}, 13 November 2007.

ECTA. (2009) *Broadband Scorecards*. Retrieved October 15, 2009, from http://www.ectaportal.com/en/basic650.html.

ERG. (2007). *ERG Opinion on Functional Separation (ERG (07)44).*

FCC. (1996). *The First Report & Order on the local Competition Provisions of the Telecommunications Act* (FCC 96-325).

FCC. (2003). Review of the Section 251 Unbundling Obligations of the Telecommunications Act of 1996; Deployment of Wireline Services Offering Advanced Telecommunications Capability; Appropriate Framework for Broadband Access to the Internet over Wireline Facilities. *(FCC03-36), 20 February 2003.*

FCC. (2004). *FCC Releases New Rule for Network Unbundling Obligations of Incumbent Local Phone Carriers*, Order on remand (FCC 04-290), 15 December 2004.

FCC. (2005). *FCC Eliminates Mandated Sharing Requirement on Incumbents' Wireline Broadband Internet Access Services, Report and order and Notice of Proposed Rulemaking*, (FCC 05-150), 23 September 2005.

FCC. (2010). *Internet Access Services: Status as of June 30, 2009.* Retrieved October 15, 2010, from http://www.fcc.gov/Daily_Releases/Daily_Business/2010/db0902/DOC-301294A1.pdf

Fuke, H. (2000). *Structural changes and deregulation in the telecommunications industry*. Tokyo: NTT Publishing. (in Japanese)

Fuke, H. (2003). The Spectacular Growth of ADSL in Japan and its Implications. *Communications & Strategies*, *52*, 175–191.

Fuke, H. (2005). Structural Changes in Telecommunications and Regulatory Policy in Japan. *Communications & Strategies*, *59*, 113–139.

Fuke, H. (2007). *Info-communications policy in the broadband era*. Tokyo: NTT Publishing. (in Japanese)

Joskow, P. L. (2006). *Vertical Integration*. Prepared for the American Bar Association Antitrust Section's "Issues in Competition Law and Policy Project." Retrieved on October15, 2009 from http://econ-www.mit.edu/files/1191

Kamino, A. (2009). Transition of the US telecommunications mega merger reviews and problems-Balancing of efficiency and public interest [in Japanese]. *Journal of Information and Communication Research*, *26*(4), 51–66.

Kamino, A., & Fuke, H. (2010). Diffusion of the Broadband Internet and the Structural Separation. In Gentzoglanis, A., & Henten, A. (Eds.), *Regulation and the Evolution of the Global Telecommunications Industry*. Northampton, MA: Edward Elgar Pub.

MIC. (2006a). *Report of Panel on Frameworks of Communications and Broadcasting* (in Japanese).

MIC. (2006b). *Agreement between the Cabinet and Ruling Parties on Frameworks of Communications and Broadcasting* (in Japanese).

OECD. (2003). *The benefits and costs of structural separation of the local loop, DSTI/ICCP/TISP (2002)*. Final.

OECD. (2008). *Broadband subscribers per 100 inhabitants, by technology, December 2008*. OECD Broadband Portal. Retrieved October 15, 2009 from http://www.oecd.org/document/54/0,3343,en_2649_34225_38690102_1_1_1_1,00.html

Ofcom. (2004a). *Strategic Review of Telecommunications Phase 1 consultation document,* 28 April 2004.

Ofcom. (2004b). *Strategic Review Telecommunications Phase 2 consultation documents*, 18 November 2004.

Ofcom. (2005d). *Final statements on the Strategic Review of Telecommunications, and undertakings in lieu of a reference Under the Enterprise Act 2002,* Statement, 22 September 2005.

OPTA. (2004). *Vertical integration: efficiency & foreclosure*. Economic Policy Note, no.3, May 2004.

Reding, V. (2006a). The Review 2006 of EU Telecom rules: Strengthening Competition and Completing the Internal Market (Speech/06/422)

Reding, V. (2006b). From service Competition to Infrastructure Competition: the Policy Options Now on the Table (Speech/06/697)

Rey, P., & Tirole, J. (2007). A Primer in Foreclosure in Armstrong. In M. & R. H. Porter (Eds.), *Handbook of Industrial Organization* (Vol. 3, pp. 2145-222). Amsterdam: North-Holland.

Riordan, M. H. (2008). Competitive Effects of Vertical Integration. In Buccirossi, P. (Ed.), *Handbook of Antitrust Economics* (pp. 145–182). Cambridge, MA: The MIT Press.

Temin, P. with L. Galambos (1987). *The fall of the Bell System*. New York: Cambridge University Press.

Williamson, O. E. (1985). *The Economic Institutions of Capitalism*. New York: Free Press.

ADDITIONAL READING

Carlton, D. W., & Perloff, J. M. (2005). *Modern Industrial Organization* (4th ed.). Boston, MA: Pearson.

Coase, R. H. (1937). The Nature of the Firm. *Economica N. S., 4*, 386–405. doi:10.1111/j.1468-0335.1937.tb00002.x

Crandall, R. W. (2005). *Competition and chaos, U. S. Telecommunications since the 1996 Telecommunications Act*. Washington, D.C.: The Brookings Institute.

de Jong, H. W., & Shepherd, W. G. (Eds.). (2007). *Pioneers of Industrial organization: How the economics of Competition and Monopoly Took Shape*. Cheltenham, U.K.: Edward Elgar Publishing.

EC. (2002). *Directive 2002/21/EC of the European parliament and of the council of 7 March 2002 on a common regulatory framework for electronic communications networks and services*. Framework Directive.

Howell, B. (2009). *Separating New Zealand's Incumbent Provider: A Political Economy analysis*. Paper presented at the 4th Africa- Asia-Australia Regional Conference of the International Telecommunications Society, August 16-18, 20

Martin, S. (1993). *Advanced Industrial Economics*. Oxford, UK: Blackwell Publishers.

Nikolinakos, N. T. (2006). *EU Competition Law and Regulation in the Converging Telecommunications, Media and IT Sectors*. Alphen, The Netherlands: Kluwer Law International.

Shepherd, W. G., & Shepherd, J. M. (1997). *The Economics of Industrial Organization*. Long Grove, Illinois: Waveland Press.

Spulber, D. F., & Yoo, C. S. (2009). *Networks in Telecommunications: Economics and Law*. New York: Cambridge University Press.

Stigler, G. J. (1968). *The Organization of Industry*. Chicago, Illinois: University of Chicago Press.

Teppayayon, O., & Bohlin, E. (2009). *Functional Separation: Next Step of Improving Broadband Competition?* Paper presented at the 4th Africa-Asia-Australia Regional Conference of the International Telecommunications Society, August 16-18, 2009

Tirole, J. (1988). *The Theory of Industrial Organization*. Cambridge, MA: The MIT Press.

Vogelsang, I., & Compaine, B. M. (Eds.). (2000). *The Internet Upheaval: raising questions seeking answers in communications policy*. Cambridge, MA: The MIT Press.

Walden, I., & Angel, J. (2005). *Telecommunications Law and Regulation*. New York: Oxford University Press.

Williamson, O. E. (1975). *Markets and Hierarchies; Analysis and Antitrust Implications*. New York: Free Press.

Williamson, O. E. (Ed.). (1990). *Industrial Organization*. Cheltenham, U.K.: Edward Elgar Publishing.

Williamson, O. E., & Scott, S. E. (Eds.). (1999). *The Economics of Transaction Costs*. Cheltenham, U.K.: Edward Elgar Publishing.

Williamson, O. E., & Winter, S. G. (Eds.). (1993). *The Nature of the Firm: Origins, Evolution, and Development*. New York: Oxford University Press.

KEY TERMS AND DEFINITIONS

Bottleneck Facilities: Facilities owned by an upstream firm and essential for the operation of downstream competitors.

Broadband: High-speed Internet access. The definition of broadband differs from country to country. The FCC in the U.S.A. includes access lines that deliver services at speeds exceeding 200 kbps. In the case of Japan, access lines faster than 1.5Mbps are regarded as broadband.

Conduct Regulation: Regulations introduced to prevent abuse of market power in the upstream market. Obligation to interconnect, control on interconnection charges and accounting separation are some of the conduct regulations implemented in the telecommunications industry.

Functional Separation: Structural separation of vertically integrated firms into upstream and downstream businesses that does not accompany ownership unbundling.

NGN: Abbreviation of Next Generation Network that aims to secure the safety and service quality of IP based networks. NTT in Japan introduced commercial NGN services in March, 2008.

Structural Separation: Separation of vertically integrated firms. When narrowly defined, only separation with ownership unbundling is included. In some cases functional separation is also included.

Transaction Costs: Costs incurred by the coordination between firms. Vertical integration eliminates transaction costs.

Vertical Integration: Firms' business operation that participates in both upstream and downstream markets. In the case of telecommunications incumbents, carriers offering both local and long-distance services have been regarded as vertically integrated.

ENDNOTES

1. Refer to Joskow (2006), Rey & Tirole (2007), and Riordan (2008) for details.
2. Although the leading Telecommunications Company in the U.S.A., the old AT&T had been a private company since the incunabula of the industry, it de facto monopolised the industry and the vertical integration had been the focus of regulation.
3. Refer to Temin (1987) for details.
4. Sweden, Italy and New Zealand implemented a kind of functional separation.
5. Refer to Fuke (2000) and Fuke (2007) for details.
6. FCC (2003) and FCC (2004)
7. EC Commissioner Viviane Reding suggested the introduction of structural separation in abroad sense including structural separation with ownership unbundling and functional separation (Reding (2006a)). She later clarified that she meant just functional separation (Reding (2006b)).
8. Refer to Fuke (2003) and Fuke (2005)
9. Recently, KDDI, the second large cellular operator in Japan, announced its acquisition of J: Com. Changes in competitive landscape is expected.
10. Translation by the author.
11. It is true that NTT became receptive to conduct regulation as it was exposed to pressure to structurally separate.
12. Riordan briefly summarises the development of academic theory on structural separation (Riordan (2008))
13. Refer to Carlton & Perloff (2005), and Williamson (1985, 1975) for the academic theory in this section.
14. Even in the case of POTS, the scope of bottleneck facilities become limited to subscriber lines with the development of digital technology.
15. The switch to FTTx from ADSL has progressed in Japan. The number of users of FTTx overtook that of ADSL in the second quarter of 2008 and the ratio of FTTx users exceeded more than half of broadband users in the second quarter of 2009.
16. NTT local companies are also offering shared FTTX services that are cheaper than dedicated services. A certain number of users will share a fibre core with neighbouring users. Competitors are arguing that they may use one of the branches of a shared core.

Chapter 6
Mobile Phones: Like any Other ICT?
The Case of Greece and its Adoption of Mobile Phones from a Socio-Cultural Perspective

Panayiota Tsatsou
Swansea University, UK

ABSTRACT

This chapter explores mobile phones and how they have been received in juxtaposition with the Internet and in close association with socio-cultural contexts of life. By examining the Greek case and its particularities, the chapter provides some sense of why different Information and Communication Technologies (ICTs), such as mobile phones and the Internet, might be received and adopted differently by people living in the same socio-cultural context (at the national level). In the case of Greece, statistical and historical data confirm the contrasting receptions of mobile phones and the Internet but empirical evidence is lacking to explain the exceptionally high adoption rates of mobile phones in the country. Thus, the chapter reports on original empirical evidence obtained in elite actors' interviews and focus groups of ordinary people to explain the contrasting ways mobile phones and the Internet have been received in the country. On the basis of these empirical findings, the chapter finds that certain socio-cultural contexts, such as that of Greece, favour mobile phones more than the Internet, thus making mobile telephony a distinctive case of ICT.

INTRODUCTION

In general, Internet indicators are relied on more than those of mobile telephony for evaluating the progress of the information society (OECD, 2002). On the other hand, mobile telephony is approached as an ICT that can be positioned in more or less the same framework as any other ICT, with Internet, mobile telephony and other information society indicators being commonly viewed from an economic and market perspective (OECD, 2002).

This chapter argues that mobile telephony is a distinctive ICT due to its particular technical features and its exceptionally high degree of social acceptance (penetration) compared to

DOI: 10.4018/978-1-60960-011-2.ch006

the Internet in societies such as Greek society. Socio-cultural indicators in particular allow one to note how people negotiate the meaning, value and use of particular technology (-ies), making ICTs look like an extremely diverse and often uneven technological and communication domain. In this sense, it is argued that the distinctiveness of mobile telephony can not only be indicated in technocratic terms but also on the basis of its establishment and penetration in socio-cultural contexts which do not equally favour generally popular ICTs such as the Internet.

The chapter justifies the interest in comparing mobile telephony to other ICTs along with the focus on the particular context of Greece by examining the exceptionally high penetration rates of mobile telephony in contrast to the low penetration of Internet technologies in the country and by attempting to explain the dissimilar way mobile phones have been received on the grounds of socio-cultural factors. In what follows, an account of the literature examining mobile phones and other ICTs from a socio-cultural perspective is offered, thus highlighting relevant literature gaps and developments. Then, a brief account is provided of the Greek information society and the divergent dissemination directions that mobile telephony and the Internet have taken in the country. This case-focused discussion paves the way for a historically grounded exploration of the socio-cultural parameters that seem to have played an important role in how Greek people have received and integrated these two different and simultaneously so similar communication technologies into their lives. The empirical findings obtained in qualitative research are then reported and discussed. These findings were obtained from interviews with elite actors and focus groups of ordinary people in Greece, and provide some concrete and diverse insights into how and why mobile telephony in the country differs from other ICTs such as the Internet. The chapter concludes by summarising the findings and their significance as well as noting areas where more work and richer conclusions are required.

BACKGROUND: MOBILES PHONES AND OTHER ICTS FROM A SOCIO-CULTURAL PERSPECTIVE

A significant number of studies has looked at the role of socio-cultural milieus in ICTs (Cathelat, 1993; Mediagruppe, 2000; Klamer et al., 2000; SevenOneMedia, 2002) and stressed their importance for explaining not only the particular shapes and aspects of ICTs but also their appropriation by users in concrete socio-cultural contexts. More specifically, the literature has attempted to examine the positive role of social values, such as values of openness, in how ICTs are used, adopted and integrated into people's lives (Hofstede, 1980; Trompenaars, 1993; Rogers, 1995; Thomas, 1995; Thomas and Mante-Meijer, 2001; Smoreda and Thomas, 2001; Mante-Meijer, 2002). The literature has also looked at the opposite role that resistant elements of social culture and associated introversion and backwardness can play, leading to the slow and difficult development of particular technological artefacts and the limited appropriation of technology in people's lives (Mokyr, 1990 & 1992). Media technologies and ICTs are often among the technologies (e.g. biotechnology, nuclear technology etc) which are opposed by dominant resistant cultures in particular socio-cultural contexts (Bauer, 1995).

Mobile telephony is an ICT that is increasingly analysed in a socio-cultural framework and from various perspectives (e.g. texting, mobiles at work, mobiles and health, mobiles in public and private spaces etc). It draws the attention of a growing number of researchers who examine it as a socially and culturally embedded technology, which entails significant implications for social life in general (Castells, 2007; Goggin, 2006; Haddon 2003 & 2004; Haddon et al., 2002; Harper et al., 2005; Katz and Aakhus, 2008; Ling, 2004;

Ling and Donner, 2009; Ling and Pedersen, 2005; Taylor and Harper, 2002). Mobile telephony has a highly flexible (e.g. mobile, text, audio and oral) and rapidly developing (e.g. the increasing convergence of other media platforms in mobile telephony) nature which allows people to fit it into their lives and select or reject certain elements of it on the grounds of the socio-cultural context(s) where they live and experience this technology. Thus, while the trait of mobility is generally acknowledged as a significant feature to be passed on to the functions and traits of other ICTs, mobile telephony is ever more viewed as a distinctive technology to be analysed from a socio-constructivist perspective.

Research has attempted to disentangle the possible influence of mobile telephony on certain aspects of social and cultural living such as the impact of electronic and mobile communications on time and space settings of social life (Tsatsou, 2009), the influence of mobiles on timekeeping and the co-ordination of everyday life, as well as the socially and individually intrusive character and operation of mobile telephony (Ling, 2004). At the same time, the literature expresses strong interest in examining the ways society and social life shape and in turn change mobile technology. On one hand, Katz and Aakhus (2008, p. 9) recognise that established approaches to mobile phones 'tend to fix either praise or blame on the purpose and consequence of mobile communication'. On the other hand, the literature (Haddon, 2004) argues that mobile phones are not used in the same way in people's daily lives and finds that the usage of mobile phones is a subjective experience (Haddon et al., 2002). Culture sets constraints which are in turn open to cross-cultural variations, setting different parameters on people's practices with and on mobile phones. This is leading to what is called the domestication of mobile telephony (Haddon, 2003) and which takes place in more or less different terms in different socio-cultural environments. This can be considered a quite prevalent thesis today, although early studies of mobile telephony in Europe (Bakalis et al., 1997) had failed to find a clear correlation between social culture and the adoption of mobile telephony. In a way, the social dimension or impact of mobile telephony (Ling, 2004) as well as its embeddedness in diverse socio-cultural and everyday life settings (Castells, 2007) are all issues which have now drawn the interest of a significant number of scholars who aim to make sense of mobile telephony distinct from or in conjunction with other ICTs.

Thus, it is presently argued that mobile phones have their own 'culture' (Goggin, 2006), being embedded in broader socio-cultural settings on one hand, and generating a fascinating set of particular cultural elements through the production and consumption of mobile products and services on the other. This 'mobile culture' is arguably significant not only for today's mobile technology and communication but also, and more importantly, for broader socio-cultural structures in communications and beyond: '…how existing and new social structures, relationships, and behaviors have incorporated and been changed by cell phones' (Goggin, 2006, p. 5). However, in this chapter it is argued that this 'mobile culture' is largely shaped by broader socio-structural trends while being in a complex and continuing interaction with them. That is to say mobile phones and their culture cannot be looked at in isolation from other forms of culture in communications, while the specific characteristics of 'mobile culture' (as defined by production and consumption processes) can only be disentangled when one considers how broader socio-cultural structures establish mobile telephony and its culture as a particular case of ICT-mediated communication today.

Further, this chapter challenges research that attempts to identify culture and the cultural dependency of the mobile telephony with the concept of 'domestication'. Haddon (2004) speaks about the domestication of mobile telephony within the framework of the domestication tradition. In his approach, ICTs are treated as objects with a symbolic value whose consumption in domestic spaces has a bearing on the present and future of those technologies. This chapter avoids the 'domestication' label, instead viewing the broader socio-cultural system as a more significant parameter to account particularly for mobile telephony. In support of this view there are national studies looking at the role and value of national values (or cultures) in the reception and adoption of mobile phones and seeking to explain certain daily practices with this ICT (see, for instance, Part I of Katz and Aakhus, 2008). Such studies reach diverse conclusions concerning the role of social or national in the adoption of mobile telephony and reject any rigid definitions of culture and its scope, thus being in line with the approach this chapter takes in order to answer the question: *does mobile telephony differ from other ICTs such as the Internet in the particular socio-cultural context of Greece?*

MOBILE TELEPHONY AND THE INTERNET IN THE CASE OF GREECE

Looking at the case of Greece in general, one can conclude that Greece has for long been slow in its diffusion of new media technologies and services and part of the North-South digital gap in the European information society (Servaes, 2003, p. 23). Quantitative figures about ICT adoption in Europe and internationally (OECD Outlook 1999; Forbes and ESIS-ISPO, 1998) in the late 1990s and early 2000s provided evidence of the delayed introduction and distribution of ICTs in Greece, thus pointing early on to the existence of serious barriers to the development of the Greek information society.

In the first decade of the new century things seem to have changed significantly with respect to mobile telephony, albeit not equally positively for other ICTs such as computers and the Internet. More specifically, the 2005 Eurobarometer (EB) survey concluded that Greece was last in Internet usage in the EU-25 with a mere 24% of the population using the Internet in 2005 (from 15% in 2003/2004) (European Commission, 2006, p. 14). The same data illustrated that mobile telephony in Greece is more widely used than the Internet, even among the younger groups of the population, with 30% of children owning a mobile phone and 26% using the Internet in 2005 (36% and 50%, respectively, in the EU-25) (European Commission, 2006, p. 19). One year later, in 2006, the EB E-Communications Household survey (European Commission, 2007, p. 12) found that only 16% of households in Greece had a fixed phone line but not mobile telephone access, a lower figure than the respective percentages in many other counties with higher Internet access rates, such as Germany and France. In addition, according to the 2007 EB E-Communications Household survey (European Commission, 2008, p. 11), the percentage of households in Greece with no mobile access was even lower in 2007 (14%), while a higher percentage of households in Greece (86%) than in the average EU member state (83%) had at least one mobile telephone access (European Commission, 2008, p. 28). On the other hand, 41% of Greek households had a computer and 19% had Internet access in 2007, positioning Greece below the EU-27 (57% and 42%, respectively) (European Commission, 2008, p. 49 & 54). More recently, the national survey of the Observatory for the Greek information society (2009) found that even if a higher percentage of Greek households had access to the Internet (39.4%) in 2008 than in previous years, the majority of the Greek population were still offline. In contrast to the Internet, the great majority (84%) of the Greek population used a mobile phone in 2008 (Observatory for the Greek information society, 2008, p. 54), with Greece overall providing a relatively unique picture of the adoption of mobile phones and Internet technologies.

In order to provide a sense of the framework of the forces leading to the enthusiastic reception of mobile telephony in the country and the simultaneous reluctance of Greek people to use other ICTs such as computers and the Internet, it is useful to briefly present a historical and literature-based perspective on Greece. This perspective can shed some light on the striking history of technology diffusion in Greece as presented by relevant analyses of the 'semi-periphery' paradigm of the country (Mouzelis, 1986).

Greece has been categorised as a country of the 'semi-periphery' since it has been marked by 'early parliamentarism' and 'late-late industrialisation' (Mouzelis, 1986). These historical conditions have determined the country's overall course and specifically contributed to the creation of a weak and state-dependent civil society. The weakness, inactivity and clientelistic state-dependency of the Greek civil society have led to the closely associated traits of individualism and strong patriotism in Greek society (i.e. a dismissive attitude to anything non-Hellenic) (Lyrintzis, 1984; Mouzelis, 1995; Mouzelis and Pagoulatos, 2002; Patmesidou, 2000 & 1996; Sotiropoulos, 1996). Put together, this blend of cultural traits and elements of 'collective' living in the country has created genuine negativism regarding anything new, technological and other (Voulgaris and Sotiropoulos, 2002) which might upset the established order of life and culture in the country.

As regards new technologies and ICTs in particular, the only exception seems to be mobile telephony since mobile telephony has been appropriate for the extroverted lifestyle of Greek people, allowing them to socialise and demonstrate their communicative and sociability skills. Mobile telephony appears to be more appealing than technologies which do not contribute to sociability and social acceptance. In a way, mobile technology is considered as an alternative for the Internet in Greece, while in countries like the UK e-mail and written speech can be regarded as the prevalent means of communication and exchange. In this sense, Greece differs, for instance, from the Finnish information society model (Castells and Himanen, 2002) and other models especially in the West and North of Europe where technological development and diffusion have been firmly encouraged by a strong sense of national identity, collectiveness and citizenship, driving ordinary people to adopt a generally positive attitude to new technologies.

Although this chapter does not allow space for a detailed presentation of these history-based claims of the Greek context, it is worth saying that so far no empirical research tests these claims and makes them more comprehensible and tangible, especially with regard to the dissimilar reception of mobile telephony and the Internet in Greece. In other words, there is a lack of research into the ways in which mobile telephony in Greece can be distinguished from other ICTs and how its high popularity can be concretely explained by considering that Greek society is resistant to the development and integration of new technologies into everyday life.

SOCIAL CULTURE AND MOBILE PHONES IN GREECE: EMPIRICAL REFLECTIONS

This section attempts to demonstrate how the above statistics- and history-based insights can be construed and translated in today's socio-cultural context of Greece. This section identifies the reasons driving the popularity of mobile telephony and the contrastingly low diffusion rates of Internet and computer technologies in Greece by reporting on interviews with elite actors and focus groups of ordinary people in the country.

Research Design

The elite actors' interviews and focus groups constituted the first and third steps, respectively, of a project that examined the drivers of digital divides in Greece.

First, 12 elite actors (e.g. policy-makers, regulators, researchers, market players etc) were interviewed and reflected on the low adoption of the Internet versus the high popularity of mobile telephony in the country. These were in-depth, semi-structured interviews of a small-scale and purposefully selected sample of elite actors in the Greek information society. Although with small-scale freestyle interviewing it is not necessary to include representativeness, I attempted to achieve a good spread of respondent characteristics so as to cover the main spectrum of key actors in the Greek information society. Specifically, the interviewees were selected after making a list of key policy, regulatory, market, civic and research bodies in Greece. Members of those bodies who play a relevant role concerning the subject matter of the research were contacted and, after a first round of telephone communication, 12 of them were selected.

The interviews were categorised according to the study's conceptual framework and on the basis of the professional status of each interviewee. Thus, the study resulted in three interview perspectives, each being addressed by a separate thematic guide and with each interviewee discussing issues derived from one interview perspective: scoping, bottom-up and theory-driven. The scoping perspective digged deeper into the scope and focus of the research and consisted of interviews with four experts in policy and regulation. These experts reported on the Greek information society and echoed the voice of policy and regulatory authorities within and outside the country. The bottom-up perspective aimed to examine issues that derive from the web of social actors. Three interviewees associated with Greek society in general and with the community of Greek users of ICTs in particular were selected to inform this interview perspective on the views of the grassroots. The theory-driven perspective highlighted issues that derive from the literature and consisted of five interviews. All five interviews touched upon research areas that relate to the benefits, risks and implications of new technology adoption and the ways Greek society perceives them. The table below presents the names, expertise and domain of activity of the interviewees, while it categorises each interviewee in the three interview perspectives.

Then, a bottom-up perspective on the issue was taken by interviewing four socio-demographically diverse focus groups of ordinary people (i.e. Internet users and non-users). Twenty-five individuals (Internet users and non-users) formed the four focus groups. 'Internet usage' was the only criterion in the sample selection and group formation, with each group consisting of either Internet users or non-users. Socio-demographic diversity was established as shown in the table below.

The number of groups was decided on the grounds of the consideration that more than one group of Internet users and non-users should inform the study, while other examples of research and practical matters concerning time and budget constraints were also taken into account. The sample recruitment process was based on the second phase of research, a survey (not reported in this chapter), where surveyed individuals were asked to give their consent to be interviewed. About 350 surveyed individuals agreed to be interviewed at a later stage and some of them were randomly contacted by phone and asked to participate in the interviews. In the end, the 25 individuals described in the table below participated in the focus groups.

Two thematic guides were used in the focus groups: one for the groups of Internet users and another for the groups of non-users. The discussions were loosely structured and involved issues of ICT usage and reasons driving the non-usage of communication technologies. Although, the interpersonal relationship between the interview situation and any matters arising during the interviews could always alter the initial thematic framework, emphasis was placed on focus group discourses concerning 'life circumstances', 'choice', 'priorities' and 'identity' so as to trace and contextualise the range of factors that influ-

Table 1. Sample for the elite actor interviews

INTERVIEW SAMPLE	INTERVIEW PERSPECTIVES		
Policy & Regulation	**Scoping interviews**	**Bottom-up**	**Theory-driven**
Vassileios Asimakopoulos (VA), former Special Secretary, Operational Programme Information Society	✓		
Manoussos Voloudakis (MV), President of the National Committee for Electronic Commerce & General Secretary of Commerce	✓		
Costas Balictsis (CB), Director of Telecommunications, Hellenic Republic National Telecommunications and Post Commission	✓		
George Papapavlou (GP), Officer, EC DG Information Society	✓		
Research	**Scoping interviews**	**Bottom-up**	**Theory-driven**
Dr.–IngVeronica Samara (VS), SafeNetHome Project			✓
Professor Gregory Yovanof (GV), Head of Broadband Wireless & Sensor Networks, Athens Information Technology Institute			✓
Ioannis Tomkos (IT), Associate Dean, Athens Information Technology Institute			✓
Internet Bodies	**Scoping interviews**	**Bottom-up**	**Theory-driven**
Nicos Vassilakos (NV), President of the Association of Greek Internet Users		✓	
Elena Spyropoulou (ES), Legal Consultant, Association of Greek Internet Users		✓	
Nikos Frydas (NF), President of SAFENET (the Hellenic self-regulation body) and SAFELINE (the Hellenic Hotline)			✓
Athena Bourka (AB), Auditor, Hellenic Data Protection Authority			✓
Market	**Scoping interviews**	**Bottom-up**	**Theory-driven**
Sophia Parissi (SP), Officer of the Federation of Hellenic Information Technology & Communications Enterprises and Product Manager of FORTHnet (ISP)		✓	

Table 2. Sample for the focus groups

1st group: users	2nd group: users	3rd group: non-users	4th group: non-users
Stefanos: 32 yrs, male, investment analyst, single, no children	Antonios, 44 yrs, male, self-employed, married, one child	Antonia, 33 yrs, female, self-employed, married, no children	Ioannis, 25 yrs, male, civil engineer, single, no children
Myriam: 27 yrs, female, PG student, single, no children	Eirini, 32 yrs, female, accountant, single, no children	Dimitrios, 18 yrs, male, student, single, no children	Evangelia, 29 yrs, female, shop owner, single, no children
Apostolos: 44 yrs, male, civil servant, married, one child	Pantelis, 25 yrs, male, PG student, single, no children	Maria, 45 yrs, female, housewife, married, three children	Anna, 38 rs, female, teacher, married, two children
Agapi: 35 yrs, female, decorator, single, no children	Anastasia, 27 yrs, female, teacher, single, no children	Konstantinos, 62 yrs, male, plumber, married, two children	Petros, 39 yrs, male, receptionist, married, two children
Ioanna: 72 yrs, female, pensioner, married, two children	Kwnstantina, 55 yrs, female, administrator, married, two children	Andreas, 50 yrs, male, doctor, married, one child	Menios, 47 yrs, male, waiter, married, three children
Petros: 19 yrs, male, military service, single, no children	Michalis, 17 yrs, male, student, single, no children	Dionysia, 36 yrs, female, saleswoman, single, no children	The 6th member of the group did not show up

ence people's evaluation and usage of ICTs in general and mobile telephony in particular.

Empirical Reflections

Both the interviews with elite actors and focus groups of ordinary people aimed to help the study understand how mobile technology differs from other ICTs in the Greek context and the particular role that socio-cultural factors play accordingly.

Elite actors in the Greek information society: the Internet vs. mobile telephony?

The intention of the elite actors' interviews was to view Greece as part of a 'whole', as a case among others in the European region, and to explore the elements making the Greek information society diverge from other countries in this region. Nevertheless, the interviewed elite actors argued about 'Greek distinctiveness', particularly in relation to Internet adoption.

More specifically, they framed and associated the theme of 'Greek distinctiveness' with arguments concerning the existence of a non-technocratic culture in Greece. Most interviewees recognised the existence of a cultural identity and argued that this identity drives Greece to maintain traditions and dissociate the country from the increasingly powerful world of new technologies. The interviewees decoded cultural identity in a pessimistic way, referring to the existence of a non-technocratic and technophobic culture in society, policy and regulation, to social ignorance, as well as to what they generally perceived as a Greek lifestyle:

People in Greece are still attached to a traditional lifestyle because they are not informed or because they have not seen examples of people in other countries using new technologies successfully (Costas Balictsis, Director of Telecommunications at the Greek National Regulatory Authority for Telecommunications).

As regards the exceptional case of mobile telephony, the interviewees supported the contention that the socio-cultural distinctiveness of the Greek context can precisely explain why mobile telephony differs in its reception from other ICTs and how this contrast should not be viewed as a surprise. Indicative are the words of George Papapavlou, Officer at the EC DG Information Society and a Greek national, who has worked at the EC for implementation of the European telecommunications regulation in Greece. George Papapavlou argued that a particular culture exists in Greece, which he called 'Mediterranean' and linked to the country's geographical location. He argued that Greek identity and culture are illustrated by the adoption of new technologies and the contrasting adoption rates of different technologies in the country, such as mobile telephony and the Internet:

...the difference is that we [Greece] are a Mediterranean country, we live more 'outside' than 'inside'...we are fans of short- and not long-term work... I think this is why we are better with mobile phones than with the Internet... On the other hand, if we understand what exactly the Internet is about, if we realise that Internet services can do more things...because we are clever, adventurers, because...we look at the future, we have fantasia, because of all these I believe we will use to a certain degree all the things the Internet offers.

Such arguments concerning the disparity in the way mobile telephony and Internet technologies are received in Greece may support what the literature and historical analyses of the Greek context (as briefly presented above) have shown about the broader socio-cultural characteristics of the country, their complex linkages with politics and their important role in technological development. From this perspective, the elite actors precisely indicated the complex interlinkages between social culture and politics in the Greek information society and how this might explain the current picture of ICT adoption in the country:

Vassileios Asimakopoulos (former Special Secretary of the Operational Programme Information Society):...the difficulties...relate to the fact that we talk about technology in a society that is marked by technophobia...and this has influenced, in terms of delays, the efficiency of the Operational Programme Information Society, as even members of the public authorities in charge, such as Ministries...do not put as much effort into it as they should, possibly because they do not understand the benefits of technology or because they are afraid of it.

Interviewer:...it has to do with culture...

VA: (interrupts) Yes...this definitely reflects the dominant culture in society.

In addition to the socio-cultural traits of the Greek context and largely as a result of them, Nicos Frydas, President of the Greek Hotline for safe usage of the Internet, referred to 'hysterical media propaganda' to explain the varying receptions of different technologies by Greek society (e.g. mobile telephony and the Internet) and Greek people's resistance to the Internet in particular: 'regarding the Internet, the fear of fraud has affected people. This has made us suspicious and reluctant, and the hysterical media propaganda has contributed to this reluctance'. He argued that the media could and should play a more positive role as the medium through which citizens' voices are heard: '...we need co-regulation, supervision and control of the central authorities. To me, the state and the media, which reflect social concerns, are the drivers of the process'. The negative role of the (mass) media in Internet usage was also argued by Sophia Parissi, a key player in the IT market in Greece: '...we constantly see negative media representations of the Internet. For example, the media presents child pornography and occasional incidents of suicide on the Internet much more extensively than the Internet benefits. We believe that this contributes significantly to social fear, obscurantism and ignorance'.

However, elite actors such as Nicos Frydas do not specifically refer to the reasons the Greek mass media does not take an equally critical approach to mobile telephony: is this because mobile telephony raises fewer risks and fears or because the Greek context is such that it treats ICTs differently and for different reasons? This distinctive treatment of apparently not particularly dissimilar ICTs, such as mobile phones and the Internet, raises not only the question of how mobile telephony differs from other ICTs but also the ambivalent role of social culture in this respect. As a result, some elite actors made contradictory statements regarding the distinctiveness and future of the Greek information society as a whole. On one hand, they acknowledged the distinctiveness of the Greek case on the basis of the example of the Internet. On the other hand, they made contradictory remarks when issues of Greek identity and culture were raised with respect to the particularly positive reception of mobile telephony in the country:

Interviewer: Is Greek society characterised by such a culture affecting, in turn, the course of the information society?

NF: We have the example of mobile telephony that enjoyed commercial success soon after its appearance. Hence, Greek citizens accept anything new...there are indications that the Greek citizen can accept the new...it might not be the rule, it might be the exception though.

Overall, most elite actors recognised the existence of a cultural identity that drives, in their view, Greek society towards maintaining a traditional lifestyle and dissociating itself from the powerful world of new technologies. This identity was characterised as non-technocratic and technophobic, while issues of social ignorance and Greek lifestyle, in general, also seemed to matter. At the same time, the interview discourses painted a puzzling picture of Greek distinctiveness, with contradictory evidence about technologies that Greek people accept enthusiastically, such as mobile telephony, sometimes being used by the interviewees against the argument of Greek distinctiveness and technophobia. Hence, the question arising from the elite actors' interviews is whether social culture in the country is multi-dimensional and certain aspects of it are expressed in relation to certain types of technologies or whether mobile telephony is a different ICT that fits the traits and elements of Greek culture better than any other ICT. Regardless of the answer to this question, the elite actors' reflections bring us back to the theoretical discussion offered in the previous section. That is because they emphasise technophobia, 'Greekness' and an extroverted lifestyle in Greece as drivers of the usage of mainly those technologies that can be employed as an asset in people's daily social activities (e.g. a mobile handset). On the other hand, more educational and less 'exposable' technologies and usages that require training and a certain level of knowledge, such as Internet technologies, have been less popular in Greece as they do not seem to contribute to the user's social exposure and acceptance.

Ordinary people and their culture in Greece: the Internet vs. mobile telephony?

In the focus groups, ordinary users and non-users of the Internet brought up a different perspective on the role of social culture in how mobile telephony is perceived, evaluated and appropriated. They challenged elite actors' top-down claims about technophobia in Greek society, while they positioned their decision to use ICTs, such as the Internet and mobile phones, in an everyday life context where certain socio-cultural and identity trends are in place.

Regarding general media use, most focus group participants said they use a wide range of mass and new media technologies, indicating the significant role the media plays in people's lives today. As far as mobile telephony is concerned, the majority said they use a mobile phone on a frequent and even daily basis mostly to communicate with others. The interest is particularly in Internet non-users and the reasons driving them, on one hand, not to use the Internet and, on the other, to choose to use mobile telephony possibly in addition to other ICTs (e.g. computer):

Ioannis (25 yrs, male, civil engineer, single, no children): I use my mobile phone a lot, everyday, at work and for communicating with family and friends…eeemmm…I use a computer everyday as well, since I need it when I am in the office…and radio when being in the car…TV as well when I am at home relaxing and watching the news. I think this is more or less…

Even those Internet non-users who have limited familiarity with and usage of ICTs and mostly engage with the mass media, such as television, appear to own a mobile handset and to use it on certain occasions:

Anna (38 yrs, female, teacher, married, two children): I don't use my mobile that much…only when I need to contact my children or my husband…and only when it is needed…I prefer to read newspapers to get informed about the news of the day…and then I like watching TV in the evenings, the news at 8pm, or when a good programme or series is on. I very much like TV programmes that satirise people and situations…it is fun and make me think at the same time…what else?

In order to understand the factors that make people in Greece treat these two technologies, the Internet and mobile telephony, differently, it is worth looking at the explanations the focus group participants provided about society's attitudes to ICTs in general and the Internet and mobile telephony in particular. Regarding the Internet, Internet non-users referred to a lack of need, interest, information, education and awareness as the factors that make Greek society sceptical about Internet-based technologies:

Antonia (33 yrs, female self-employed, married, no children): It's the same for me...I work a lot, I'm self-employed and don't need the Internet for work...I prefer to spend my time with family or going out.

Anna (38 yrs, female teacher, married, two children): I don't have the knowledge or expertise to use the Internet, but I could easily get some training so as to start using it...I don't need it and don't want to be subject to all new technological wonders that mislead people. This is something I would like to teach my children too, but I don't think they will be in a position to resist.

Ioannis (25 yrs, male civil engineer, single, no children):...in the future everyone will be expected to use the Internet. Undoubtedly, the Internet is a technology we cannot ignore as it is becoming increasingly important for most things in life. I sometimes wonder whether it's wrong that I don't use it, but it's just that I haven't had the chance in the family and school to learn about it and how to use it.

At the same time, some participants acknowledged more bluntly the role of culture in how ICTs are perceived and adopted by referring to the established everyday life culture and customs in the country. For instance, Ioannis (25 years old, male, civil engineer, single, no children) pointed out the contrasting attitudes of Greek society to different technologies, strikingly confirming the contradictions within the Greek context that the literature has highlighted. Ioannis, an Internet non-user himself, brings up the role of learning and fashion or trend parameters as critical for explaining the divergent adoption rates of technologies such as the Internet and mobile telephony in the country:

In general...I think we are quite reactive to anything that changes our habits and lifestyle. On the other hand, we immediately adopt trends that come from abroad as long as they do not require effort and are fashionable. About the Internet...I think that some of us said that we don't have the knowledge and incentives to use it...

This socio-cultural and everyday life context can also be used as the explanatory platform for understanding the opposite treatment of mobile telephony in the country, particularly as far as Internet non-users are concerned. For instance, Evangelia, a 29-year-old woman who works as a shop owner and is a non-user of the Internet, explains how her lifestyle (i.e. extroverted) and daily work schedule (i.e. many hours outside of home) drive her to use mobile telephony on a regular basis and not the Internet: '...because of the long work hours...I always have to have my mobile with me so that people can find me and talk to me about either work-related issues or more personal matters'. Some Internet non-users seem to use mobile telephony as a necessary tool for almost every activity in their lives, with mobile telephony often being perceived as an integral element of social culture (capital) in Greece. For instance, Andreas, a 50-year-old male doctor who is married with one child, emphasises how mobile telephony is a fundamental element of his everyday routine and how nicely it fits in with his everyday culture:

...for me the mobile is necessary, I don't know how I managed things without it in the past. I need it for everything: my work, when patients call me if in need; for when my wife contacts me about family issues mainly concerning our child; for my friends to keep in touch with...for everything, I would say...it is very useful for practical and communication matters.

From an Internet-user perspective, everyday and other socio-cultural parameters still make a difference between the Internet and mobile telephony as in many cases Internet users appear to be more positive regarding mobile phones than the Internet, even though they are Internet users themselves. For instance, Eirini, a 32-year-old single female accountant, implies that the Internet is not as important for her as mobile phones:

Ok, I may be the exception; I love radio and the TV, I use mobile technology...sometimes read newspapers...but computer, Internet and other new technologies are not that much in my life...I mean, I use them whenever needed only...

Likewise, everyday culture and lifestyle in Greece seem to explain the comparatively few instances of people who do not use mobile telephony. For instance, Maria, a 45-year-old married housewife and mother of three, does not use the Internet or mobile phones. To explain this decision not to use mobile phones in particular, she points to her daily routine, time schedule and hobbies. Maria spends most of her time at home and mostly watches TV, considering TV watching relaxing and 'good company':

Interviewer: Have you all got a mobile?

Maria: No, not me (hesitation)...you see, I stay home most of the time to bring up my children. I don't work, so don't need it....everyone can find me on the landline...besides, it's expensive for me...

Interviewer: So, do you use any other media?

Maria: Of course....my landline to contact relatives and friends from time to time, and particularly TV....it relaxes me...I'm tired by the end of the day and TV gives me the chance to rest and relax...

(a few moments of silence)

Maria: ...the TV is on most of the time...it's good company...

Nevertheless, in attempting to problematise the elite actors' claims about the technophobic and traditional character of Greek society, the focus group participants not only emphasised the role of everyday culture and customs, but some of them also pointed to phenomena such as a lack of education and training, placing societal reaction to new technologies on a logical basis and as not deriving from society only. For instance, Internet users but not great supporters of the Internet (e.g. Agapi and Apostolos) questioned individual responsibility and, although they admitted the existence of social ignorance and inactivity, they emphasised the role of the state and education especially in relation to Internet penetration in the country:

Agapi (35 yrs, female decorator, single, no children): We all have so many things to deal with...I don't know how we can be aware and keep up with everything new that is coming up. Yes, there is a lack of awareness that makes us inactive...but who is responsible for this?

Stefanos (32 yrs, male investment analyst, single, no children): Sorry Agapi, but why should we always expect everything to be offered to us? Aren't we, as individuals, at all responsible? Besides, if the authorities force us to keep up with new technologies, a huge public reaction will occur.

Apostolos (44 yrs, male civil servant, married, one child): Stefanos, I don't agree with you. I mean, how am I expected to use such technologies when I haven't been taught anything about them?

Petros(19 yrs, male, military service, single, no children): ...but I see how negative my parents are when I ask them to learn to use computers and other gadgets... I'm giving them free training and they do not take it (laughs).

CONCLUSION

The elite actors' reflections and the focus group discourses have illustrated the role of socio-cultural elements of life in the striking popularity of mobile telephony in Greece. Although elite actors and ordinary people in Greece take a different perspective and approach social culture in a relatively divergent way, they have enabled this study to answer the question: *does mobile telephony differ from other ICTs such as the Internet in the particular socio-cultural context of Greece?*

The elite actors argued that a technophobic identity is in place, driving people in Greece to maintain a traditional lifestyle that dissociates them from the increasingly powerful world of new technologies. They considered that this non-technocratic and technophobic culture leads to social ignorance about technologies and to a particular Greek lifestyle that resists the incorporation of new technologies. In a way, these arguments indicated a lack of citizenship in the country and its importance not only for how each individual treats new technologies but also for how individuals might 'collectively' prevent beneficial technological developments for the whole society due to fears and insecurities related to individual interests. On this basis, the elite actors considered the element of culture and relevant evidence important for explaining not merely why Greek people treat generally popular technologies such as the Internet with great scepticism, but also why society in the country enthusiastically accepts some technologies only, such as mobile telephony. As regards the latter, mobile telephony seems to lie outside the scope of Greek society's technophobia. This is because it does not require any training or learning, mostly facilitates oral communication and does not, at least apparently, challenge people's values, traditions or security, while it provides Greek people with an extra asset (e.g. handset, accessories) to improve their social profiles and further develop their extroverted lifestyles.

On the other hand, the focus groups shed more and a different light on culture and the adoption of mobile telephony in the country. Ordinary people confirmed that mobile telephony is far more popular and more socially integrated than the Internet and other ICTs in Greece. Unlike the elite actors, they contextualised claims of technophobia in an everyday life framework and construed Greek society's attitudes to ICTs in a broader context in which diverse systemic, socio-cultural and other parameters operate. Thus, although they acknowledged the importance of culture, they attempted to disentangle the various cultural elements and their role in the adoption of mobile telephony by looking at two different issues. First, they emphasised how ordinary people appreciate the role and usefulness different technologies can have in everyday life. Hence, they brought up the role of everyday culture, customs and lifestyle and viewed the parameters of 'interest' and 'need' as determining technology usage. In a sense, they argued that mobile telephony is more needed and interesting than the Internet. Second, they raised broader issues, also touched on by the elite actors, such as a lack of social awareness, education and skills that prevent people from using Internet and other technologies. Some focus group participants argued, however, that this is not society's but rather the state's responsibility, also related to the culture and practices of the broader system (e.g. politics) in Greece. As regards mobile telephony, they argued that this technology is not challenged by the current socio-cultural and systemic structures since no training or particular knowledge is required to use mobile phones and adjust their usage in various everyday life activities.

Although these empirical insights offer a far from complete and exhaustive socio-cultural analysis of the adoption of mobile technology in Greece, they move beyond assumptions about the general and floating role of socio-cultural factors involved in ICT adoption, while pointing to specific reasons why mobile telephony is a distinctive ICT, at least in the case of Greece. These insights also challenge accounts of 'mobile culture' (Goggin, 2006) as being partly autonomous from socio-cultural contexts, while moving beyond the over-emphasis on the social impacts of mobile telephony in general (Ling, 2004). At the same time, these insights extend socio-cultural accounts of mobile telephony and particularly 'domestication' approaches (Haddon, 2003 & 2004), showing that culture is something up to the individual and his or her domestic space as well as dependent on broader cultural structures and historical parameters. As a result, the chapter attempted to go deeper into the socio-cultural traits affecting the adoption of new technologies in Greece, while it illustrated the somewhat contested character of those traits and the contrasting impacts they have on different technologies.

Hence, the contribution of this work lies in:

- Its effort to translate social culture and its role in the adoption of mobile telephony into specific trends and values in life, while indicating the need to contextualise culture in a historical and everyday life framework.
- Its attempt to provide a somewhat subtle picture of the diverse and often contrasting effects of culture on ICTs, using the example of the distinctiveness of mobile telephony from the Internet in Greece.

However, this paper contains a relatively limited range of findings which must be elaborated and further developed in order to provide more broadly applicable and insightful evidence and to advance the discussion. In these terms, further empirical research in Greece is suggested coupled with the development of a comparative framework to allow Greek culture to be compared with other cultures in relation to the course and adoption of mobile telephony.

REFERENCES

Bakalis, S., Abeln, M., & Mante-Meijer, E. (1997). The Adoption and Use of Mobile Telephony in Europe. In L. Haddon (Ed.), *Communications on the Move: The Experience of Mobile Telephony in the 1990s* (pp. 19-28). Farsta: COST248 Report.

Bauer, M. (Ed.). (1995). *Resistance to New Technology – Nuclear Power, Information Technology, Biotechnology*. Cambridge: Cambridge University Press. doi:10.1017/CBO9780511563706

Castells, M. (2007). *Mobile Communication and Society*. Cambridge, MA: MIT Press.

Castells, M., & Himanen, P. (2002). *The Information Society and the Welfare State: The Finnish Model*. Oxford: Oxford University Press.

Cathelat, B. (1993). *Socio-Lifestyles Marketing*. Chicago: Probus.

European Commission. (2006). *Safer Internet*. Special Eurobarometer 250/Wave 64.4. Luxemburg: European Commission. Retrieved October 1, 2006, from http://europa.eu.int/information_society/ac tivities/sip/docs/eurobarometer/euro barometer_2005_25_ms.pdf.

European Commission. (2007). *E-Communications Household Survey*. Special Eurobarometer 274/Wave 66.3. Luxembourg: European Commission. Retrieved November 3, 2008, from http://ec.europa.eu/public_opinion/ archives/eb_special_280_260_en.htm

European Commission. (2008). *E-Communications Household Survey*. Special Eurobarometer 293/Wave 68.2. Luxembourg: European Commission. Retrieved November 3, 2008, from http://ec.europa.eu/public_opinion /archives/ebs/ebs_293_full_en.pdf.

Goggin, G. (2006). *Cell Phone Culture: Mobile technology in everyday life*. New York: Routledge.

Greek Information Society Observatory. (2008). *Project: "Study for measuring the indicators of the i2010 initiatives for the year 2008". Research Findings amongst Individuals-Households*. Retrieved October 20, 2009, from http://www.observe.gr/page/defaul t.asp?la=2&id=2101&pk=161&return=183

Greek Information Society Observatory. (2009). *Measurement of eEurope/i2010 indicators*. July 2009. Retrieved October 25, 2009, from http://www.observatory.gr/page/default.asp?id=4

Haddon, L. (2003). Domestication and Mobile Technology. In Katz, J. E. (Ed.), *Machines that Become Us: The Social Context of Personal Communication Technology* (pp. 43–56). New Brunswick, New Jersey: Transaction Publisher.

Haddon, L. (2004, October). *Cultural Differences in Communication: Examining Patterns of Daily Life*. Paper presented at the Conference Mobile Communication and Social Change, Seoul, South Korea.

Haddon, L., de Gournay, C., Lohan, M., Östlund, B., Palombini, I., Sapio, B., & Kilegran, M. (2002). *From Mobile to Mobility: The Consumption of ICTs and Mobility in Everyday Life*. The COST269 Mobility Workgroup. Retrieved October 5, 2008, from http://www.cost269.org

Harper, R., Palen, L., & Taylor, A. (Eds.). (2005). *The Inside Text: Social, Cultural and Design Perspectives on SMS*. Dordrecht: Springer.

Hofstede, G. (1980). *Culture's Consequences: International Differences in Work-Related Values*. London: Sage.

Katz, J. E., & Aakhus, R. (Eds.). (2008). *Perpetual Contact: Mobile Communication, Private Talk, Public Performance*. Cambridge: Cambridge University Press.

Klamer, L., Haddon, L., & Ling, R. (2000). *The Qualitative Analysis of ICTs and Mobility, Time Stress and Social Networking*. Heidelberg: A Report to EURESCOM P-903.

Ling, R. (2004). *Mobile connection: The Cell Phone's Impact on Society*. San Francisco, CA: Morgan Kaufmann.

Ling, R., & Donner, J. (2009). *Mobile Phones and Mobile Communication*. Cambridge, MA: Polity.

Ling, R., & Pederson, P. (Eds.). (2005). *Mobile Communication: Re-negotiation of the Social Sphere*. London: Springer.

Lyrintzis, C. (1984). Political Parties in Post-Junta Greece: A Case of Bureaucratic Clientelism. *West European Politics*, *7*(2), 99–118. doi:10.1080/01402388408424473

Mante-Meijer, E. (2002). The Netherlands and the US Compared. In Katz, J., & Aakhus, R. (Eds.), *Perpetual Contact: Mobile Communication, Private Talk, Public Performance* (pp. 110–125). Cambridge: Cambridge University Press. doi:10.1017/CBO9780511489471.010

Mediagruppe München (2000). *Online-Nutzung in den Sinus-Milieus*. @facts Newsletter.

Mokyr, J. (1990). *The lever of riches. Technological creativity and economic progress*. Oxford: Oxford University Press.

Mokyr, J. (1992). Technological inertia in economic history. *The Journal of Economic History*, *52*(2), 325–338. doi:10.1017/S0022050700010767

Mouzelis, N. (1986). *Politics in the Semi-Periphery: Early Parliamentarism and Late Industrialisation in the Balkans and Latin America*. London: Macmillan.

Mouzelis, N. (1995). Modernity, late development and civil society. In Hall, J. (Ed.), *Civil Society: Theory, History, Comparison* (pp. 224–249). Cambridge: Polity.

Mouzelis, N., & Pagoulatos, G. (2002). *Civil Society and Citizenship in Postwar Greece*. Discussion Paper. Athens: Athens University of Economics and Business.

OECD. (1999). *Communications Outlook*. Paris: OECD.

Petmesidou, M. (1996). Social protection in Greece: a brief glimpse of a welfare state. *Social Policy and Administration*, *30*(4), 324–347. doi:10.1111/j.1467-9515.1996.tb00564.x

Petmesidou, M. (2000). Social protection in Greece in the nineties: reforming the 'weak' welfare state. In Mitsos, A., & Mossialos, E. (Eds.), *Contemporary Greece and Europe* (pp. 303–330). Aldershot: Ashgate.

Rogers, E. M. (1995). *Diffusion of Innovations* (*Vol. 4*). New York: Free Press.

Servaes, J. (2003). The European information society: a wake-up call. In J. Servaes (Ed.), *The European information society: A reality check* (pp. 11-32). Bristol, Portland, Or: Intellect.

SevenOneMedia (2002). *Die Sinus-Milieus 2001*. Das neue Gesamtdeutsche Modell, Unterföhring: SevenOneMedia.

Smoreda, Z., & Thomas, F. (2001). *Social Networks and Residential ICT Adoption and Use*. Paper presented at the EURESCOM Summit, Heidelberg.

Sotiropoulos, D. (1996). Civil society and the Greek state in the Third Hellenic Republic. In Lyrintzis, C., Nikolakopoulos, E., & Sotiropoulos, D. (Eds.), *Society and Politics* (pp. 113–138). Athens: Themelio. (in Greek)

Taylor, A., & Harper, R. (2002, April). *Age-old practices in the 'new world': a study of gift-giving between teenage mobile phone users*. Paper presented at Changing Our World, Changing Ourselves, the SIGCHI Conference on Human Factors in Computing Systems. Minneapolis, Minnesota.

Thomas, F. (1995). *Telefonieren in Deutschland. Organisatorische, Technische und Räumliche Entwicklung eines Grosstechnischen Systems* (*Vol. 21*). Frankfurt, New York: Campus.

Thomas, F., & Mante-Meijer, E. (2001, June). *Internet Haves and Have Nots in Europe*. Paper presented at the e-Usages, Paris.

Trompenaars, F. (1993). *Riding the Waves of Culture*. Antwerp, Amsterdam: Economist Books.

Tsatsou, P. (2009). Reconceptualising Time and Space in the Era of Electronic Media and Communications. *PLATFORM: Journal of media and communication, 1*, 11-32.

Voulgaris, Y., & Sotiropoulos, D. (2002). *Information Society, Sociology and Technology*. Athens: Operational Programme for the Information Society. (in Greek)

KEY TERMS AND DEFINITIONS

Greece: A Southern European country with a population of 11 million. It is world renowned for its culture, philosophy and science in ancient times.

Information Society: The development and adoption of Information and Communication Technologies and related socio-economic, political and cultural changes linked to the establishment of new forms of information and means of communication.

Information and Communication Technologies: An umbrella term referring to the combination of informatics technology with other, related technologies, specifically communication technology. Thus, it refers to technologies that provide access to information through telecommunications. It is similar to Information Technology (IT), but focuses primarily on communication technologies such as the Internet, wireless networks, mobile phones, and other communication mediums.

Mobile Telephony: A non-stationary telephone service which takes place over a cellular network of specialised base stations and offers a telephone service beyond a fixed geographical distance. It also supports many additional services and accessories, such as SMS for text messaging, email, packet switching for access to the Internet, gaming, Bluetooth, a camera with a video recorder and MMS for sending and receiving photos and video, an MP3 player, radio and GPS.

Social Culture: A complex set of meanings, habits, values and behaviours adopted by one or more social formations.

Everyday Life: The context and experience of living that all human beings are subject to. An instance of how and where social culture is shaped,accommodated and developed, making itself visible.

Section 2

Cross-Country Analysis and Global Perspectives on Policies and Strategies

Chapter 7
The Effectiveness of Government Policies in Broadband Deployment:
An Assessment of Singapore, Hong Kong SAR and South Korea

Elizabeth Fife
University of Southern, USA

Francis Pereira
University of Southern, USA

ABSTRACT

While the potential economic and social benefits of broadband internet use are significant, adoption levels vary greatly among countries around the world. Many governments, particularly those in Southeast Asia, have adopted aggressive policies to deploy broadband networks and to encourage the use of applications. Governments are motivated to promote broadband adoption in order to realize both economic and social benefits. This paper argues that the generally higher levels of broadband adoption rates witnessed in many Asian economies, is attributable in part to the aggressive policies pursued by these governments. There is some evidence to suggest that these governmental policies have been successful in achieving their stated goals.

INTRODUCTION

Despite the fact that many countries around the world today have modern fiber optic telecommunications networks that support broadband access to the home, broadband penetration rates still vary dramatically (Figure 1). A key factor explaining this difference in penetration rates are government initiatives, particularly in the more developed Asian economies, to establish national "information superhighways. Specifically, the governments of South Korea, Hong Kong, Japan, Malaysia and Singapore, have launched "Version 4.0" of the National Information Infrastructure (NII) plan first deployed in the early 1990s.

While governments may deploy national broadband networks for different reasons, a com-

DOI: 10.4018/978-1-60960-011-2.ch007

Figure 1. Broadband penetration rates per 100 households for select countries, December 2008 (OECD, 2009)

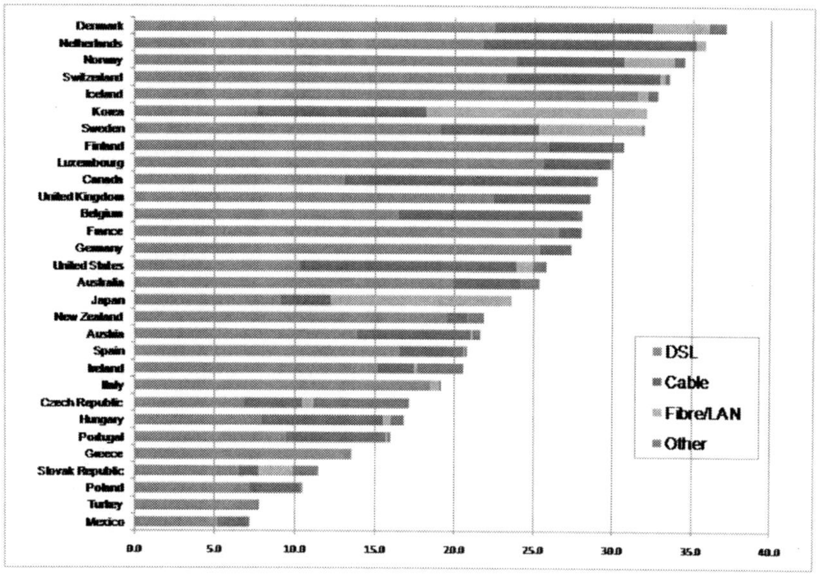

mon significant reason is that efficient information infrastructures, theoretically, can enhance productivity by providing intelligent networks that can accommodate converging voice, data and electronic commerce applications (Frieden, 2005). These infrastructures can provide a competitive advantage in the knowledge-based industries that include data processing, insurance, management, customer relationship management and logistics and distribution.

Such a competitive advantage in the area of information and telecommunications technology (ICT), when combined with a stable economy and favorable regulatory system, should translate to higher levels of Foreign Direct Investment (FDI). In this respect, endogenous growth theory argues that it is technology and human capital, when endogenously present, that contribute to continuous economic growth and therefore play an essential role in a country's development (Easterly, King, Levine, & Rebelo, 1994; Barro, 1997). Specifically, in developed countries, existing ICT infrastructures have been found to causally attract FDI; a higher level of ICT investment leads to a higher level of FDI inflows. This suggests that ICT contributes to productivity and economic growth indirectly by attracting more FDI (R. Gholami et. al., 2006).

FDI increases domestic capital formation, augments host country stocks of technology and managerial expertise, improves access to export markets, and provides a comparatively stable source of external financing (Matthews, 1999; Lehman, 2002).

The deployment of such networks, either directly through the use of tax dollars or indirectly by use of appropriate policies, such as spectrum allocation by fiat, thus would allow national governments to exploit the benefits of e-government. This could encourage end-user adoption, by effectively lowering the usage cost of applications for the consumer, thus improving living standards and enhancing productivity. Specifically, telemedicine would allow governments, particularly in developing countries where "transactions costs are high because of logistical problems (Sein and

Harindranath, 2004), to extend universal healthcare services to remote areas and provide, especially for developing countries, an economically viable means to increase the quality of medical services. Similarly, tele-education could provide governments with the means to increase the level of educational services to resource strapped inner-cities and the means to extend the learning experience to the home.

This paper assesses the effectiveness of these government NII policies, in South Korea, Hong Kong, SAR, and Singapore to achieve the espoused goals of broadband development. Specifically, some of the questions that this paper addresses include the following:

- Is there any empirical evidence to suggest that governments have been able to achieve their espoused economic goals through their national policies?
- Have some national governments been more successful in achieving their stated goals? What are the factors explaining the differences in achievements?
- Are there any social or cultural factors that can explain differences, if any, in the rate of adoption of these ICT applications and services?

AN OVERVIEW OF NATIONAL INFORMATION INFRASTRUCTURE POLICIES

Hong Kong, Special Administrative Region (SAR)

Hong Kong's Special Administrative Region (SAR) has one of the world's most sophisticated and advanced telecommunications networks, which is crucial for Hong Kong to maintain its status as a leading business, financial and industrial center. Sizeable investments in communications technology have made the telecom market in Hong Kong highly efficient and cost effective for the manufacturing industry. An advanced network, together with a wide range of professional and high value-added manufacturing support services, an ample supply of skilled and low-cost workers from China and other Asian-Pacific countries, all coupled with the cross-border production competence of Hong Kong companies, has positioned Hong Kong as a control center in global production and sourcing (Chan et al., 2000, p.288).

Hong Kong boasts what is widely considered the world's freest economy, as well as a highly skilled workforce fluent in English, the lingua franca of the business, legal and technology industries. The government of Hong Kong has been one of the leading investors in its IT sector through its resource allocation, e-government program, its implementation of information systems and procurement arrangements, as well as through administrative adoption. The government's investment in this industry has averaged HK$ 4.6 billion per year (Commerce, Industry and Technology Bureau, 2004).

The city launched its bid to privatize and demonopolize local and long-distance telephone services in 1995 and 1998, respectively. By 2001, all telecom services and infrastructures were liberalized, which resulted in wider choices and a higher quality of services at more competitive prices. Today, Hong Kong has one of the most advanced and competitive telecom markets in the world, with six major mobile operators competing to serve Hong Kong's market, none of them dominating (Xu, 2006). In fact, with a penetration level of 123 mobile phones per 100 inhabitants, it may not be an exaggeration to claim that Hong Kong has the highest level of telecom competition as well as adoption of mobile technology, in the world (Xu, 2006).

The government's policies and the drive toward market liberalization created a great deal of pressure for innovation. As a result, Hong Kong's former telecom monopolist became the first provider in the world to launch Interactive Television (Video-On-Demand) service commercially in 1998, and was also the first to offer Internet Provider Television (IPTV) service. In 2006, they launched a mobile real-time TV broadcasting service as well (Yan, 2006). The people of Hong Kong have benefited in numerous ways from the government's promotion of competitive policies: Hong Kong was the world's first city to fully digitize its telecommunications networks. It was also the first economy in the world to incorporate fixed-line number portability, and third in the world with fixed-line mobile number portability.

Against this background, the government's first Digital 21 IT Strategy was published in 1998. It was revised and updated in 2001, and then again in 2004. The plan is a comprehensive strategy to bring Hong Kong into the 21st Century as a global leader in e-business, new technologies, and the information economy. The government's vision is that it should be an effective facilitator to promote innovation and the development of industry, as well as to ensure that the entire community will benefit from IT development, by taking measures to bridge the digital divide.

Two flagship programs of the Government's Digital 21 Plan are the Cyberport and the Science and Technology Park. Both of these industrial clusters provide high technology-focused valued-add to Hong Kong's overall economy, with the larger concept of attracting leading IT companies and a critical mass of professional talent into Hong Kong.

The Cyberport is a $2 billion high-technology multimedia hub designed to attract the research and development sectors of major high-tech companies. Cyberport's Phase 1 occupancy rate has already reached 80 percent, attracting such big name tenants as Microsoft, and GE Information Services. It is designed to be more than a workplace, incorporating both a world class living and working environment, and is meant to attract businesses from many industries, including entertainment, communications, advertising, sales, trading, and finance (MAIT, 2005). Another similar initiative is the Science and Technology Park which focuses on four sectors: ICT, electronics, biotechnology and precision engineering. It is equipped with fully equipped laboratories and engineering services that can be rented by the hour or on a project basis. Over 200 companies that range from start-ups to multinational conglomerates are housed here.

Hong Kong's economy benefits from its status as a major global financial services center as well as a trade, commerce, transportation and logistics hub. All of the sectors comprising these industries are major users of IT, which has helped provide the impetus for Hong Kong to be at the forefront of the IT development and innovation, as well as for the government to facilitate and augment this process. Hong Kong SAR's banking sector is the second largest in Asia, after Japan, and in terms of total bank assets to GDP it is only matched by Singapore.

Hong Kong's trade and financial community is extremely "IT oriented." The Hong Kong Trade Development Council (HKTDC) promotes Hong Kong's goods and services by serving as an information broker and matchmaker among traders, exporters, manufacturers, buyers, and service providers. With the advent of the World Wide Web, the HKTDC introduced a host of electronic advertising and publishing activities to take advantage of the Web's instantaneous, low-cost information distribution to a wide on-line audience.

Besides using the Internet for web browsing, there is an increasing trend by Hong Kong corporations to use the Internet as the backbone to develop intranets for mission critical applications, including dynamic access to data bases, and functional applications such as purchasing, trading, and inventory control. The rapid growth of the Internet and intranets is generating rapid expansion in LANs (IMF, 2002). Hong Kong has a high level of LAN penetration - approximately 70% of all corporations.

Hong Kong's economy has witnessed remarkable growth in FDI inflows over the past few decades. Global FDI inflows rose substantially in 2005, up 29% from 2004, as Hong Kong maintained is number two position (behind China) as East Asia's leading regional FDI target.

Hong Kong's culture is characterized by industriousness, and by individuals who desire to better themselves (MAIT, 2005). As was mentioned in Korea's case below, such characteristics may promote the adoption and acceptance of new technologies, as these are seen as methods for increasing efficiency, tools for learning, and as an opportunity to show that one owns and knows how to use the latest device.

A great proportion of Hong Kong's population has realized the benefits of the government's pro-technology efforts. Broadband is now available to over 95 percent of households and to more than 98 percent of business buildings. Thc PC penetration rates in 2006 stood at 71 percent for households and 88 percent for businesses, while internet penetration rates held at 65 percent for households and 83 percent for businesses (CEPAL, 2008). The mobile penetration rate of 123 percent was mentioned above, while 70 percent of Internet users have used e-government services.

Despite this level of technological progress, the government still acknowledges that it must go further in order to bridge the digital divide within its borders. In order to ensure that the entire community will benefit from IT development, the government has begun collaborating with both industry and NGOs in order to introduce a wide range of measures to further enhance residents' quality of life.

The "IT Honk Kong" campaign was designed as a result, to promote IT adoption in the greater community. It consists of free courses providing IT training, free talks in libraries, district promotional activities and infotainment programs for citizens. Other initiatives include providing public computers with Internet access at convenient locations, computer recycling for the needy, financial assistance to people with disabilities for the purchase of computer facilities to enable working from home, and the installation of devices in public computers so that they may be accessed by the blind and visually impaired (Commerce, Industry and Technology Bureau, 2004).

Singapore

The government of Singapore has been particularly aggressive in promoting the deployment of new technologies and broadband access, as manifested in several initiatives, which are discussed below. Given Singapore's lack of natural resources, its government has pursued an international-oriented business focus for a number of years. One result of this plan is that the total production of U.S. MNC affiliates constituted some 13% of Singapore's Gross Domestic Product (GDP). Altogether, there over 7,000 MNCs located in Singapore (Lim, 2006).

The Singapore government has been most aggressive in promoting the deployment of new technologies and broadband access, as manifested in several initiatives. Relatively high labor costs in Singapore, as shown in Figure 3, coupled with high employee turnover rates, particularly in the IT sector, a manpower shortage and a high overall cost of doing business in Singapore is motivating the government, in this respect, as some have observed, to eventually encourage firms to adopt teleworking (Lim and Teo, 2000).

One of the first two Singapore government initiatives was the deployment of TradeNet, and the Singapore ONE program. TradeNet, which facilitates trade documentation online, processes 99% of all trade permit declarations, and is used by over 2,400 companies. This service has reduced processing time from several days to five minutes. TradeNet Plus, a more complete and efficient version that incorporates other online features, such as payment and insurance, is expected to further reduce processing time to between one and three minutes. Furthermore, it is estimated that it will generate annual savings of some S$2.8 billion (Shih, 1999, p.2).

Table 1. Indexes of hourly compensation for production costs in manufacturing for select countries (U.S. Bureau of Census, 2009)

	1990	2000	2001	2002	2003	2004	2005	2006
U.S.	100	100	100	100	100	**100**	100	100
Canada	111	85	80	79	88	**95**	101	108
Japan	85	112	98	88	91	**96**	92	85
Hong Kong	22	28	28	27	25	**24**	24	24
Singapore	25	38	36	33	34	**32**	31	36
Korea	25	42	38	41	43	**47**	54	62
Taiwan	26	32	30	27	27	**26**	27	27
Portugal	24	23	23	24	28	**31**	31	32
Mexico	11	11	12	12	11	**11**	11	12
Sri Lanka	2	22	2	2	2	**2**	2	n.a.

The Singapore ONE (One Network for Everyone) program, is a high-speed optical fiber network that manifests the government's IT2000 master-plan, a blue-print to deploy IT in almost every government department with the end goal of transforming the state into an "intelligent island" (Economist, 2000). The IT2000 master-plan was designed to allow all of Singapore's over 800,000 households to connect to a hybrid-fiber co-axial network, irrespective of whether they intended to subscribe to the various services (Cable and Satellite Asia, 1997). The Singapore ONE initiative strives to develop a nationwide broadband network to link businesses, homes and schools, to deliver interactive, multimedia applications and services (Santiago, 2000). It is estimated that by 2005, some forty-three percent of all households had connected to this broadband network (Burton, 2005). In this respect, the Singapore government's eCitizen Center, an integrated service delivery system, was designed to ensure that the public sector operated, and was seen to operate, as a single entity. This particular project is aimed at bringing together useful services and delivering them to Singaporeans in convenient and easily accessible packages. The Singapore government hopes that eCitizen will enhance the ability of the public to be increasingly familiar and comfortable with IT, which has become a critical component in the knowledge economy (Economist, 2000).

In June 2006, the Singapore government launched its latest 10 year masterplan, dubbed iN2015 (Intelligent Nation 2015,) which established ambitious goals for the country: to establish Singapore as first in the world in harnessing ICT to add value to the economy; a two-fold increase in value-add of the ICT industry to S$26 billion; a three-fold increase in ICT export revenue to S$60 billion; the creation 80,000 additional jobs; a 90 per cent rate of home usage for broadband; and the achievement of 100 per cent computer ownership for all homes with school-going children (Business Wire, 2006). The government hopes to accomplish these goals through programs and initiatives that include for example, the CXO programme to facilitate strategic discourse among key decision makers, CEOs, CIOs and CFOs and the iEnterprise Challenge Programme focuses on developing innovative products and solutions as reference projects, which is intended to have an impact across various sectors, while simultaneously creating new intellectual property (Lim, 2006).

A third government initiative seeks to raise the percentage of small and medium-sized enterprises (SMEs) conducting e-commerce to twenty-five percent within three years' time. According to a survey conducted late in 1999 by the Singapore IT Federation, while 91 per cent of the top 1,000 companies already had Internet access, only 4 per cent of the 92,000 SMEs were conducting e-commerce (Raj, 2000, p.7). Sixty-six per cent of SMEs with sales below $1 million cited lack of in-house IT skills and infrastructure, inadequate knowledge of e-commerce, and low budgets as barriers to implementing e-commerce (Koh, 2000, p.14). As such, the Economic Development Board (EDB) will assist businesses in their e-commerce strategy by supporting up to 50 per cent of their external consultants' costs in formulating a business plan or business collaboration venture, conducting feasibility studies, or providing assistance in implementation, up to a maximum limit of S$250,000 (Lee, 2000, p.71). The government also unveiled a three-year, $11.7 million plan to prod the city's growing logistics industry into doing more with the Internet. Singapore's transport and logistics industry contributed more than 7 percent to gross domestic product last year (Bangsberg, 2000, p.10).

The Singapore government's GeBIZ initative is an attempt to create a one-stop, round-the-clock centre for the government's business dealings. The first phase was launched in April 2000. GeBIZ enables the financial systems of ministries and the procurement applications to work together. Trading partners can find invitations to tender and purchase orders on the site. Suppliers can also submit invoices, check payment status, post their catalogues and bid for contracts. Currently, purchases are capped at S$30,000 (US$17,341), but the Ministry of Finance estimates that once new payment and security systems are introduced, 80% of all government procurement will transfer to GeBIZ. In 2005, some 41,000 small-value purchases worth some $36 million were transacted over GeBiz (Business Times, 2006). As with other online B2B trading networks, the benefits come in the form of more competitive bidding, easy access to suppliers round the world, time saved by online processing of orders, lower stocks and automated collection of high-quality data (Economist, 2000).

South Korea

The South Korean Government has taken a strongly proactive approach to promoting an information society for all of its people, businesses and industries. In all of its endeavors, the government's overarching objective has been to utilize ICT to enhance the quality of life in Korea.

Korea's focus on promoting information and communication technologies began in the mid-1990s, with the establishment of the Ministry of Information and Communication (MIC) in 1994, and the announcement of its overall broadband strategy, the Korea Information Infrastructure (KII) in 1995. Korea's broadband market is currently an open market, free of regulation and controls over licensing and pricing. Its incumbent telecom provider, Korea Telecom, has been completely privatized as of May 2002 (Lee, 2002).

One aim of the KII initiative was to stimulate private sector investment and competition in the broadband market, by lowering the regulatory barriers to entry. A second aspect of the initiative was for the Government to provide a high-speed network that would serve as a public backbone, providing broadband services to 30,000 government and non-profit organizations, as well as to research institutes, and to around 10,000 schools. The government had another ambitious goal: to provide high-speed internet to every household in Korea. Not only has this goal nearly become reality, but the government's targeted internet speed (threshold) is continually revised upward: the current target is to provide 50-100 Mbps to all homes by the end of 2010 (Lee, 2002).

The Korean Government is continually revising its strategies, due in part to rapid changes in technological innovation and market demands (Lee and Chan-Olmsted, 2004), and putting forth new initiatives in its wide-ranging telecommunications efforts. In December of 2002, Korea announced its e-Korea Vision 2006 plan. This initiative, with a focus on improving the quality of Korean citizens' lives, had five objectives (NIA, 2008).

- *"Efficient Knowledge Government" to Serve the People"*

The first was to maximize the opportunity for and ability of all Korean citizens to utilize the information society and communication technologies in order to actively participate in the information society (or, to be "prosumers"—both producers and consumers of information.) The government took action by providing increased opportunities for Internet access for all, and by establishing a lifelong education system available through online learning (Ministry of Information and Communication, 2002). Citizens' ability to utilize IT was intended to lead to the creation of added value in all aspects of society. The objective was to establish an intelligent administrative procedure by providing one-stop customized services, such as birth registration, sharing administrative information and providing public information in a much larger scale to help achieve a creative and efficient government.

- *"People Prospering Through Digitalization" to support a Facilitated Market Economy*

The second objective was to strengthen the country's level of global economic competitiveness by promoting "informatization" within all Korean industries, as the IT industry strengthens the competitiveness and productivity of most all other industries. This objective also had an environmental component: the use and application of ICT would enable improvements in the ecological system and ensure the conservation of a clean and natural environment by strengthening environmental policy functions, while its use in the realm of transportation would relieve traffic problems and reduce logistics costs (Ministry of Information and Communication, 2002).

- "Trusted Information Society" to Actively Support Welfare System

The third objective was to realize a more efficient and transparent government structure through "informatization," by offering more online public services in the areas of education, culture, and social welfare.

- "Creative Soft-Power" to become A Brain Power Nation

The Ministry's fourth objective was to facilitate continued economic growth by promoting the IT industry in general, and both core and new technologies in particular. The government aimed to achieve this through the building of a value-creating knowledge infrastructure that would serve as a nation-wide base for utilizing knowledge.

- "Cutting-edge Infrastructure for Digital Convergence" to Become a Mature Country

Finally, the Ministry also aimed to become a leader in regional technological cooperation, by establishing a system that promotes ICT collaboration within the Asia-Pacific region (e-Korea, 2002).

In 2004, the MIC produced a new and ever further-reaching policy plan: the IT 839 Strategy. One of the overarching economic goals announced in this document was to see Korean GDP per capita grow to (US)$20,000. The exploitation and utilization of ICT is seen as crucial to achieving this goal (Ministry of Information and Communication, Republic of Korea, 2004).

The IT 839 strategy also sought to establish and promote new and cutting-edge technologies, including a Broadband Convergence Network, Internet Protocol version 6, and new services, such as WiBro, Digital Multimedia Broadcasting, Home Network Service, Telematics Service, and Internet Telephony (VoIP). These services, available over a converged network, were intended to ensure Internet access at any time in any place (Ministry of Information, 2004).

In terms of competition, both DSL and Cable have been unbundled (de-monopolized and made competitive). This has had the result the making of Korea into one of the most truly competitive nations in terms of the levels of domestic ICT competition. In turn, this high level of competition has made for comparatively low costs and high rates of subscription by Korean consumers. Indeed, Korea is the world's leading nation in household broadband penetration, with 94.5 percent of households having access in 2008 (OECD, 2009) and has the lowest costs for broadband access in the OECD (Mathiason, 2008).

The importance of foreign direct investment in the Korean economy can be seen in the table below. With policy reforms to allow increased foreign participation in the Korean economy, including the purchase of troubled corporations, the value of FDI between 1997 and 2000 quadrupled and its share increased from 2% of business financing to 13%. In 1998, FDI was equal to some 14.8% of Total Gross Domestic Investment in the Korean economy. Although there has been some fluctuation in investment levels over recent years, 2004 data show a strong resurgence in foreign interest in Korea's economy.

In addition to the MIC's overall policy plans, the Korean government has launched a number of specific broadband-related initiatives, many of them with an eye to bridging the digital divide. Among these has been the Public Fund Program, wherein the government provided low-interest loans to providers to ease the financial burden of bringing access networks to small and medium sized cities.

An additional initiative the government has taken to close the digital divide was to provide 3,000 free internet access points in public places, such as post offices and community centers. The government has also provided PCs and Internet infrastructure to schools. Further, within the context of its "Plan for Promoting ICT Use and Distributing PCs to Children of Low Income Families," the government provides discounted ICT access and a PC to those who otherwise could not afford it (ITU, 2003).

Further, the government has launched IT literacy programs, targeted at social groups that traditionally would not have access to computers or the Internet, such as housewives, the elderly, and prisoners. More than eight million people were trained between 2000 and 2002. The government has also started offering high school education programs that are broadcast over the Internet, increasing the demand for PCs and Internet connectivity in family homes (ITU, 2003).

Beyond government initiatives, Korean culture has played a large role in determining ICT adoption. Some cultural characteristics prevalent in Korea include a high valuation placed on family relationships, social advancement, (as well as on social status in general,) and on learning and education (Davies, 2006).

These cultural values promote hard work and personal development, and education is seen as a key means for achieving economic and social success. Another cultural characteristic at play is the "copycat syndrome" or "keeping up with the Joneses" scenario. If one household subscribes to broadband Internet service, other households in the neighborhood will subscribe to an equal or superior service in order to maintain their social standing (Davies, 2006). Both of these social characteristics have promoted the adoption of ICT within the Korean society.

One downside of the cultural propensity for ICT uptake has been the widespread addiction to gaming. Most see this addiction spreading in reaction to the high level of societal pressure to excel in one's performance: gaming offers an attractive—and addictive—form of escape.

Table 2. Korea: FDI by source country (millions of U.S. Dollars) (Ministry of Commerce Industry and Energy, 2004)

	1995	1996	1997	1998	1999	2000	2001	2002	2003	2004
Japan	418	255	266	504	1749	2448	776	1403	541	2258
US	643	875	3190	2971	3739	2921	2885	4500	1240	4717
EU	475	1058	2409	2968	6262	4396	3064	1680	3062	3009
Total	1947	3203	6971	8852	15541	15217	11292	9103	6468	12785

This phenomenon has been abetted by the increasing prevalence of PC Baangs (PC rooms). These are inexpensive internet cafes that, over the past few years, have morphed mostly into gaming rooms.

These Baangs, and gaming in general, are becoming so popular that addiction to gaming is currently considered one of the biggest social problems facing Korea. The situation has grown so acute that 10 South Koreans—mostly teenagers and people in their twenties—died in 2005 from game addiction-related causes; from sitting in a single, cramped position for too long. To combat this problem, the government launched a game addiction hotline in 2006, after having opened a treatment center in 2002. Numerous private hospitals and psychiatric clinics have also opened units to treat this addiction (Faiola, 2006).

In terms of the benefits of ICT, however, as mentioned above, one of the government's key objectives is to ensure that ICT will make a significant difference in the daily lives of its citizens. To this end, the government has implemented numerous ICT-oriented public services, such as e-Health, eTransport, eLearning, with positive results (Davies, 2006, p.8). The government's promotion of ICT development is also expected to stimulate job creation.

EVALUATING THE EFFECTIVENESS OF GOVERNMENT POLICIES

Various theoretical approaches have been developed to try and measure the effect of telecommunications and ICT on development (ITU, 2006; OECD 2007, ENECA, 2007). Specifically, one of the earliest efforts to explore the relationship between telephone service and economic development by Hardy, found that greatest impact of the telephone was on information diffusion and organizational efficiency (Hardy, 1980; Thompson, 2007). Other studies suggest the possible effects of telecommunications on economic development (Saunders et al.,1983; Leff, 1984), but also warn of the problems and some possibly negative aspects (Thompson, 2007). In a 2001 study, Roller and Waverman (2001) estimated a structural model that included telecommunications investment as an endogenous variable. They conclude that for OECD countries, telecommunications infrastructure growth has a positive effect on economic growth, especially when a threshold of universal telecommunications service was achieved sometime during the 1970–1990 period. However Jacobsen, in a 2003 study, re-estimated the Roller and Waverman model for a larger set of countries, including poor countries, and concluded that the threshold effects were no longer statistically significant, and that the estimated growth rate impacts of telecommunications are implausibly large (Thompson, 2007). Garbacz and Thompson show for the United States that both untargeted and targeted universal service policies for households from 1970 to 2000 were ineffective, inefficient and generally counter-productive (Garbacz and Thompson, 2005; Thompson, 2007). In 2005, Waverman tested a version of the 2001 Roller and Waverman model on data from Africa but was unable to rely on the results as the model was not robust to either changes in sample size or changes in model specification (Waverman, 2005).

The ambiguity in empirical evidence to demonstrate the effect of ICT, in general, and government policies, specifically on macroeconomic variables, to date, have led several international agencies to conclude that the casual link between ICT and development have yet to be established. The measurement of ICT impacts is challenging for a number of reasons (CEPAL, 2008B) including:

- Absence of a clear definition of impacts, and these have been variously characterized, for instance, as strong or weak influences, direct or indirect impacts, positive and negative impacts, short or long term impacts, intended and unintended impacts, intermediate and final impacts;
- The range of possible statistical approaches to impact measurement, combined with a lack of both comparable measurement models and data;
- The general challenge in measuring effects of any kind as demonstrating the impact of one factor on another can be difficult because a positive correlation cannot readily be attributed to a cause and-effect relationship; and
- The nature of ICT itself, as an enabling technology, implies that its effects will be indirect. As such, the use and impacts of ICT are ubiquitous yet difficult to measure. It is not ICTs per se that have an effect on the economy and society but how they are used to transform organization, processes and behaviors. (OECD, 2007).

There are various approaches to assessing the impact of ICT on productivity at the firm level, as well as a variety of data sources used. While this diversity has some benefits, it also limits cross-country comparison (CEPAL 2008B). Some generalized findings from firm level studies are that ICT affects productivity in a positive way– for example, through hardware and software investment, through firm level use of multiple electronic business links, and through greater employee engagement with computers and the Internet. However, the extent of the benefits differs according to the type of firm – for example across industries (with some service industries showing a particularly strong effect), between young and established firms (with the former showing greater gains from IT investment), and between firms with different ownership and geographic scope. There are also relationships with other firm attributes, for instance, there is a greater positive impact where firms have more fixed investment, a higher number of skilled employees and a greater level of innovative activity (Eurostat, 2007).

For example a study of Thai manufacturing firms found that the use of ICT (computers, the Internet and web presence) was associated with significantly higher sales per employee. More importantly, the study illustrates that the use of even the most basic of these ICTs – computers – accounts for large differences in labor productivity between firms. Furthermore, variations in the intensity of computer use resulted in larger productivity differentials, for example, a 10 per cent increase in the share of employees using computers was associated with a 3.8 per cent rise in labor productivity (UNCTAD, 2007).

While the causal effects of government ICT policies on macro-economic variables may be difficult to establish, there are some indications that there is a high correlational relationship between broadband/Internet penetrations rates and growth. This is shown in Figure 2.

Figure 2. Internet penetration, broadband penetration and per capita GDP (OECD, 2009)

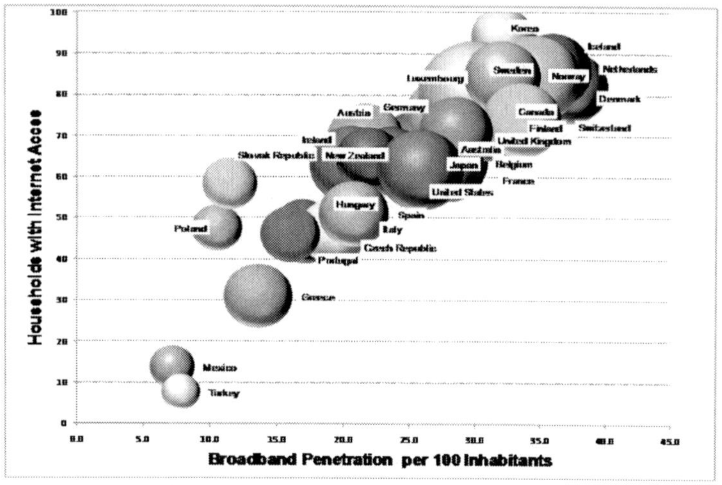

Figure 3. The effects of "informatization" on Korean society (National Information Agency, Korea, 2008)

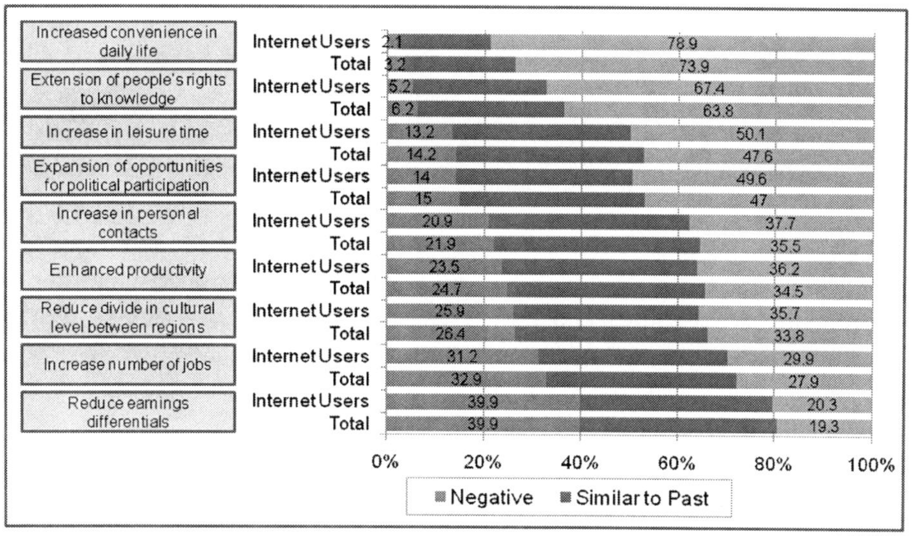

Singapore

In a recent survey of businesses (IDA, 2007) on the effectiveness of the government's policy on Singapore, it was found that:

- 76% of businesses obtained information from the Government in the past 12 months.
- 95% of businesses that obtained information from the Government did so electronically (via the Internet, automated kiosks, email, telephone via the Interactive Voice Recognition System, SMS, etc.), at least once in the past 12 months.
- 74% of businesses that had visited the EnterpriseOne website were satisfied with the overall quality of the Government in-

formation found on the website. The result is similar to the 75% achieved in FY2005.

- 95% of businesses transacted with the Government in the past 12 months.
- 94% of businesses that transacted with the Government did so electronically (via the Internet, automated kiosks, email, telephone via the Interactive Voice Recognition System, SMS, etc.), either on their own or with help from intermediaries, at least once in the past 12 months.
- 82% of businesses were satisfied with the overall quality of Government electronic services. This implies an increase in general satisfaction levels with regard to businesses using Government electronic services, as compared to FY2005.
- 80% of businesses that transacted with the Government were satisfied with the overall support provided for electronic services usage. The result was 71% in FY2005, reflecting a significant 9%-point increase in general satisfaction levels.

Korea

The Korean government's overarching ICT objective: to create an information society, has been marked by its strong and continually renewed commitment. The results have been remarkable: one decade after the announcement and implementation of its initial ICT plan, the country now ranks at or near the top of almost any global-level survey measuring broadband use, uptake, or readiness, or of ICT penetration

A recent survey of households in Korea similarly reflects the effectiveness of the government's ICT policies on their everyday lives as shown in Figure 4. Figure 5 shows the Internet usage in Korea. Over 50 percent of Koreas use the Internet for education and learning services and some 36 percent use it for financial services.

Hong Kong and FDI

To the extent that one of the stated goals of the NII initiatives of the governments of Singapore, Korea and Hong Kong were to increase FDI in their respective economies, then there is some evidence that they have been successful, as shown in Tables 5 and 6. In fact, generally, the investment flows to many of the East Asian economies have been directed at creating export-oriented industries (Fan and Dickie, 2000). As Table 5 illustrates, the role of FDI in the economies of Hong Kong and Singapore are extremely significant, and the FDI stock as a percentage of the GDP of these countries is much higher than most other countries.

Table 3. Global awards for e-government systems in Korea (National Information Agency, Korea, 2008)

SYSTEM	MINISTRY	AWARDING BODY	AWARD	YEAR
UNI-PASS	Korea Customs Service	UN	Best Practice in Anti-Corruption Forum	2001
		WCO	WCO Trophy for Intellectual Property Protection	2006
KONEPS	Public Procurement Service	UN	Public Service Award	2003
		OECD	Best Practice for Improving Transparency	2004
		WCIT	Global IT Excellence Award	2006
Home Tax Service	National Tax Service	OECD	Best Practice for e-Tax Administration	2006
e-People	Ministry of Public Administration and Security	OECD	Top 10	2006

Figure 4. Purpose of use of Internet in Korea (Korea Internet and security agency, 2009)

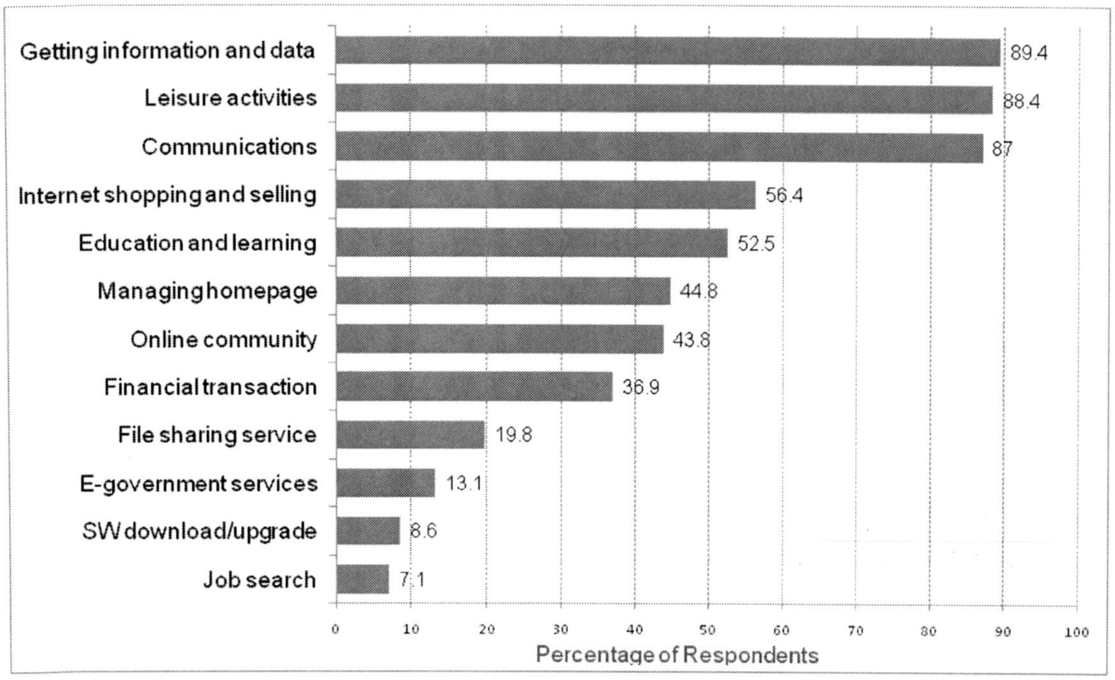

Table 5. Inward FDI stock as percentage of gross domestic product (unctad, 2008)

	1980	1990	2000	2001	2002	2003	2004	2005	2006	2007
Korea, Republic of	1.8	2.0	7.4	11.0	11.5	10.9	12.9	13.3	13.4	12.3
Hong Kong, China	616.8	262.3	269.3	251.7	205.3	240.5	273.1	294.3	390.7	573.0
Singapore	45.7	82.6	121.5	140.5	153.7	158.8	160.3	164.1	165.1	154.7
Chinese, Taiwan	5.7	5.9	6.1	11.9	10.1	12.2	11.6	12.1	13.7	12.7
Ireland	167.9	79.4	131.9	128.1	149.2	141.7	111.1	81.4	71.4	73.6
Malaysia	21.1	23.4	56.2	36.6	37.2	37.4	34.5	32.4	34.5	41.1
European Union	6.3	10.6	25.9	28.0	31.1	33.0	34.3	33.0	39.0	40.9
Caribbean	10.6	14.3	86.7	94.2	96.1	100.0	120.9	103.2	109.8	111.5
West Asia	3.1	10.1	9.7	11.5	11.0	12.5	12.6	15.7	17.6	21.5
East Asia	41.9	25.9	32.1	31.4	26.2	26.2	26.5	26.0	28.6	
South Asia	1.7	1.5	4.5	4.6	5.1	5.5	5.5	5.4	5.6	6.5
South-East Asia	9.5	18.2	44.9	42.5	41.0	41.7	42.8	45.0	44.1	43.0
South America	7.1	9.6	23.6	27.4	28.6	31.6	29.7	27.6	26.4	27.7
Central America	0.9	9.7	17.7	23.1	25.4	28.2	29.7	26.9	26.4	30.6

Table 6. Employment, total assets and gross value-add of U.S. majority owned foreign affiliates as percentage of gross domestic product of selected home countries (Mataloni, 2008)

	Gross Value-Add						Employees ('000)	Assets (million $)
	1989	2001	2002	2003	2004	2005	2005	2005
Ireland	12.4	16.6	18.9	18.4	19.5	18.5	89.8	381,943
Singapore	7.8	12.0	11.3	13.2	13.2	15.0	123.6	150,696
Canada	9.5	10.3	9.7	9.8	10.1	9.5	1,106.8	695,945
Honduras	5.6	6.2	6.1	6.6	6.3	5.8	91.2	n.a.
Nigeria	5.7	10.0	9.7	9.8	7.9	6.7	1	n.a.
United Kingdom	6.2	7.1	6.7	6.5	5.9	6.2	1,251.9	2,377,098
Malaysia	4.6	5.8	6.1	6.3	5.3	5.3	120.8	33,098
Hong Kong	4.6	4.9	4.4	5.3	4.8	4.7	110.9	168,928
Thailand	2.5	3.5	3.0	3.2	4.0	3.9	143.1	38,676
Chinese Taipei		1.3	1.2	1.5	1.6	1.7	88.1	80,360
China		0.5	0.6	0.6	0.6	0.7	316.7	44,997
Japan		0.9	0.9	0.9	1.0	1.1	554.8	574,655
South Korea	0.3	1.0	0.9	0.9	1.0	1.0	114.2	56,106

CONCLUSION

In sum, although it difficult to completely account for the differences in the deployment and adoption rates of broadband access throughout the world, a few provisional conclusions can be made. First, due in large part to the high costs for the deployment of advanced networks, thus far the most extensive examples of implementation have occurred due to strong government involvement. The governments that have taken an aggressively pro-ICT approach are motivated to implement high-speed networks largely for clear economic reasons: to encourage foreign direct investment and to realize multiple benefits from transforming their economies to technological, information-based ones. Secondly, while it is difficult to show empirically, particularly through regressional analyses that the ICT policies of the governments have led to a sustained level of FDI in these economies, there are some evidence to suggest that the governments of Hong Kong, SAR and Singapore have been successful in their policies.

REFERENCES

Bangsburg, P.T. (2000). Singapore to meld Internet logistics. *Journal of Commerce*, (March),10.

Barro, R. J. (1997). Economic growth in a cross-section of countries. *The Quarterly Journal of Economics*, (106): 407–443.

Barro, R. J. (1997). *Determinants of Economic Growth*. Cambridge: The MIT Press.

Burton, J. (2005, March 21). High speed links to the Internet encounter consumer slowdown. *Financial Times London*.

Business Times (2006, March 25). New GEBiz features lets firms post electronic catalogue.

Business Wire (2006, June 26). Singapore iN2015 masterplan offers digital future for everyone

Cable and Satellite Asia (1997). No couch potatoes? *Cable and Satellite Asia,* July, 18.

CEPAL (2008a, April). *The global information society: A statistical view*. Chile: United Nations.

CEPAL (2008b, June). *Digital Review 2007 of Latin America and the Caribbean*. Chile: United Nations.

Chan, J. W. K., Yung, K. L., & Burns, N. D. (2000). Environment-strategy fit: A study of Hong Kong manufacturing logistics. *Logistics Information Management*, *13*(5), 286–300. doi:10.1108/09576050010378513

Commerce, Industry and Technology Bureau (2004). *Digital 21 strategy: Sustainability and opportunities*. Hong Kong: Commerce, Industry and Technology Bureau, The Government of the Hong Kong Special Administrative Region.

MAIT Country Intelligence (2005, June). *Hong Kong's ICT Industry, 48*.

Davies, J.R. (2006, July). Korea broadband market December 2005. *eCommerce Innovation Centre*. Cardiff University.

Easterly, W., King, R., Levine, R., & Rebelo, S. (1994). Policy, Technology Adoption and Growth. *NBER Working Paper No. 4681*. Cambridge: National Bureau of Economic Research.

Economist, (2000, June 24). Government and the Internet: A survey. *The Economist*, 17.

Eurostat (2007). *ICT Impact Assessment by Data Linking*. Luxembourg: Working Group on Information Society Statistics (ISS), Doc.F6/ISS-WG/Oct07/08.

Faiola, A. (2006, May 27). When escape seems just a mouse-click away: Stress-driven addiction to online games spikes in S. Korea. *Washington Post*.

Fan, X., & Dickie, P. (2000). The contribution of foreign direct investment to growth and stability: A post crisis ASEAN-5 review. *ASEAN Economic Bulletin*, *17*(3), 312–323. doi:10.1355/AE17-3F

Frieden, R. (2005). Lessons from broadband development in Canada, Japan, Korea and the United States. *Telecommunications Policy*, *29*, 595–613. doi:10.1016/j.telpol.2005.06.002

Garbacz, C., & Thompson, H. G. Jr. (2005). Universal telecommunications service: a world perspective. *Information Economics and Policy*, *17*, 495–512. doi:10.1016/j.infoecopol.2005.03.001

Gholami, R., Sang-Young, T. L., & Heshmati, A. (2006). The Causal Relationship Between Information and Communication Technology and Foreign Direct Investment. *World Economy*, *29*(1), 43–62. doi:10.1111/j.1467-9701.2006.00757.x

Hardy, A. P. (1980). The role of the telephone in economic development. *Telecommunications Policy*, *4*(4), 278–286. doi:10.1016/0308-5961(80)90044-0

Indjikian, R., & Siegel, D. (2005). The Impact of Investment in IT on Economic Performance: Implications for Developing Countries. *World Development*, *33*(5), 681–700. doi:10.1016/j.worlddev.2005.01.004

InfoComm Development Authority. Singapore (2007). *Annual e-Government Perception Survey, 2007*. Retrieved from http://www.ida.gov.sg/Publications /20071001170637.aspx

International Monetary Fund. (2002). People's Republic of China- Hong Kong Special Administrative Region: Selected Issues. *IMFCountry Report*, 02/99.

International Telecommunications Union. (2003). Broadband Korea: Internet Case Study. Retrieved February 1, 2007 from http://www.itu.int/ITU-/ict/cs/korea/ material/CS_KOR.pdf

International Telecommunications Union. (2006). *World Telecommunication/ICT Development Report 2006: Measuring ICT for social and economic development*, http://www.itu.int/ITU-D/ict/publications/wtdr_06/index. html.

Kim, F. F. (1997). Government IT Infrastructure. International Market Insights. *National Trade Data Base*. U.S. Department of Commerce, December 16 1997.

Koh, J. (2000, July). E-biz yet to touch SMEs. *Business Times* (p. 14).

Korea Internet and Security Agency. (2009). *Survey of the Internet Usage 2009*. Seoul: Korea Internet and Security Agency.

Lee, C., & Chan-Olmstead, S. M. (2004). Competitive advantage of broadband internet: A comparative study between South Korea and the United States. *Telecommunications Policy*, *28*(9-10), 649–677. doi:10.1016/j.telpol.2004.04.002

Lee, N. C. (2002, February). Broadband Internet Service: Korea's Experience. Retrieved January 28, 2007 from eng.mic.go.kr/eng/secured.tdf?seq=5&idx=2&board_id=E_04_03

Lee, S.S. (2000, April 12). EDB 50% grant to aid switch to e-commerce. *The Straits Times* (p.71).

Leff, N. H. (1984). Externalities, information costs, and social benefit-cost analysis for economic development: an example from telecommunications. *Economic Development and Cultural Change*, *32*(2), 255–276. doi:10.1086/451385

Lehmann, A. (2002). Foreign direct investment in emerging markets: Income, repatriations and financial vulnerabilities. *International Monetary Fund Working Paper Series*, WP/02/47, 1-31.

Lim, R. (2006, July 3). Masterplan to transform InfoComm. *The Business Times*, Singapore.

Lim, V., & Teo, T. (2000). To work or not work at home: An empirical investigation of factors affecting attitudes toward teleworking. *Journal of Managerial Psychology*, *15*(6), 560–586. doi:10.1108/02683940010373392

Mathiason, N. (2008, July 27). Shaping the way we live: A Korean vision of hi-tech future. *The Observer*.

Matthews, J. (1999). Silicon Island of the East: Creating a semiconductor industry in Singapore. *California Management Review*, *41*(2), 55–78.

Ministry of Commerce. Industry and Energy, Republic of Korea (2005). *Key figures: FDI.* Ministry of Commerce, Industry and Energy, Republic of Korea. Retrieved January 28, 2007 from http://english.mocie.go.kr/ language/eng/news/news_view2.jsp

Ministry of Information and Communication, Republic of Korea. (2004). *IT 8-3-9 strategy: The road to $20,000 GDP per capita.* Ministry of Information and Communication, Republic of Korea.

Ministry of Information and Communication. (2002, December 16) Republic of Korea. *e-Korea Vision 2006*. Ministry of Information and Communication, Republic of Korea. MSC revitalised for global stage. (2006, April 7) Retrieved January 28, 2007 from http://sso.msc.com.my/index.php?option= com_content&task=view&id=114&Itemid=57

National Information Agency. Korea (2008). *Informatization White Paper*. National Information Agency, Korea 2008.

OECD. (2007*). Measuring the impacts of ICT using official statistics*, Working Party on Indicators for the Information Society, DSTI/ICCP/IIS(2007)1/FINAL, Paris, http://www.oecd.org/dataoecd/ 43/25/39869939.pdf

OECD. (2009). *OECD Communications Outlook, 2009*. Geneva: OECD.

Raj, C. (2000, April 11). Singapore to pump up SMEs doing e-commerce. *Business Times* (p. 7).

Raymond Mataloni. (2008). U.S. multinational companies: operations in 2006. *Survey of Current Business.* U.S. Department of Commerce, (November), 26-47.

Roller, L., & Waverman, L. (2001). Telecommunications infrastructure and economic development: a simultaneous approach. *The American Economic Review, 74*, 909–923. doi:10.1257/aer.91.4.909

Santiago, T. (2000, June 10). Singapore's e-commerce net strategy bears fruit. *Electronic Engineering Times.*

Saunders, R. J., Warford, J. J., & Wellenius, B. (1983). *Telecommunications and Economic Development.* Baltimore: Johns Hopkins University Press.

Sein, M. K., & Harindranath, G. (2004). Conceptualizing the ICT artefact: Toward understanding the role of ICT in national development. *The Information Society, 20*(1), 15–24. doi:10.1080/01972240490269942

Shih, T.H. (1999, April 8). Singapore at home and abroad. *Business Times* (p. 2).

Tan, C. S. (2006, September 1). Steady flows of FDIs into ASEAN. *The Star Online.* Retrieved January 28, 2007 from http://sso.msc.com.my/index.php?option= com_content&task=view&id=184&Itemid=57

Thompson, H., & Garbacz, C. (2007). Mobile, Fixed Line and Internet Service Effects on Global Productivity Efficiency. *Information Economics and Policy, 19*, 189–214. doi:10.1016/j.infoecopol.2007.03.002

UNCTAD. (2008). *World Investment Report, 2008.* Geneva: United Nations.

UNECA. (2007). *Project on Information Society (IS) indicators – Current situation and prospects.* Geneva: United Nations.

United States Bureau of Census. (2008). *Statistical abstract of the United States, 2008.* Washington, D.C.: Bureau of Census.

United States Department of Commerce. (2001, October 10). NTT Launches world's first 3G cellular service. *International Trade Administration, International Market Insights*, Department of Commerce.

Waverman, L., Meschi, M., & Fuss, M. (2005). The impact of telecoms on economic growth in developing countries in Africa: The Impact of Mobile Phones. *Vodafone Policy Paper Series, 2*(March), 10–24.

Xu, Y. (2006). The regulatory environment for future mobile multimedia services: The case of Hong Kong SAR and China. *ITU New Initiatives Workshop on the Regulatory Environment for Future Mobile Multimedia Services.* Geneva: International Telecommunication Union.

Chapter 8
Best Practices and Strategies for Broadband Deployment: Lessons Learned from Around the World

Christos Bouras
University of Patras and Research Academic Computer Technology Institute, Greece

Apostolos Gkamas
Research Academic Computer Technology Institute, Greece

Thrasyvoulos Tsiatsos
Aristotle University of Thessaloniki, Greece

ABSTRACT

Broadband deployment is a necessity nowadays. It could help each country, municipality and region to grow and offer better quality of life to the citizens. Today, the emphasis on the development of broadband networks is on fixed Fibre To The Home solutions The lessons learned from countries that are leaders in broadband penetration and Fibre To The Home deployment could be proven very useful for under-served communities, regions and countries where the broadband penetration is low. Therefore, this chapter summarises the lessons learned from implementing (a) country-wide strategies formulated at the national level, and (b) local strategies formulated by the municipalities. Concerning the role of national and local governments, it should be noted that nowadays it is very urgent the involvement of government in the development of broadband infrastructure. Proposed noteworthy remarkable cases are Japan, South Korea and Singapore.

INTRODUCTION

Broadband technology has emerged as the natural next step in Internet evolution and diffusion (Papacharissi & Zaks, 2006). With dialup connections limiting bandwidth and therefore, Internet applications, broadband technology promises high speed and opens up a seemingly limitless gamut of possibilities (Langdale, 1997).

Broadband networks will be as critical to the 21st century as roads, canals, and railroads were to the 19th Century and the Interstate Highway System and basic telephone networks were to the 20th Century (Copps, 2001; OECD, 2001). In addition, broadband is a key element of the

DOI: 10.4018/978-1-60960-011-2.ch008

developments taking place in the electronic communications markets. Prestona & Cawleyb (2008) refer that the European Commission (EC) has pushed broadband to the fore of social and economic policies in recent years. The same direction adopted by many many national governments in Europe. It has aligned broadband developments with furthering information society and knowledge economy developments. To this direction, EC has been particularly active in promoting broadband developments. In particular, the EC adopted an initiative supporting the Lisbon 2010 goals, i2010 (EU (2005).), where broadband take-up is considered an important factor for the emerging digital economy and competitiveness.

The main focus of this chapter is to summarize lessons learned from implementing (a) country-wide strategies formulated at the national level, and (b) local strategies formulated by the municipalities. The policies adopted by these countries for supporting the broadband growth could be proved very beneficial for countries with very low broadband penetration rate.

The rest of the chapter is structured as follows: the next section presents the current status and trends concerning the broadband deployment worldwide. The third section describes the impact of broadband in the economy of a country. The fourth section presents the advantages of broadband technologies and their positive impact concerning Green Information Technology (IT) infrastructures. The fifth section presents the role of governments in the deployment of broadband as well as the advantages of the broadband services for a country. The sixth section presents broadband policies adopted and applied in various countries as well as these policies proposed by European Union (EU) and Organisation for Economic Co-operation and Development (OECD). The next section presents best practices concerning the deployment of broadband networks. The eighth section presents suggestions for broadband deployment, while the last section (i.e. "Conclusions") summarises the results of this chapter.

BACKGROUND

This section presents the current status and trends concerning the development of broadband networks. According to Berkman Center for Internet and Society (2009) these networks offer the highest available speed, by using fixed line, fixed wireless, or mobile infrastructure. These networks are deployed in households and business places. Therefore they are mainly deployed at high population areas and business centres.

The emphasis concerning the development of these networks is on fixed Fibre To The Home (FTTH) solutions.

However, there are still operators adopting VDSL or VHDSL (Very High speed DSL) technology (ITU-T, 2004). Examples are Tele2 in Netherlands (Telecompaper, 2009) and Vodafone in Heilbronn, Germany (Telegeography, 2009). VDSL is a DSL technology providing faster data transmission over a single flat untwisted or twisted pair of copper wires. Therefore VDSL is capable of supporting high bandwidth applications such as High Definition Television (HDTV), as well as telephone services (such as Voice over IP) and general Internet access, over a single connection. VDSL is deployed over existing wiring used for POTS (Plain Old Telephone Service) and lower-speed DSL connections. This standard was approved by ITU in November 2001. The second-generation of VDSL, namely VDSL2 (ITU-T G.993.2 approved in February 2006), supports bandwidth of up to 30 MHz and provides data rates exceeding 100 Mbit/s simultaneously in both the upstream and downstream directions. The maximum available bit rate is achieved at a range of about 300 meters. However, performance degrades as the loop attenuation increases. VDSL2 is relatively inexpensive solution, in case that the operator has already a fibre backhaul and ADSL2+ network. Table 1 compares ADSL, ADSL2+, VDSL1 and VDSL2.

However, a pure FTTH solution is the more promising solution for many countries. As re-

Table 1. Comparison of ADSL, ADSL2+, VDSL1 and VDSL2

	ADSL	ADSL2+	VDSL1	VDSL2
Downstream (Mbps)	8	25	100	100
Upstream (Mbps)	1	1	50	100
Range (feet)	>15000	>15000	5000	12000

ported by Schrier (Government Technology, 2009) for the case of United States, the core of a national broadband strategy is fibre-optic cable in every home and business place in the country. According to Schrier's report, such a network is an investment that would last 50 years or more. The network would support many value added services, such as high-definition video streaming, video conferencing station, teleworking, etc. Furthermore, fibre networks would also support high-speed wireless connections, because wireless access points can be added at any place the fibre terminates (such as home and business place).

There are many variations concerning the business model concerning the management of the passive and active layers of the networks as well as their services. The main emphasis is on open access to in-building, last drop, last mile fibres. Also there is a preference for point-to-point topologies focused on competitive access to passive components.

Furthermore, there is a technological debate concerning optical networks. More specifically, the telecommunications industry, having ten years' experience with both active and passive optical networks, debates about the pros and cons of each solution. Optic FTTH can be implemented with both Passive Optical Networks (PONs) and Ethernet Point-to-Point (PtP) fibre optic networks. A PON (Kramer, 2005) is a point-to-multipoint, Fibre To The Premises (FTTP) network architecture. This architecture uses unpowered optical splitters to enable a single optical fibre to serve multiple premises. The typical amount of premises is from 32 to 128. A PON consists of an Optical Line Terminal (OLT) at the service provider's central office and a number of Optical Network Units (ONUs) near end users. A PON configuration uses less fibre and central office equipment than point to point architectures. The standards in this area are the following:

- BPON (Broadband Passive Optical Network).
- EPON or GEPON (Ethernet PON).
- GPON (Gigabit PON).

Table 2 compares BPON, EPON and GPON (Marcheck, 2006; Li, 2005).

Except the above standards, presented in Table 2,there are two new standards, namely IEEE 802.3av-2009 and SCTE IPS910. IEEE 802.3av-2009 (10G-EPON, 10 Gigabit Ethernet PON) has been approved at September 2009. This standard has been created by IEEE P802.3av 10G-EPON Task Force (http://www.ieee802.org/3/av/). It is backward compatible with 802.3ah EPON. 10GigEPON uses separate wavelengths for 10G and 1G downstream traffic. 802.3av will continue to use a single wavelength for both 10G and 1G upstream with TDMA separation. SCTE IPS910 (RFoG, RF over Glass) is an SCTE (Society for Cable Telecommunications Engineers, http://www.scte.org/) Interface Practices Subcomittee standard in development for Point to MultiPoint (P2MP) operations. Its proposed wavelength plan is compatible with data PON solutions including EPON, GEPON and 10G-EPON. RFoG offers an FTTH PON like architecture for MSOs without having to select or deploy a PON technology.

Table 2. Comparison of BPON, EPON and GPON

	BPON	EPON	GPON
Standard	ITU (G.983)	ITU-T (G.984)	IEEE (802.3ah)
Max. Throughput	622 Mbps	2.488 Gbps	1.25 Gbps
Bandwidth Support	Symmetrical (622/622)	Asymmetrical (2.4/1.2)	Symmetrical (1.25/1.25)
Network Interfaces	ATM	ATM, TDM, GbE	GbE
Network Reach	Up to 20 km	Up to 60 km	Up to 20 km
Pros	• Multiservice (TDM, Data, Video) • Fully defined carrier-class management • Widely deployed technology • Supported by major vendors • Carrier-Class TDM (low jitter, low delay synchronous transport for structured and unstructured TDM traffic)	• Leverages Ethernet cost efficiencies • Easy interface with carrier Ethernet networks • More mature than GPON • Maintains benefit of faster development cycles • Lower cost optics • Simplicity of IP management	• Low delay for TDM traffic in mixed mode only • Mixed traffic mapping
Cons	• Lower upstream bandwidth	• "Best Effort" syndrome • Overhead reduces useable bandwidth • Higher delays and jitter due to store and-forward architecture • Undefined Multiservice • Poor QoS without per-flow queuing • Lack of Carrier-Class management	• More expensive than EPONs • System complexity hampers cost reduction efforts • Complex implementation for mixed mode traffic • Higher cost optics

Ethernet Point-to-Point (PtP) systems are also often deployed. Examples are Metro Ethernet switches and IP edge routers (KEYMILE, 2008). However, optical Ethernet technology can also be used economically for direct subscriber connection via fibre optics. To date, existing systems in the IT world were simply converted and therefore not cost effective. Current FTTx systems offer very high fibre optic port densities, low power consumption and functions specifically for FTTH/FTTB applications. In contrast to PON systems, today's FTTx systems are cost effective and designed specially for use in harsh environmental conditions in public networks. Ethernet PtP systems use standard Ethernet technology for transmitting services to the end subscriber. Commonly available components, such as switches and routers with optical interfaces, can therefore be used for the subscriber as CPE (Customer Premises Equipment).

BEST PRACTICES AND STRATEGIES FOR BROADBAND DEPLOYMENT

Broadband and Economy

Broadband infrastructure is fundamental for the efficient participation of companies and organisations in today's economic. According to a study from Strategic Networks Group (SNG, 2009) local economy development and the secondary investment which is done through broadband is ten times the initial broadband investment. Broadband today is so vital like the electricity in 1930 and increase the participation of digital economy and improve the quality of life (Copps, 2001; OECD, 2001).

Broadband is needed as a complementary investment to other infrastructure such as buildings, roads, transportation systems, health and electricity grids, allowing them to be characterized as "smart". Broadband networks increase the impact and efficiency of public and private investments which depend on high-speed communications. Some governments have realised the importance of broadband and they have invested public funds to address the limitations of the communication market. These investments fall into two general categories: (a) these for extending access to unserved or under-served areas and (b) these for upgrading networks with optic fibre technologies capable of supporting competitive services in regions and municipalities. Announced government stimulus spending on communication infrastructure will largely target these two types of investment. For example the American Recovery and Reinvestment Act of 2009 (Recovery Act) was signed into law on February 17, 2009. The Broadband Initiatives funded in the Act are intended to accelerate broadband deployment across the United States (AAR, 2009). In addition, as part of Canada's Economic Action Plan (BC, 2009), $225 million was provided to Industry Canada over three years to develop and implement a strategy to extend broadband coverage to as many unserved and underserved households as possible, beginning in 2009-2010. Moreover the UK Government has issued "The Digital Britain Report" (DB, 2009) which describes the Government's strategic vision for ensuring that the UK is at the leading edge of the global digital economy. The report introduces policies to maximize the social and economic benefits from digital technologies.

Economic crisis presents serious challenges and opportunities for structural reform and targeted investment in strategic areas such as broadband. The above raises various issues concerning the best manner that governments could accomplish these goals without displacing or disrupting private investment.

According to Organisation for Economic Co-operation and Development (OECD) (OECD, 2009) the recent economic crisis has led policy makers in OECD countries to consider policies for helping their economies. Most of these policies involve large government expenditures to support demand for goods and services while simultaneously increasing the longer-term productive capacity of the economy. Investments in public infrastructures such as electricity, gas, water, transportation and communications are key elements of most policies because they have immediate impact on demand and employment as well as their strong potential to expand future supply.

The strongly pro-cyclical nature of communication network investment also means that skilled labor and equipment may be left idle and planned projects shelved until the economy improves. This labor and equipment could be quickly shifted to government-funded projects. At the same time, governments must ensure that interventions do not interfere with properly functioning markets or displace private investment.

Broadband and Green IT

Fibre optics could be characterized as "green" technology. It dissipates much less energy than copper based cables and. Furthermore, it saves a lot of materials because one single strand of glass can carry as much data as many thousands copper cables, and it can do over a longer distance without using electronic equipment to regenerate it.

For example the fibre-optic submarine cables provide an efficient approach for supporting an ever growing need for international communications.

Studies supported by, or associated with, telecommunications companies in various countries, suggest broad potential for broadband technologies in supporting activities relative to energy consumption reduction and consequence carbon emissions reduction. When we talk about broadband, connectivity comes to mind. When we talk about broadband technology it shouldn't be limited to technology that supports broadband. The broadband technology includes also technology which is supported by broadband.

The information and communications technology (ICT) industry contributes to 2 per cent of total global CO2 emissions (EU, 2008) but ICT industry can significantly contribute to control and reduce the 98% of CO2 emissions caused by other activities and industries.

Importance of Government Intervention in the Development of Broadband

Nowadays it is very urgent the involvement of government in the development of broadband infrastructure. This involvement can take place either directly with the finance of broadband development projects or indirectly with the creation of an attractive investing and regulatory environment. The creation of an attractive environment aims at private investments for broadband development.

The main reasons concerning the importance of broadband development are summarized in the following paragraphs.

First of all, broadband could be regarded as the new public utility. In other words Broadband is our generation's infrastructure challenge. Broadband is as important as electricity and highways were for past generations. Broadband can facilitate:

- Flexibility on education and training: Distance learning could not substitute the entire educational experience, but it can be used in conjunction with online meetings or weekly face-to-face meetings to create a rich flexible and collaborative educational environment. Broadband services make that possible today, and faster Internet access will enrich these online environments supporting a more flexible and accessible learning environment.
- Medical care improvements: Broadband also can save medical costs and improve access to health care. Delivering radiological scans via broadband requires fat pipes and rapid speeds, but the benefit to patients, insurers and doctors would be many: fewer scans, faster delivery of images where they are needed and lower costs associated with this process.
- Telecommuting expansion: Another benefit of better broadband would be the ability for people to telecommute. This has far-reaching benefits, from fewer cars on the roads to increasing a family's resilience in the face of economic uncertainty.

The above are some examples about the necessity to boost broadband access through governmental policies. Broadband can help promote an educated citizenry, could help lower the costs of providing health care and could increase workforce flexibility and decrease traffic.During the last years it is common the opinion that broadband can contribute to overcome the latest economy crisis (EC, 2009). According to OECD studies (OECD, 2009), broadband networks increasingly contribute to the development of the economy and the society. Broadband networks are increasingly recognised as fundamental for economic and social development. They serve as a communication and transaction platform for the entire economy and can improve productivity across all sectors. Advanced communication networks are a key component of innovative ecosystems and support economic growth.

Broadband Policies

The following paragraphs provide information about some representative broadband policies. More specifically we provide brief description of EC broadband policies, OECD broadband policies as well as national broadband policies of leader broadband countries such as Japan, Korea and Singapore.

EU Policies

Information and communication technologies are a powerful driver of growth and employment. A quarter of EU GDP (Gross Domestic Product) growth and 40% of productivity growth are due to ICT (EC 2005). Differences in economic performances between industrialised countries are largely explained by the level of ICT investment, research, and use, and by the competitiveness of information society and media industries. ICT services, skills, media and content are a growing part of the economy and society.

In recent years, ICT developments have presented a massive growth in information society and media, due to widespread fast communications, connecting multiple devices. Multimedia content (such as films, video, music) is now available in digital formats, and new digital services, such as interactive television, are emerging. The digital convergence of information society and media services, networks and devices is finally becoming an everyday life: ICT will become smarter, smaller, safer, faster, always connected and easier to use.

Proactive policies are needed to support the technological evolution. Digital convergence requires policy convergence and a willingness to adapt the current regulatory frameworks in order to be consistent with the emerging digital economy.

EC proposes a strategic framework, i2010 – European Information Society 2010 (EU, 2005), laying out broad policy orientations. It promotes an open and competitive digital economy and emphasizes ICT as a driver of inclusion and quality of life. i2010 will build towards an integrated approach to information society and audio-visual media policies in the EU.

Drawing on a comprehensive analysis of information society challenges as well as on wide stakeholder's consultation on previous initiatives and instruments, the EC proposes three priorities for Europe's information society and media policies:

- the completion of a single European information space, which promotes an open and competitive internal market for information society and media
- strengthening innovation and investment in ICT research to promote growth and more and better jobs
- achieving an inclusive European information society that promotes growth and jobs in a manner that is consistent with sustainable development and that prioritizes better public services and quality of life.

OECD Policies

OECD argues that policy makers should evaluate the costs and benefits of any public investment in telecommunication infrastructure. Therefore they should fund projects, which can deliver both strong immediate aggregate demand effects and strong longer term aggregate supply-side effects.

In most cases, governments try to create suitable environments for supporting innovative and robust participation by the private sector. Over the past three decades, the role of governments in OECD countries has been to increasingly support market-led innovation and investment in the ICT sector. At the same time, governments recognise that competitive broadband communication networks are crucial for supporting economic and social development. Broadband networks are viewed as a general purpose technology that will not only support critical services but are necessary for innovation, competitiveness and economic growth. In this direction some governments wish to address areas of potential market failure (e.g. reaching under-served areas) through their stimulus packages.

Other Policies

The following paragraphs present significant national policies adopted for broadaband deployment.

Japan has one of the fastest broadband services in the world due to an increase in government support (DOGC, 2009). Recent investments have been directed towards (optic) FTTH network installations which have resulted in broadband speeds of up to 1Gbps. In Japan FTTH is not only fast and reliable, but it is also offered with one of the best prices to speed ratio in the world. Moreover the Japanese government regularly provides subsidies of up to 33% for the maintenance and upkeep of the entire network infrastructure with the aim of propelling Japan to the forefront of the fast growing global digital economy. The Japanese government has introduced a new "zero broadband areas elimination" policy, which will spread broadband coverage to the entire country within two years, to confront this issue. Japan remains the world's leader in connecting homes to optic fibre networks with over 30 million people connected to high-speed networks. Other developed countries around the world are now embracing optic fibre technology so as not to be left behind. It is however not certain if any country can compete with the Japanese, who have already begun investing in high-speed wireless network (Wimax) and 4G mobile transmissions which will be closely followed by the commercialization of 10 Gbps optic fibre networks and eventually 160 Gbps high speed optical connection.

The Korea Communications Commission (KCC, 2009) has announced that about 34.1 trillion won (which is composed of about 32.8 trillion won of private funds and about 1.3 trillion won of government funds) will be invested to change the domestic wired, wireless and broadcasting communication network into an Internet network. They plan to make the world's best converged infrastructure of wired, wireless and broadcasting systems with a guaranteed speed of 1 Gbps for wired networks and 10 Mbps for wireless networks. When this Ultra Broadband convergence Network (UBcN) is built, customers can use all converged services including phone calls, TV, online shopping, and interactive TV finance with UDTV. Furthermore, customers will also be able to use high quality and interactive services like tutoring, medical care, civil affairs solutions, and e-commerce. When people are outside of the house, they can use multi-converged services like Internet access, phone calls, and IPTV 10 times faster than now, using their mobile phones or private portable devices. The Singapore Government has decreed the structural separation of new national broadband network infrastructure from the operating company that will operate its switches and routers (IDA, 2009). The Singapore Government would provide up to S$750m of funding for the network.

The Singapore government in the corresponding Request for Proposal for the passive infrastructure defines that a Network Company, or NetCo, will be selected to design, build and operate Layer 1 passive infrastructure that will carry the NGN traffic, expected to range between 100Mbps and 1Gbps to the end user, with a minimum upload speed of 50Mbps. A separate Layer 2/3 Operating Company or Opco will deploy the electronics such as switches and routers to manage the flow of traffic on the passive infrastructure. This same entity will offer wholesale broadband access to downstream Retail Service Providers, or RSPs. There must be operational separation between the Opco and the RSPs. The network is expected to be complete by 2015 with 50% coverage by 2012.

Best Practices

This section presents best practices concerning the deployment of broadband networks from two viewpoints. The first one regards the business models adopted by several regions or municipalities. The second viewpoint regards the technological solutions adopted by regions concerning their broadband networks.

Business Models

International experience records various business models (OECD, 2003) on broadband infrastructures exploitation, and a few indicative ones are mentioned in the following paragraphs. In general, significant efforts in Europe can be recorded in Ireland, Sweden, The Netherlands and Spain. Furthermore, remarkable examples could be recorded in Canada, New Zealand and United States

- **The Irish development model.** Metropolitan networks are designed as general-purpose networks and not as a technical solution, at least, for a limited number of service suppliers. The networks have been selected and designated by the local authorities, in the perspective of social and economic development. The operation of the metropolitan networks is effectuated centrally, through the establishment of a specific service that operates and administrates the networks. The channel network is planned and constructed so that it can support both fibre optic and copper wire infrastructures. In this particular case the open access model has been adopted.
- **The Swedish model**. The Stockholm case (Stokab): Stokab constitutes a business plan, which is being applied in the wider region of Stockholm, Sweden, aiming at the development and operation of fibre optic communication networks, as well as at leasing of fibre optic connections. The Stokab company was established for this purpose and belongs to the Stockholm Stadshus AB group, which, in turn, belongs entirely to the Municipality of Stockholm. The company was supposed to achieve this goal by providing the market with network infrastructures that are going to permit both operators and different service providers to offer their services to the end-users.
- **The Netherlands' model**. The case of Amsterdam: In the case of the Netherlands, the policies and goals concerning the implementation of broadband technologies regard also the municipalities of the country. The most indicative example is the case of Amsterdam: The Municipality of Amsterdam opted for fibre optics, considering that copper and coaxial cables are soon going to be obsolete. The model selected consists in the creation of a utility service in which the Municipality participates with a 20% percentage, while it owns and exploits the passive network (fibre optic). The other sides participating in the Public Private Partnership (PPP) are private carriers and companies. The network's active part belongs to a private company.
- **The case of Catalonia**. The LocalRet (or Local Network for these Catalonian Municipalities) was formed in 1998 and its intention is to connect 300 municipalities. The main concept of LocalRet network is that the service providers of the network will be its final operators and among its major stake holders. LocalRet will design a homogenous network for all Catalonia, at least at passive infrastructure level. This will be open and parallel to Telefonica's network. LocalRet will start integrating the small parts of the networks that the municipalities and Generalitat (government) control, as part of their ownership of railways, highways, streets, and so on. LocalRet will design Metropolitan Area Networks (MANs) in every city with more than 10.000 inhabitants.
- **United States (US).** In the US, the cases of the State of Utah (UTOPIA network) and the city of Philadelphia are of great interest, concerning the successful application of business models for exploiting broadband metropolitan area networks. In the State of Utah, eighteen (18) cities have committed

to ensure that citizens and enterprises will keep on being successful and competitive in the 21st century, by securing access to advanced telecommunication services. For this reason, they established the Utah Telecommunication Open Infrastructure Agency (UTOPIA). UTOPIA will develop the infrastructure and will install fibre optic connections to every house and enterprise. Private contractors will undertake the construction of the network and will participate in its operation. All important parts of the network have been funded by government contributions. The UTOPIA business model is classified as an open access/wholesale provider model, which becomes feasible through the interlocal agreement between the participating communities. Apart from the Utah project, in the USA, municipal fibre optic networks are being developed in fifteen (15) states, while wireless networks have been realised in 30 states. The attempt of creating a huge wireless network in the city of Philadelphia presents special interest. The Wireless Philadelphia Project received both public and private funding and its target was the provision of wireless access in the whole city, so that the digital gap would be surpassed and the quality of living of all residents would be improved. Steps have been taken towards the direction of establishing an organisation that assigns the network's planning, development and administration to private companies. The same organisation will provide network access to service suppliers, who will, in turn, offer services to subscribers.

- **The case of Canada (CANARIE)**. The great majority of the current and proposed initiatives for the encouragement of broadband infrastructure development in Canada may be classified into two broad strategies: the demand aggregation model and the public infrastructure support model. The former promotes collective regulations that aggregate the demand for broadband services, on a community or even larger scale, so that the scale economies required for the development and support of a broadband network are achieved. The latter is based on direct public funding for the construction of broadband infrastructures. Although these two models can be applied independently, it has to be noted that, in many cases, they are often combined.
- **New Zealand: The case of Wellington**. The CityLink company was established by the Municipality of Wellington, New Zealand, aiming at the development of a low cost telecommunication network, intending to offer a comparative advantage to the local enterprises and governmental organisations. This network provides the citizens of Wellington with a variety of services, but it also gives massive company users and Internet Service Providers (ISPs) the possibility to lease part of the network (dark fibre) at cost-oriented prices. One of the innovative and most interesting network features, which contributes enormously in its viability, is that it uses low-cost passive and active equipment (i.e. Zebra on Linux), and not expensive commercial network devices.

FTTH Deployment

The following list presents noteworthy efforts concerning FTTH deployment:

- **OneCommunity** (Ohio, US, www.onecommunity.org): The communities in northeastern Ohio have long suffered from deindustrialization and loss of population. In order to revitalize the local economy, OneCommunity, a nonprofit organisation, was created in Cleveland in 2003 to serve a variety of public and nonprofit organisations in 22 counties across Northern Ohio. The fibre network deployment has been leveraged by government and nonprofits to jumpstart new investment, improve health-care, education, and engage thousands of area leaders in collaboration over regional economic development.

- **BVU OptiNet** (Bristol Virginia, US, www.bvu-optinet.com): Optinet is recognised as Broadband Fibre Network of Year in US by the National Association of Telecommunications Officers and Advisors (NATOA). OptiNet, which extends from Bristol Virginia approximately 850 miles through eight Southwest Virginia counties, was recognised for "pioneering community fibre to the home and for demonstrating how true broadband can bring jobs and economic development to rural America." Bristol Virginia Utilities (BVU) is a city-owned public utility that provides electric, water, wastewater, cable and advanced fibre-optic broadband services to customers in a 125-square-mile area that includes Bristol, Abingdon and Washington County. In 2003, BVU became the first municipal entity in the United States to build and successfully maintain a fibre-to-the-premises broadband network offering phone, cable and data services.
- **blizznet** (Vienna, Austria, http://www.wienenergie.at): In the field of telecommunications, the Wien Energie subsidiary Wien Energie Wienstrom offers Vienna's extra broadband network (blizznet), which is one of the most modern and fastest broadband fibre optic networks in Europe. Around 12,800 households spread across nine districts in Vienna are currently connected to blizznet. Consequently, the current network consists of more than 1,400 km of fibre optic cables - a network which is constantly being expanded. Moreover, there are 2,200 km of existing empty piping available to expand the network. blizznet offers top-level quality for all (Internet access, TV, telephone, security, data protection and gaming) supporting high-speeds (100 mbps, upload and download).
- **TRE-FOR** (East Jutland, Denmark): TRE-FOR is a multi-utility (for water supply, district heating, electricity and broadband telecom) serving an area of about 28km radius in East Jutland, Denmark. There are three principal towns – Fredericia, Kolding, and Vejle – and about 300,000 customers. The company is organized as a cooperative and is essentially owned by its customers, who can be elected to the 115-member Stakeholders' Committee, which, among other things, helps to set strategic priorities and directions. TRE-FOR has crafted a role for itself as a middleman between its customers and service providers. It's rolling out fibre-to-the-home (FTTH) and Ethernet infrastructure, offering customers a self-service portal for them to order services, and providing a platform that makes it easy for third parties to provide those services. TRE-FOR is not taking on any of the risky business of guessing what applications will prove a hit with its customers. It is still in the utility business, where investments can be amortized over a long period, and thus it's able to keep a lid on costs. At the same time, it's created an environment that encourages innovation and competition among service providers so that its customers (who are also its shareholders) get a wide choice of offerings at low prices. The TRE-FOR network is based on a core-plus-aggregation layer. The core is a dual 5-node redundant IP/MPLS backbone using routers with 10-Gigabit Ethernet interfaces.

FUTURE RESEARCH DIRECTIONS

Concerning the broadband deployment the European Broadband Portal has created a thorough check list for public authorities that covers all steps to be taken in the complex process of developing structural actions and interventions in broadband, mainly for under-served territories but valid also for any other territories (European Broadband Portal, 2009). The guide describes aspects to be taken into account in 5 main aspects of the process:

- **The Mapping and the Coordination**: concerning this aspect the guide proposes actions for gather information about existing infrastructures in the territory; mapping the territory and identifying gaps in provision and opportunities to exploit other infrastructure; identifying the key players (suppliers and users) in the territory; promoting the benefits of broadband; creating a demand aggregation strategy; reviewing best practices in similar territories worldwide.
- **Developing the Business Case**: concerning this aspect the guide proposes actions for developing the action plan and obtaining the support of all key players in the territory; elaborating feasibility study; comparing alternative technological solutions etc
- **Funding rise and find**: concerning this aspect the guide proposes actions for identifying available sources of funding
- **Business models available**: concerning this aspect the guide proposes actions for developing an appropriate business model; ensuring economic/business sustainability dependent on commercial success, etc,
- **Implementation**: concerning this aspect the guide proposes among others actions for preparing open calls for procurement under the "neutral technology approach" for suppliers to build the network and for operators to operate; elaborating plans for providing the appropriate training for public sector employees; designing and establishing systems for monitoring and evaluation

Furthermore, Intel Corporation (Intel, 2009) has identified five essential best practices that can help countries successfully expand broadband penetration:

- **Adopt supportive regulations that embrace innovation and competition**: A wide range of regulatory reforms can be made to create an enabling environment. Hundreds of countries now have national regulatory bodies. Market liberalisation and privatisation can increase private sector investment and healthy competition, and strengthening regulations and making them more transparent can build the trust necessary to spur greater investment and encourage infrastructure sharing.
- **Form mutually beneficial public/private partnerships**: successful broadband deployments begin with a strong commitment from the government that is then carried out through extensive collaboration and partnerships among government, industry and private groups.
- **Invest in infrastructure and the latest innovative technology**: Particularly in developing nations, where the infrastructure is not likely to be fully developed, investment is necessary to support the intermediate infrastructure — including everything from utilities to computers — that in turn enables successful diffusion of broadband networks.
- **Encourage competitive ecosystems**: When policy and regulatory reforms are considered, the encouragement of competition should be a priority, as competition is

critical to the market growth seen in nearly 80% of countries worldwide (ITU, 2008).

- **Release spectrum suitable for sustained broadband deployment**: Releasing spectrum offers a clear benefit: It enables governments to place bandwidth in the hands of private businesses that can then develop its use through emerging industries and technologies, thus offering a significant source of sustained revenue. The question is whether to release spectrum now or later, and based on basic economic realities, the clear answer is that the time is now.

According to Intel, when these best practices are combined with a firm governmental and national commitment and with private and public partnerships, developing nations can achieve the many benefits of broadband networks.

CONCLUSION

Broadband infrastructure is fundamental for the efficient participation of companies and organisation in today's economic. Broadband today is so vital like the electricity in 1930 and increase the participation of digital economy and improve the quality of life.

Furthermore, broadband could have positive impact not only because it offers value added services and it is a vital infrastructure in a country, but also because it can support the vision of many countries for Green IT. Fibre optics is a very "green" technology. It dissipates a lot less energy than copper based cables and it saves a lot of materials.

Concerning the role of governments, it should be noted that nowadays it is very urgent the involvement of government in the development of broadband infrastructure. The involvement of government could take place either directly, with the finance of broadband development projects, or indirectly, with the creation of an attractive environment in order to attract private investments for broadband development. The policies adopted by each country and its government are vital for the broadband deployment. Both EU and OECD are trying to support their members by proposing such kind of policies. Furthermore, almost each country with high broadband penetration elaborates broadband strategies as well as policies in order to support their strategies. Proposed noteworthy cases are Japan, South Korea and Singapore.

Concerning the process of broadband deployment there are many lessons learned from the leading countries in broadband penetration worldwide. However, it is vital to map the process of broadband deployment. Such a map has been elaborated by the European Broadband Portal in the form of a thorough check list for public authorities. This list covers all steps to be taken in the complex process of developing structural actions and interventions in broadband, mainly for under-served territories but valid also for any other territories (European Broadband Portal, 2009). The main aspects to be taken into account are (i) the mapping and the coordination; (ii): the developing of the business case; (iii) the funding rise and find; (iv) the elaboration of business model; and (v) the implementation.

In addition, Intel proposes a list of best practices that can help countries successfully expand broadband penetration: (i) adoption of supportive regulations that embrace innovation and competition; (ii) formation of mutually beneficial public/private partnerships; (iii) investment in infrastructure and the latest innovative technology; (iv) encouragement of competitive ecosystems; and (v) releasing of spectrum suitable for sustained broadband deployment.

To conclude, it could be said that broadband deployment is a necessity nowadays that could help each country, municipality and/or region to grow and offer better quality of life to their citizens. The lessons learned from countries that are leaders in broadband penetration and FTTH deployment could be proven very useful for under-served communities, regions and countries where the broadband penetration is low.

REFERENCES

AAR. (2009). The American Recovery and Reinvestment Act of 2009. Retrieved January 14, 2010, from http://www.broadband.gov/

BC. (2009) Broadband Canada: Connecting Rural Canadians, Retrieved January 14, 2010, from http://www.ic.gc.ca/eic/site /719.nsf/eng/home

Berkman Center for Internet and Society. (2009). *Next Generation Connectivity: A review of broadband Internet transitions and policy from around the world.* Harvard University.

Copps, M. (2001). Remarks of Michael J. Copps, Commissioner Federal Communications Commission to National Association of State Utility Consumer Advocates 2001 Annual Meeting.

DB. (2009). Digital Britain: The Final Report, Retrieved January 14, 2010, from http://www.culture.gov.uk/ what_we_do/broadcasting/6216.aspx

DOGC. (2009). Digital opportunities for global community. Retrieved November 11, 2009, from http://www.dosite.go.jp

EC. (2005) i2010 – A European Information Society for growth and employment Retrieved January 14, 2010, from http://europa.eu/rapid/pressReleases Action.do?reference=MEMO/05/184&format=HTML&aged=0&language=EN&guiLanguage=en

EC. (2009) Digital economy can lift Europe out of crisis, says Commission report, European Commission, Retrieved January 14, 2010, from http://europa.eu/rapid/pressReleases Action.do?reference=IP/09/1221& format=HTML&aged=0&language=EN&guiLanguage=fr

EU. (2005). i2010 - A European Information Society for growth and employment. European Commission. Retrieved November 11, 2009, from http://ec.europa.eu/information_society/eeurope/i2010/index_en.htm

EU. (2008). Commission casts ICT in green role. Reference IP/08/733, European Commission. Retrieved November 11, 2009, from http://europa.eu/rapid/press ReleasesAction.do?reference=IP/08/733&format= HTML&aged=0&language=EN&guiLanguage=en

European Broadband Portal. (2009). Check-list of actions for public authorities considering broadband interventions in under-served territories. Draft (1st Version, April 2009). Retrieved November 11, 2009, from http://www.broadband-europe.eu/ Documents/EBP-CHECK-LIST-Broaband%20Actions% 20of%20Public%20Authorities%20v%201-4-2009.pdf

Government Technology. (2009). The Future Is Fibre. Retrieved November 11, 2009, from http://www.govtech.com/gt/692922

IDA. (2009). Infocomm Development Authority of Singapore. Retrieved November 11, 2009, from http://www.ida.gov.sg/

Intel (2009). The Economic Impact of Broadband in Developing Nations, Best practices enable developing nations to reap economic benefits of broadband. White paper, Intel Corporation, 2009. Retrieved November 11, 2009, from http://download.intel.com/intel/ worldahead/pdf/economic_impact.pdf

ITU. (2008). Report on the World Summit on the Information Society Stocktaking. World Summit on the Information Society, Geneva 2003 - Tunis 2005. Retrieved November 11, 2009, from http://www.itu.int/wsis/stocktaking/ docs/2008/WSIS-Stocktaking2008-e.pdf

ITU-T. (2004). Very high speed digital subscriber line transceivers. ITU-T Recommendation G.993.1. Retrieved January 15, 2010, from http://www.itu.int/rec/ T-REC-G.993.1/en

KCC. (2009). Korea Communications Commission. Retrieved November 11, 2009, from http://eng.kcc.go.kr

KEYMILE. (2008). Ethernet Point-to-Point vs. PON – A comparison of two optical access network technologies and the different impact on operations. Retrieved November 11, 2009, from http://www.eurocomms.com/asset/ 922/White%20Paper%20EP2P%20vs%2 0PON%20(E%202009%2008%2017).pdf

Kramer, G. (2005). *Ethernet Passive Optical Networks (professional Engineering)*. McGraw-Hill Professional Publishing.

Langdale, J. (1997). International competitiveness in East Asia: Broadband telecommunications and interactive multimedia. *Telecommunications Policy, 21*(3), 235–249. doi:10.1016/S0308-5961(97)00004-9

Li, M. (2005). Passive Optical Networks (PON) Protocols. Hakko-opto, presentation. Retrieved January 15, 2010, from http://www.hakko-opto.com/pdf/ Comparison_PON_Protocols.pdf

Marcheck, J. (2006). FTTx: The Many Roads to Last Mile Fiber. Current-Analysis, presentation, Retrieved January 15, 2010, from http://www.currentanalysis.com /t/fiberpremise2006/files/CurrentAnalysis-fttx.pdf

OECD. (2001). *The development of broadband access in OECD countries*. Paris: OECD.

OECD. (2003). *Policies for Broadband Development. Recent OECD work on Broadband Committee for Information*. Computer and Communications Policy.

OECD. (2009). The Role of Communication Infrastructure Investment in Economic Recovery. DSTI/ICCP/CISP(2009)1/FINAL, Organisation for Economic Co-operation and Development. Retrieved November 11, 2009, from http://www.oecd.org/dataoecd /4/43/42799709.pdf

Papacharissi, Z., Zaks, A. (2006). Is broadband the future? An analysis of broadband technology potential and diffusion. Telecommunications Policy, *30*(2006), 64–75 70.

Philadelphia, Pennsylvania November 14, 2001 Retrieved January 15, 2010, from http://www.fcc.gov/Speeches /Copps/2001/spmjc108.html

Prestona, P., & Cawleyb, A. (2008). Broadband development in the European Union to 2012—A virtuous circle scenario. *Futures*..doi:10.1016/j.futures.2008.07.035

SNG. (2009). Measuring the impacts of broadband. *Strategic Networks Group*. Retrieved November 11, 2009, from http://www.govtech.com/gt/692922

Telecompaper (2009). Tele2 to introduce VDSL2 in Netherlands. Retrieved November 11, 2009, from http://www.telecompaper.com/news /article.aspx?cid=688321

Telegeography (2009). Vodafone to launch VDSL pilot in Heilbronn. Retrieved November 11, 2009, from http://www.telegeography.com /cu/article.php?article_id=29221

Chapter 9
The European Research and Education Networks:
Ensuring Europe's Leadership in e-Science

Navonil Mustafee
Swansea University, UK

Simon J. E. Taylor
Brunel University, UK

ABSTRACT

e-Science is the future of Science. It necessitates the use of powerful computing resources, massive data sets, remote instruments, scientific/visualisation software and expertise that are distributed around the world. e-Infrastructures refer to the underlying computing technologies that facilitate e-Science. To weave these distributed resources into a cohesive entity, and thereby enabling such large-scale science to be conducted in highly distributed environments, needs the utilisation of high speed communication networks. The European Research and Education Networks provide the connectivity required to ensure Europe's leadership in e-Science. These networks not only provide the communication resources required to conduct e-Science, but they also offer high-speed Internet access to Universities and educational institutes to facilitate teaching and learning. The objective of this book chapter is to inform the readers of the organisation of the underlying European networks that are used by millions of researchers, academics and students. The importance of these networks are highlighted by presenting six e-Science and e-Infrastructure projects that are being funded by the European Commission.

INTRODUCTION

e-Science (and arguably e-Research) can be defined as Science that may necessitate the utilisation of non-trivial amounts of computing resources and massive data sets to perform scientific enquiry; Science that requires access to remote scientific instruments and distributed software repositories; Science that generates data that may demand analysis from experts belonging to multiple organisations and specialists in different knowledge domains. Such Science is usually carried out in highly distributed environments by exploiting advanced hardware and software technologies like grid computing and high-speed networking.

DOI: 10.4018/978-1-60960-011-2.ch009

Grid computing has been defined as a system that coordinates distributed resources using standard, open, general-purpose protocols and interfaces with the aim of delivering non-trivial qualities of service (Foster and Kesselman, 2004). The nature of resources being shared in the grid environment not only includes file exchange but rather direct access to computers, software, data, and other resources, e.g., sensor networks and scientific instruments (Foster et al., 2001). e-Infrastructure enable such sharing of distributed resources. The term encompasses communication networks that bind these dispersed computing resources and thereby facilitate e-Research. These networks also provide high speed Internet access to Universities and educational institutes all over the world. In European context, these networks are collectively referred to as the European Research and Education Networks, or EREN for short.

The EREN presently comprises of 36 regional National Research and Education Networks (NRENs) that are interconnected through the high-bandwidth pan-European GÉANT (GÉANT, 2009a) network. The NRENs are usually funded through national initiatives. GÉANT, on the other hand, is funded not only by the NRENs but also by the European Commission through the GÉANT & e-Infrastructure Unit. For example, the third generation GÉANT project – *GÉANT3* (launched in December 2009 and funded till 2013) – has been awarded funding of €93 million from the European Commission under the EU's 7th Framework Programme, and a matching amount from Europe's NRENs (GÉANT Press Release, 2009).

The ERENs collectively provide the networking backbone that enable e-Research in Europe. There are presently over 40 European e-Infrastructure and e-Science projects that utilise the ERENs. These projects range from those focussing on computing (supercomputing, grid computing, desktop computing), networking, data repositories, scientific experimentation to projects that are involved in creating global virtual research communities, formulating e-Infrastructure policies and promoting standards. Although the focus of this book chapter is on e-Science projects that are supported by ERENs, it is important to note that the same networks are also used for providing high speed Internet connectivity to further and higher education institutes in Europe.

The focus of this book is on broadband. There have been several studies on the adoption, diffusion and usage of broadband with regard to household consumers and in the context of several countries (Choudrie & Dwivedi, 2004; 2005; 2006abc; 2007; Dwivedi & Irani, 2009; Dwivedi & Lal, 2007; Dwivedi & Weerakkody, 2007; Dwivedi & Williams, 2008; Dwivedi et al., 2006abc; 2007ab;2008ab; 2009; Irani et al., 2009; Khoumbati et al., 2007). In this chapter we focus on yet another consumer of this technology, namely, further and higher education institutes and research centres in Europe. These institutions are usually connected to ERENs which enable them to provide high-speed broadband access to the employees, students, etc. affiliated to such organisations. The objective of this book chapter is to provide the reader with an understanding of the organisation of the ERENs and to highlight the importance of these networks with regard to the European e-Science and e-Infrastructure projects that are being funded by the European Commission. The Commission's GÉANT & e-Infrastructure Unit plays a vital role in this.

The rest of the chapter is organised as follows. An overview of the pan-European GÉANT network is presented in section 2. This is followed by Section 3 that deals with the European NRENs. Section 4 describes six current EU funded e-Science and e-Infrastructure projects that utilises the networking services provided by GÉANT and the NRENs. Section 5 is the concluding chapter of this book chapter and highlights the importance of the ERENs in ensuring Europe's leadership in e-Science.

THE PAN-EUROPEAN GÉANT NETWORK

GÉANT, launched in the year 2000 with a funding €200 million and with the objective of upgrading Europe's research and education networking infrastructure, is the world's largest multi-gigabit pan-European computer network dedicated to research and education (EC Press Release, 2008). GÉANT is presently in its third phase and is a successor to GÉANT2. The Delivery of Advanced Network Technology to Europe (DANTE, <http://www.dante.net/>) is responsible for planning, building and managing the GÉANT network. The Trans-European Research and Education Networking Association (TEREN, < http://www.terena.org/>) is responsible for managing GÉANT outreach activities and supports the co-ordination of the research and development effort amongst the 36 GÉANT project partners.

GÉANT provides advanced interconnectivity between Europe's NRENs and operates at data transfer speeds of up to 10 Gbps across approximately 50,000 kilometres of network infrastructure, thereby enabling nearly 40 million research and education users in over 8,000 institutions across 40 countries to collaborate across multiple disciplines throughout Europe and the rest of the world (GÉANT Press Release, 2009). The GÉANT project seeks to develop all aspects of European research and education networking, including, shaping the Internet of the future through the network and a portfolio of advanced services for the user community; tackling the digital divide of research and education networking across Europe; fostering technological research to assure Europe's role at the forefront of networking and e-Science. The portfolio of services offered includes connectivity, virtual private networks with reserved bandwidth capacity for specific project requirements, network performance monitoring tools and secure roaming and authorisation services (GÉANT Press Release, 2009).

GÉANT is commonly referred to as the "network of networks" since it provides high speed backbone network that presently interconnects 36 European NRENs (refer to Table 1 in section 3). Furthermore, GÉANT directly connects to other non-European networks with links to, for example, North America (*ESnet*, <http://www.es.net/>; *Internet2*, <http://www.internet2.edu/>), the Mediterranean (*EUMEDCONNECT2*, < http://www.eumedconnect2.net/>), Latin America (*ALICE2*, < http://alice2.redclara.net/>), the Asia-Pacific (*TEIN3*, <http://www.tein3.net/>), India (*ERNET*, < http://www.ernet.in/ >), China (*ORIENT*, <www.dante.net/orient>), Central Asia (*CAREN*, < http://caren.dante.net/>), and Southern and Eastern Africa (*UbuntuNet Alliance*, < http://www.ubuntunet.net/>) (GÉANT, 2009b). This enables European scientists to undertake e-Research with their counterparts across the world.

GÉANT is utilised for a wide range of collaborative research projects ranging from astronomy, biotechnology, particle physics, seismology to archaeology, medicine and climate change. A few of these projects are discussed in section 4. The next section provides an overview of the organisation of NRENs in Europe.

NATIONAL RESEARCH AND EDUCATION NETWORKS

Most of the countries in Europe have publicly-funded national initiatives that provide networking services geared towards their research and educational institutes. These national networking initiatives are called National Research and Education Networks, or NRENs for short. NRENs are responsible for delivering networking services to Universities, institutes engaged in further and higher education and government funded research institutes. For example, JANET is UK's NREN and it manages the network on behalf of Joint Information Systems Committee (JISC) for the UK Further and Higher Education Funding Councils.

Table 1. List of European NRENs (adopted from GÉANT (2009c))

NREN	NREN - Name	Country	Website
ACOnet	Austrian Academic Computer Network	Austria	http://www.aco.net
BELnet	Belgian National Research Network	Belgium	http://www.belnet.be
BREN	Bulgarian Research and Education Network	Bulgaria	http://www.acad.bg
CARNet	Croatian Academic and Research Network	Croatia	http://www.carnet.hr
CYNET	Cyprus' Academic and Research Network	Cyprus	http://www.cynet.ac.cy
CESNET	Czech National Research and Education Network	Czech Republic	http://www.cynet.ac.cy
EENet	Estonian Educational and Research Network	Estonia	http://www.eenet.ee
RENATER	Réseau National de télécommunications pour la Technologie l'Enseignement et la Recherche	France	http://www.renater.fr
DFN	Das Deutsche Forschungsnetz	Germany	http://www.dfn.de
GRNET	Greek Research and Education Network	Greece	http://www.grnet.gr
NIIF	National Information Infrastructure Development Institute	Hungary	http://www.niif.hu
HEAnet	Ireland's National Education and Research Net-work	Ireland	http://www.heanet.ie
IUCC	Inter-University Computation Center	Israel	http://www.iucc.ac.il
Consortium GARR	The Italian Academic & Research Network	Italy	http://www.garr.net
SigmaNet	Sigmanet Academic Network and Data Centre	Latvia	http://www.sigmanet.lv
LITNET	Academic and Research Network of Lithuania	Lithuania	http://www.litnet.lt
RESTENA	Réseau Téléinformatique de l'Education Nationale et de la Recherche	Luxembourg	http://www.restena.lu
MARNet	Macedonian Academic and Research Network	Macedonia	http://dns.marnet.net.mk
University of Malta	University of Malta	Malta	http://www.um.edu.mt
MREN	Montenegrin Research and Education Network	Montenegro	http://www.mren.ac.me
SURFnet	Education and Research Network of Netherlands	Netherlands	http://www.surfnet.nl
NORDUnet	Nordic Infrastructure for Research and Education	Nordic region (includes Norway, Sweden, Finland, Denmark and Iceland)	http://www.nordu.net
PSNC	Poznan Supercomputing and Networking Center	Poland	http://www.man.poznan.pl
FCCN	Fundação para a Computação Científica Nacional	Portugal	http://www.fccn.pt
AARNIEC/ RoEduNet	Romanian National Research and Education Network	Romania	http://www.nren.ro
AMRES	Academic Network of Serbia	Serbia	http://www.amres.ac.rs
SANET	Slovak Academic Network	Slovakia	http://www.sanet.sk
ARNES	Academic and Research Network of Slovenia	Slovenia	http://www.arnes.si/english
RedIris	Spanish Academic and Research Network	Spain	http://www.rediris.es
SWITCH	SWITCH Telematikdienste für Lehre und Forsc-hung Foundation	Switzerland	http://www.switch.ch

continued on following page

Table 1. Continued

ULAKBIM	Turkish Academic Network and Information Centre	Turkey	http://www.ulakbim.gov.tr
JANET	UK's Education and Research Network	UK	http://www.ja.net
BASNET	Education and Research Network of Belarus	Belarus	www.bas-net.by
JSCC	Joint SuperComputer Center	Russia	www.jscc.ru
RENAM	Research and Educational Networking Association of Moldova	Moldova	www.renam.md
URAN	Ukrainian Research and Academic Network	Ukraine	www.uran.net.ua

Some of the services offered by Janet are as follows (JANET, 2009):

- *Primary connections* are provided to all government funded further and higher education institutions.
- *Sponsored connections* are connections of a third-party organisation's network to JANET via a Regional Network Operator or an organisation that itself has a Primary Connection.
- *Domain registration* of *ac.uk* and *gov.uk* domains.
- *Video services* like JANET Videoconferencing Service (JVCS) and JANET Video Technology Advisory Service (VTAS).
- *Certificate service* that provide users with some assurance of the identity of the server they are accessing.
- *Training courses* pertaining to services offered by JANET. Some of the courses provided by JANET are introductory courses, technical courses, JANET service courses, wireless courses, security courses and videoconferencing courses.

Services similar to JANET are provided by NRENs of other European countries. Table 1 provides a list of the NRENs and their respective website URLs for reference.

GÉANT provides connectivity between the aforementioned NRENs, thus enabling researchers all across Europe to collaborate together in e-Science projects. Some of the leading EU funded research projects that utilise the networking infrastructure provided by the NRENs and GÉANT are described in the next section.

EC FUNDED RESEARCH PROJECTS SUPPORTED BY NRENS AND GÉANT

This section focuses on six European e-Infrastructure and e-Science projects that are presently being funded by EU's 7th Framework Programme. The common denominator of these projects is that they rely on the NRENs and GÉANT. The projects referred to in this section relate to supercomputing (DEISA-2), grid computing (EGEE-III), particle physics (LHC), experimental network infrastructure (FEDERICA), access to remote instruments (EVALSO) and expansion of the GÉANT network (BSI).

EGEE-III: The World's Largest Multi-Disciplinary Production Grid Infrastructure

Enabling Grids for E-sciencE - III (EGEE-III) will expand and optimise the existing EGEE grid infrastructure that was deployed during the previous EGEE-I and EGEE-II projects. The EGEE grid infrastructure currently processes up to 330,000 computation jobs per day and supports thousands of researchers from around the world who conduct research in diverse scientific domains. The networking backbone of EGEE is provided by GÉANT and the NRENs. In this section we use the acronym EGEE to refer to EGEE-III and its predecessors - EGEE-I and EGEE-II.

The achievements of this project are aptly demonstrated by statistics pertaining to the EGEE infrastructure and its users. As of July 2009, 290 sites in over 50 counties are connected to the EGEE infrastructure and they provided approximately 144,000 processing cores to its users. It provides a disk storage capacity and long-term tape storage of approximately 25 petabytes and 38 petabytes respectively. In terms of the users of the EGEE infrastructure, approximately 17,000 researchers belonging to around 120 virtual organisations (organisations comprising of multiple stakeholders that virtually collaborate on an e-Science project) use this resource. The application domains making use of the EGEE infrastructure includes archaeology, astronomy & astrophysics, civil protection, computational chemistry, computational fluid dynamics, computer science/tools, condensed matter physics, earth sciences, finance, fusion, geophysics, high-energy physics, life sciences, multimedia and material sciences (EGEE, 2009).

The success of EGEE is paving the way for a sustainable pan-European grid infrastructure. EGEE has been instrumental in implementing a hierarchal structure for efficient management of grid resources in member counties. This has been achieved through the EGEE consortium and the National Grid Initiatives (NGIs). The NGIs use their respective NRENs for providing interconnectivity between the grid resources deployed at the national level. The EGEE consortium consists of 42 beneficiaries from academia and industry. However, in real terms, these 42 beneficiaries represent more than 120 partners since the member countries have grouped their academic partners on a national level through the NGIs. These NGIs are an important milestone in the transition of the EGEE infrastructure to a sustainable operational model – the European Grid Initiative (EGI). The EGI is a partnership between the NGIs and a coordinating body, the EGI Organisation (EGI.eu). It is expected that by the year 2010, the resources currently coordinated by EGEE will be managed through the EGI and each country's grid infrastructure will be run by their respective NGIs (EGEE, 2009).

DEISA-2: The Pan-European Supercomputer

The Distributed European Infrastructure for Supercomputing Applications (DEISA) is a consortium of leading national supercomputing centres that aims at fostering the pan-European world-leading computational science research. DEISA-1, and now DEISA-2 (funded by EU FP7), is paving the way towards the deployment of a cooperative European High Performance Computing (HPC) ecosystem (DEISA, 2009). In the remainder of this section, we use the acronym DEISA to refer to these projects.

The DEISA infrastructure is based on a tight coupling of eleven national supercomputing centres from seven European countries, using dedicated network 10 Gbps interconnections of GÉANT and the NRENs. The aggregated power of the DEISA supercomputing resources exceeds one PetaFlop per second. DEISA is structured as a layer on top of the national supercomputing services and provides generalised interfaces and services that allow users to access and utilise this pool of computing resources in a consistent and efficient manner. The HPC infrastructure and services offered by DEISA thus combines, for its users and the user communities it serves, the advantage of having access to a variety of supercomputing architectures for different computing purposes with the advantage provided by consistent interfaces to these different resources and services. Accordingly, the profile of using the DEISA HPC infrastructure and services can be regarded as being similar to the usage of a monolithic supercomputing system of single HPC centre (DEISA, 2009). DEISA incorporates several different platforms and operating systems, for example, Cray XT4/5 (Linux), IBM Power5 (AIX), IBM Power6 (AIX, Linux), IBM BlueGene/P (Linux), IBM PowerPC (Linux), SGI ALTIX 4700 (Linux), NEC SX8/9 vector systems (Super UX). The middleware deployed on the DEISA resources enable transparent access to distributed resources, high performance data sharing at European scale, and transparent job migration across similar platforms.

Above and beyond the research DEISA has empowered and hopes to stimulate, its most important contribution may well be the extent to which it has prepared the ground for the future development of Europe's supercomputing capabilities. Since DEISA started just four and a half years ago, the aggregated peak computing power it can offer has multiplied by a factor of 300, from 30 Teraflops (30 thousand billion floating point operations per second) to over a Petaflop (a million billion operations per second) (ICT Results, 2009).

Large Hadron Collider (LHC) Project: The World's Largest Scientific Experiment

The LHC project features the world's largest and the most powerful particle accelerator and four state-of-the-art particle physics collision detectors (ALICE, ATLAS, CMS, LHCb). The four LHC experiments, named after the four collision detectors, are designed to be able to study particle physics under conditions well beyond any other previous experiment (Lamanna, 2004). The LHC experiments are the single largest scientific experiment ever undertaken. Its purpose is to recreate the conditions just after the Big Bang, by colliding two beams of hadrons (subatomic particles) head-on at very high energy. "There are many theories as to what will result from these collisions, but what's for sure is that a brave new world of physics will emerge from the new accelerator, as knowledge in particle physics goes on to describe the workings of the Universe" (CERN LHC, 2009).

The Worldwide LHC Computing Grid (WLCG) provides the computation and storage resources for the four LHC particle physics experiments. WLCG is a global collaboration of more than 170 computing centres in 34 countries and aims to harness the power of 100,000 CPUs distributed among the computing centres to process, analyse and store data produced from the LHC, making it equally available to all partners, regardless of their physical location (CERN WLCG, 2009). GÉANT provides global communications support for WLCG. The experiments at CERN depends on the high speed GÉANT links around the globe for seamless transmission of unprecedented amounts of data (15 million Gigabytes/year) to 5,000 scientists working in 500 institutes worldwide.

FEDERICA: Implementing Experimental Network Infrastructure

The Federated E-infrastructure Dedicated to European Researchers Innovating in Computing network Architectures (FEDERICA, 2009) is a networking project funded by the EU's 7th Framework Programme that aims to implement an experimental network infrastructure for trialling new networking technologies. The project started on 1st January 2008 and the project duration is 30 months. The FEDERICA infrastructure is composed of contributions of Gigabit circuits, transmission equipment and computing nodes from ten participating NRENs, namely, *Consortium GARR* (Italy), *CESNET* (Czech Republic), *DFN* (Germany), *FCCN* (Portugal), *GRNET* (Greece), *NIIF* (Hungary), *PSNC* (Poland), *RedIris* (Spain), *SWITCH* (Switzerland) and *HEAnet* (Hungary), and the pan-European GÉANT network.

FEDERICA uses virtualisation in computing and network systems to create a technology agnostic and neutral infrastructure that makes it is possible to create "slices" of the infrastructure as per the user's request, wherein each slice comprising of a set of virtual network and computing resources (Campanella, 2008). Each virtual network is a "slice" of the infrastructure and many slides can be active in parallel. The virtual networks will permit disruptive experiments in new Internet architectures and protocols to be conducted over existing networks, without affecting the production use of the underlying networks by other non-disruptive network applications. On a more holistic level, the FEDERICA project is dedicated to the future Internet research (Campanella, 2008).

The target users of the FEDERICA infrastructure are the researchers who use networks not just as the "tool" but primarily as the "subject' of their work, and thus it is expected that those utilising the FEDERICA infrastructure will include EC-funded projects, telecommunication research centres and manufacturers of network equipment. In addition to supporting research conducted by the aforementioned entities through provisioning of virtual network slices, FEDERICA also aims to provide solutions for monitoring, management and control of parallel virtual networks.

EVALSO: Providing Access to Remote Astronomical Observatories

The Enabling Virtual Access to Latin-American Southern Observatories (EVALSO) project aims to create a physical infrastructure, and the tools for its exploitation, to efficiently connect the European Sothern Observatory (ESO) Paranal and the Cerro Armazones Observatories in Latin-America to Europe. EVALSO will use the international networking infrastructures created with the EC support - RedCLARA (RedCLARA, 2009) and GÉANT - to provide European research a competitive edge by enabling faster access to the collected data and use of facilities in an ever more efficient way (EVALSO, 2009).

The main objectives of this project are:

- *Link upgrade* through the creation of the physical networking infrastructure (where nonexistent) or procurement of services in order to upgrade the connectivity to the observing facilities.
- *Fast data access* by enabling a drastic improvement in the time needed for making the data available from the moment of the physical observation.
- *Virtual presence* by planning and implementing the tools that could be used to make possible the virtual presence of scientists, engineers, and experts at remote facilities and the possibility to perform remote observations. The virtual presence system is described next.

EVALSO aims to build a set of tools to implement a "virtual presence system" (VP system) for the scientists and the experts. The VP system will re-create the observatory environment at a remote site in order to allow the needed expert to interact with the local operators. Such a system will optimize the results from actual observations, allowing the observer to interact with the local staff. A survey of the already available GPL software has been performed in order to choose the tools and the software packages that satisfy the requirements of the VP system architecture (Cirami et al, 2008).

Black Sea Interconnection: Expanding the GÉANT network

Black Sea Interconnection (BSI, 2009) project is funded by the EU's 7th Framework Programme and spans from March 2008 to February 2010. The BSI project intends to bridge the digital divide that exists between the South Caucasus countries and Europe by establishing a regional Research and Education Network in the South Caucasus (Armenia, Azerbaijan and Georgia) and connecting it to GÉANT network. This would allow the Caucasian research communities to participate effectively in joint research and educational activities with the rest of Europe, and this would be an important leap towards the integration of the scientific potential of Armenia, Azerbaijan and Georgia with the European Research Area (incrEAST, 2009).

Although NRENs already exist in the South Caucasus countries (namely, *GRENA* in Georgia, *AzRENA* in Azerbaijan, *NAS RA* in Armenia), there is need for consolidation of these existing networks and then connecting them to the pan-European Education and Research Network - GÉANT. The primary objective of this project, therefore, is to design interconnection strategies, implementing those strategies through networking solutions, and to maintain the established network. Like the *100Mbps GÉANT-ERNET London-Mumbai link* that makes it possible for scientists in India and Europe to collaborate in disciplines like high-energy physics, the BSI project will enable scientists in South Caucasus to join force with their European counterparts to solve challenging problems. Furthermore, dissemination activities will encourage scientists in both the geographical regions to exploit BSI for collaborative research.

On the local level, the BSI project will develop a high-speed backbone network among the NRENs of the South Caucasus countries and this will supply the infrastructure needed by the academic and scientific communities in this geographical region. The interconnection will provide an impetus to the Governments in the South Caucasus to implement projects on e-Learning, e-Health, e-Democracy, e-Society. Finally, the BSI project will advance collaboration between scientific communities based in South Caucasus and working in disciplines like earthquake sciences, astronomy, and high-energy physics. It will provide them access to resources like online libraries and experiments results (BSI, 2009).

CONCLUSION

This book chapter has provided an overview of the organisation of the European regional NRENs and the over arching pan-European GÉANT network. It has highlighted the importance of the ERENs by showcasing six e-Science and e-Infrastructure projects that are presently being funded by the Commission. This is but a very small subset of such projects that are being undertaken both at the European and the national level, but almost all such e-Research projects utilise the ERENs. Moreover, the European further and higher educational institutes rely on the very same networks for their day to day operations. It is therefore imperative that the European governments continue to build capacity into the existing networks and to extend its reach. Furthermore, it is essential that strategic collaborations with non-European countries are established so that the pan-European GÉANT network can be extended and does not remain confined only to Europe. This would enable researchers not only from Europe but from around the world to collaborate together to undertake scientific enquiry in issues affecting all of humanity, for example, climate change, response to natural disasters, finding cure for diseases, etc.

For more information on EU-funded e-Science and e-Infrastructure projects, the reader is referred to the BELIEF-II website <http://www.belief-project.org/>. The acronym BELIEF stands for *"Bringing Europe's eLectronic Infrastructures to Expanding Frontiers"* and it is a global access point to European e-Infrastructures. BELIEF-II is a co-ordination and support action project and is funded by the EU's 7th Framework Programme under the Research Infrastructures part of the FP7 Capacities Programme. BELIEF-II works with approximately 50 EU-funded e-Infrastructure projects on supercomputing, grid computing, scientific repositories, high-speed networking, remote instrumentation, e-Science applications, etc., to ensure effective communication of results and best practices among the various projects and their users.

Finally, we conclude with a remark on the importance of computer networks in today's age. In some cultures light symbolises knowledge and darkness denotes ignorance. In auspicious occasions therefore, a traditional lamp is lit. The lighting of the lamp symbolises dispelling of ignorance through knowledge. In the present century this largely symbolic gesture of "lighting the lamp" is concretely realised through lighting the fibre!

REFERENCES

BSI. (2009). Black Sea Interconnection (BSI) project homepage. Retrieved December 2009 from http://www.blacksea-net.eu/

Campanella, M. (2008). The FEDERICA project - A federated infrastructure for Future Internet research. EURESCOM mess@ge, issue 2/2008. Retrieved December 2009 from http://www.eurescom.eu/ message/messageJul2008/The_FEDERICA _project_A_federated_infrastructure-%20for_Future_Internet_research.asp

CERN LHC. (2009). The Large Hadron Collider: Our understanding of the Universe is about to change. Retrieved December 2009 from http://public.web.cern.ch/ public/en/LHC/LHC-en.html

CERN WLCG. (2009). Worldwide LHC Computing Grid (WLCG) project homepage. Retrieved December 2009 from http://lcg.web.cern.ch/LCG/public/

Choudrie, J., & Dwivedi, Y. K. (2004). Towards a Conceptual Model of Broadband Diffusion. *Journal of Computing and Information Technology, 12*(4), 323–338. doi:10.2498/cit.2004.04.06

Choudrie, J., & Dwivedi, Y. K. (2005). The Demographics of Broadband Residential Consumers of a British Local Community: The London Borough of Hillingdon. *Journal of Computer Information Systems, 45*(4), 93–101.

Choudrie, J., & Dwivedi, Y. K. (2006a). A Comparative Study to Examine the Socio-Economic Characteristics of Broadband Adopters and Non Adopters. *Electronic Government -. International Journal (Toronto, Ont.), 3*(3), 272–288.

Choudrie, J., & Dwivedi, Y. K. (2006b). Broadband Diffusion: Usage and Impact on the Consumer-User Behaviour. *Journal of Intelligent Systems, 15*(1-4), 389–408.

Choudrie, J., & Dwivedi, Y. K. (2006c). Investigating Factors Influencing Adoption of Broadband in the Household. *Journal of Computer Information Systems, 46*(4), 25–34.

Choudrie, J., & Dwivedi, Y. K. (2007). Broadband Impact on Household Consumers: Online Habits and Time Allocation Patterns on Daily Life Activities. *International Journal of Mobile Communications, 5*(2), 225–241. doi:10.1504/IJMC.2007.011817

Cirami, R., Coretti, I., Marcantonio, P. D., Pucillo, M., & Santin, P. (2008). EVALSO: Enabling Virtual Access to Latin-America Southern Observatories (Poster presentation). SPIE conference on Astronomical Instrumentation, Marseilles, 23-28 June 2008. Retrieved December 2009 from http://www.evalso.eu/evalso/wp- content/uploads/2008/08/jra2_spie08_poster.pdf

DEISA. (2009). Distributed European Infrastructure for Supercomputing Applications (DEISA) project homepage. Retrieved December 2009 from http://www.deisa.eu/

Dwivedi, Y. K., Choudrie, J., & Brinkman, W. P. (2006a). Consumer Usage of Broadband in British Households. *International Journal of Services and Standards, 2*(4), 400–416. doi:10.1504/IJSS.2006.010472

Dwivedi, Y. K., Choudrie, J., & Brinkman, W. P. (2006b). Development of a Survey Instrument to Examine Consumer Adoption of Broadband. *Industrial Management & Data Systems*, *106*(5), 700–718. doi:10.1108/02635570610666458

Dwivedi, Y. K., & Irani, Z. (2009). Understanding the Adopters and Non-adopters of Broadband. *Communications of the ACM*, *52*(1), 122–125. doi:10.1145/1435417.1435445

Dwivedi, Y. K., Khan, N., & Papazafeiropoulou, A. (2007a). Consumer adoption and usage of broadband in Bangladesh. *Electronic Government: An International Journal*, *4*(3), 299–313. doi:10.1504/EG.2007.014164

Dwivedi, Y. K., Khoumbati, K., Williams, M. D., & Lal, B. (2007b). Factors affecting consumers' behavioural intention to adopt broadband in Pakistan. *Transforming Government People. Process and Policy*, *1*(3), 285–297.

Dwivedi, Y. K., & Lal, B. (2007). Examining the Socio-economic Determinants of Broadband Adoption in the United Kingdom. *Industrial Management & Data Systems*, *107*(5), 654–671. doi:10.1108/02635570710750417

Dwivedi, Y. K., Lal, B., & Williams, M. D. (2009). Managing Consumer Adoption of Broadband: Examining Drivers and Barriers. *Industrial Management & Data Systems*, *109*(3), 357–369. doi:10.1108/02635570910939380

Dwivedi, Y. K., Papazafeiropoulou, A., Gharavi, H., & Khoumbati, K. (2006c). Examining the Socio-economic Determinants of Adoption of an e-Government Initiative 'Government Gateway'. *Electronic Government -. International Journal (Toronto, Ont.)*, *3*(4), 404–419.

Dwivedi, Y. K., Papazafeiropoulou, A., Willem-Paul, B., & Lal, B. (2008a). Examining the Influence of Service Quality and Secondary Influence on the Behavioral Intention to Change Internet Service Provider. *Information Systems Frontiers*. Retrieved January 7, 2010 from http://www.springerlink.com /content/aw80021039441235/

Dwivedi, Y. K., & Weerakkody, V. (2007). Examining the Factors Affecting the Adoption of Broadband in the Kingdom of Saudi Arabia. *Electronic Government: An International Journal*, *4*(1), 43–58. doi:10.1504/EG.2007.012178

Dwivedi, Y. K., & Williams, M. D. (2008). The Influence of Demographic Variables on Citizens' Adoption of e-Government. *Electronic Government: An International Journal*, *5*(3), 261–274. doi:10.1504/EG.2008.018874

Dwivedi, Y. K., Williams, M. D., Weerakkody, V., Lal, B., & Bhatt, S. (2008b). Consumer Adoption of Broadband in India: A Pilot Study. *Journal of Cases on Information Technology*, *10*(3), 39–52.

EC Press Release. (2008). Europe's GÉANT, the world's highest speed computer network, goes global. Retrieved December 2009 from http://europa.eu/rapid/pressReleases Action.do?reference=IP/08/354&

EGEE. (2009). Enabling Grids for E-sciencE (EGEE) project homepage. Retrieved December 2009 from http://www.eu-egee.org

EVALSO. (2009). Enabling Virtual Access to Latin-America Southern Observatories (EVALSO) project homepage. Retrieved December 2009 from http://www.evalso.eu/evalso/

FEDERICA. (2009). Federated E-infrastructure Dedicated to European Researchers Innovating in Computing network Architectures (FEDERICA) project homepage. Retrieved December 2009 from http://www.fp7-federica.eu/

Foster, I., & Kesselman, C. (2004). Concepts and architecture. In Foster, I., & Kesselman, C. (Eds.), *The grid: blueprint for a new computing infrastructure* (2nd ed.). San Francisco, CA: Morgan Kaufmann.

Foster, I., Kesselman, C., & Tuecke, S. (2001). The anatomy of the grid: enabling scalable virtual organizations. *International Journal of High Performance Computing Applications, 15*(3), 200–222. doi:10.1177/109434200101500302

GÉANT. (2009a). GÉANT project homepage. Retrieved December 2009 from http://www.geant.net

GÉANT. (2009b). Global Connectivity - Connecting research communities across the world. Retrieved December 2009 from http://www.geant.net/ Network/GlobalConnectivity

GÉANT. (2009c). GÉANT Partners. Retrieved December 2009 from http://www.geant.net/About_GEANT/Partners

GÉANT Press Release. (2009). Third Generation GÉANT project launched to extend Pan-European research and collaboration capabilities. Retrieved December 2009 from http://www.geant.net/Media_Centre/PressReleases/Pages/3rd Generation-ofprojectlaunched.aspx

incrEAST. (2009). Project of the month (January): Black Sea Interconnection. Retrieved December 2009 from http://www.increast.eu/en/441.php

Irani, Z., Dwivedi, Y. K., & Williams, M. D. (2009). Understanding Consumer Adoption of Broadband: An Extension of Technology Acceptance Model. *The Journal of the Operational Research Society, 60*(10), 1322–1334. doi:10.1057/jors.2008.100

JANET. (2009). JANET Services. Retrieved December 2009 from http://www.ja.net/services/index.html

Khoumbati, K., Dwivedi, Y. K., Lal, B., & Chen, H. (2007). Broadband Adoption in Pakistan. *Electronic Government: An International Journal, 4*(4), 451–465. doi:10.1504/EG.2007.015038

Lamanna, M. (2004). The LHC computing grid project at CERN. *Nuclear Instruments & Methods in Physics Research. Section A, Accelerators, Spectrometers, Detectors and Associated Equipment, 534*(1-2), 1–6. doi:10.1016/j.nima.2004.07.049

PRACE. (2009). Partnership for Advanced Computing in Europe (PRACE) project homepage. Retrieved December 2009 from http://www.prace-project.eu

RedCLARA. (2009). RedCLARA project homepage. Retrieved December 2009 from http://www.redclara.net

Results, I. C. T. (2009). Supercomputing gets its own superhero. Retrieved December 2009 from http://cordis.europa.eu/ictresults/

Section 3
Impact and Social Consequences of Broadband Diffusion

Chapter 10
Toward Understanding U.S. Rural-Urban Differences in Broadband Internet Adoption and Use

Peter Stenberg
Economic Research Service, USDA[1], USA

Mitchell Morehart
Economic Research Service, USDA, USA

ABSTRACT

The Internet has become entrenched in the U.S. economy over the last 15 years;access and use of the Internet has increased for all regions of the United States, most types of households and work places, and all income groups. In this chapter we explore how access technologies may affect household on-line activity patterns and address some of the aspects that differentiate urban and rural household Internet use. Rural households are less likely than urban households in having broadband Internet access but this varies regionally across the country. Study suggests that broadband Internet access is no longer perceived a luxury, but as a necessity and that there is pent-up demand for broadband Internet access in rural areas.

INTRODUCTION

Analysis suggests that rural economies generally benefit from broadband Internet availability. In comparing American counties that had broadband access relatively early (by 2000) with similarly situated counties that had little or no broadband access as of 2000, employment growth was higher and nonfarm private earnings greater in counties with a longer history of broadband availability (Stenberg, 2008).

Government policies that encourage deployment of broadband Internet access services have broadened broadband availability in rural America. The 2008 Farm Act (Food, Conservation, and Energy Act of 2008) reauthorized U.S. Department of Agriculture (USDA)'s telemedicine, distance learning, and rural broadband access grant and loan programs. The American Recovery and Rein-

DOI: 10.4018/978-1-60960-011-2.ch010

vestment Act of 2009 provided $2.5 billion to the USDA for loans and grants to increase broadband provision in rural areas. As much as these funds address the needs of unserved and underserved communities, rural broadband availability will increase.

In this chapter we tackle some aspects that differentiate urban and rural household use of the Internet, such as the question of what on-line behavior becomes more likely as improved access technology becomes available? And, are the differences in rural and urban household use significant?

BACKGROUND

Broadband Internet access has become viewed by many as necessary to fully utilize the potential from the Internet (Greenstein and Prince, 2006; Leamer and Storper; Parker, 2001). As the Internet economy has evolved, more and more applications have required higher data transmission rates for their use, even in the case of simple shopping web-sites.

Studies by the National Telecommunications and Information Administration and Economic and Statistics Administration (NTIA) that first appeared in 1994, have described the changing user demographics of Internet use. More recent studies describe the current, and more static, situation or examine the adoption of newer, broadband, technologies. Household studies by Choudrie and Dwivedi (2005, 2006); Stanton (2004); Stenberg and Morehart (2008); and the U.S. General Accounting Office (2001) tested socio-economic factors distinguishing adopters and non-adopters of broadband Internet use. Choudrie and Dwivedi (2005) found age, gender, and social grade were important when distinguishing between adopters and non-adopters. Their 2006 study found that characteristics such as income and education were just as important. Stanton (2004) tested for a digital divide and found it the widest for computer ownership and the narrowest for broadband Internet connections. Most studies on Internet adoption have focused on the household. Studies by the NTIA in the 1990s, for example, described differences across many demographic and geographic groupings, not only for households, but also for Internet activity in the workplace.

Rural communities have not been left out of the evolving digital economy, though from the onset there has been an issue of equal access to the Internet (Parker and Hudson, 1992; Oden and Strover, 2002). Rural households have become almost as likely as urban households to use the Internet (Stenberg, 2006). Broadband Internet access in rural areas, however, has been less prevalent than in much more densely populated areas of the country. Broadband Internet access has become the crux of today's policy debate on equal access between urban and rural communities (Malecki; Stenberg et al, 2009).

ON-LINE CONSUMER ACTIVITIES

The household, or consumption, sector is a major, perhaps the greatest, demand-side driver in the development of the Internet economy. There are two major data sources directly addressing individual household on-line activity: the Bureau of the Census and PEW (PEW Internet & American Life Project). Unfortunately, the Bureau of Census has not collected data on household on-line activity since 2003 (outside the limited "are households on-line" questions more recently) so we must rely on the small sample PEW surveys for any understanding of individual household's on-line behavior instead of the aggregate e-retail, peer-to-peer, web-page access counts, and other useful information that are commonly reported.

PEW survey samples, because of their small sample size do not allow much analysis between urban and rural household in their on-line behavior. What can be ascertained from the data suggests that on-line behavior is alike if one controls for type of Internet access. In other words rural and urban households with broadband Internet access have similar on-line behavior patterns; rural and urban households with dial-up Internet access have similar on-line behavior patterns; but broadband and dial-up households have dissimilar on-line behavior.

Overall, three-quarters of all adults have Internet access with 69 percent having access at home; 55 percent of all adults have broadband access at home. Rural households are less likely to have broadband access with only 41 percent of adults in rural households having broadband access at home. Home broadband Internet access has increased for urban and rural households (Figure 1).

On-line pursuits diverge between dial-up users and broadband users, but not between rural and urban broadband users (Horrigan, 2008). Broadband Internet users were more likely than dial-up users to take part in any specific on-line activity, such as getting news on-line (table 1). The more data intensive the on-line activity, the greater the difference between dial-up and broadband user participation rates becomes, especially such activities as watching video and downloading podcasts.

Most adults get on-line everyday (Table 2). Daily use of the Internet is mostly for e-mail, but many other pursuits also take place. Getting information is one of the most common activities, including visiting state and local government web-sites. Many on-line activities are of a personal nature, such as hobbies and web-surfing, but activities are often also related more directly to work. Twenty-three percent of adults used the Internet to conduct research for their job in the 24 hours prior to being surveyed.

In the U.S. economy one major trend with the new technology has been towards disintermediation, or eliminating or reducing the economic involvement of the broker and other business middlemen. More consumer bank transactions, for example, are taking place through ATMs or on-line instead of on-site through the teller window. According to a PEW survey in 2005 a quarter of all American adults, or 44 percent of all adult Internet users, use the Internet for on-line banking. On any given day 14 percent of all American adults perform some on-line banking activity (Table 2), including paying bills or transferring money. Internet users with broadband access are especially heavy users of on-line banking services as over 60 percent of urban and nearly 50 percent of rural broadband users conducted some on-line banking activity by 2007.

Most, if not all, on-line activities are becoming more highly developed and complex, using more data intensive methods, especially graphics. Even

Figure 1. Trends in home broadband adoption by region (Source Horrigan, 2008)

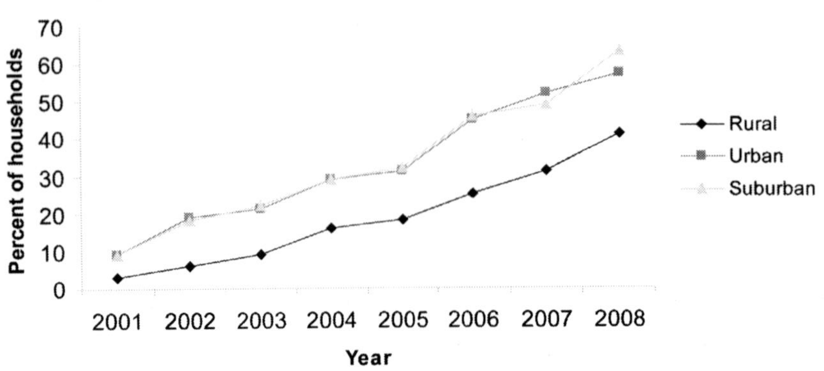

Table 1. Online activities, 2008

Selected activity that has ever been done by a user	All internet users (percent)	Dial-up at home (percent)	Broadband at home (percent)
Use an online search engine	89	80	94
Check weather reports and forecasts	80	75	84
Get news online	73	61	80
Visit a state or local government website	66	55	72
Look online for information about the 2008 election	55	37	62
Watch a video on a videosharing site like YouTube or GoogleVideo	52	29	60
Look online for information about a job	47	36	50
Send instant messages	40	38	44
Read someone else's blog	33	15	40
Use a social networking site like MySpace, Facebook, or LinkedIn.com	29	21	33
Make a donation to charity online	20	9	23
Downloaded a podcast	19	8	22
Download or share files using peer-to-peer networks such as BiTorrent or LiveWire	15	15	17
Create or work on your own blog	12	8	15

Source: Pew Internet & American Life Project Survey, April 2008.

e-mail has become more data intensive as people are more inclined to attach photos and videos to them. As data intensiveness has increased broadband has become more a necessity rather than a luxury in everyday life. Broadband Internet access, however, is not uniformly available across rural and urban America, nor is the broadband access transmission rate identical across the country.

PROCLIVITY OF RURAL-URBAN ON-LINE ACTIVITY

While data from the surveys conducted by PEW provide an understanding of what kind of adult on-line activities take place, the Bureau of the Census' Current Population Survey (CPS) data provide a better understanding of where, and a more complete understanding of who, uses the Internet. Table 3 shows the percent of all individuals irrespective of age that went on-line at home, school, work or elsewhere. A high percentage, 77 percent, of all persons went on-line during the year. The CPS data suggests a variation across the country in the occurrence of persons going on-line (Table 3). The regions are the four basic Census regions (Figure 2). Southerners were the least likely to go on-line of the four Census regions.

Nonmetropolitan areas (aka rural areas), in the aggregate, had a lower percentage of individuals going on-line. While the variation in overall on-line use was insignificant between regions outside of the south, the same did not hold for nonmetropolitan areas. Only in the Northeast was there not some drop-off in on-line activity going from metropolitan to nonmetropolitan areas.

One distinction often glossed over, and often ignored, though important if remediation is a goal of government Internet policy, is the distinction between individuals and households. Many policies have either an implicit individual or household impact, where the policy is designed to address directly one economic actor with only secondary effects going to the other economic actor.

Table 2. Percent of Internet users conducting selected activity within the last day

Selected activity	Percent	year of survey
Use the internet	69	2007
Send or read e-mail	56	2007
Use a search engine to find information	41	2006
Get on-line news	37	2007
Look for info on a hobby or interest	29	2007
Surf the Web for fun	28	2006
Do any type of research for their job	23	2007
Check the weather	22	2004
Research a product or service prior to buying it	20	2008
Research for school or training	16	2005
Get sports scores and info online	15	2006
Visit a local, state or federal government website	14	2006
Do any banking online	14	2005
Search for a map or driving directions	10	2006
Watch a video clip or listen to an audio clip	10	2004
Get financial info online, such as stock quotes or mortgage interest rates	9	2006
Play online games	9	2006
Look for information on Wikipedia	8	2007
Watch a video on a video-sharing site like YouTube or GoogleVideo	8	2006
Get travel info	8	2004
Pay bills online	7	2005
Purchase a product on-line	6	2008
Look online for info about a job	5	2005
Create content for the internet	4	2004
Make travel reservations on-line	4	2008
Create a web log or "blog"	2	2005
Take a class online just for personal enjoyment or enrichment	2	2005

Source: PEW, various surveys

Table 3. Persons going on-line at home or elsewhere, 2007

	Metropolitan (percent)	Nonmetropolitan (percent)	Total (percent)
Northeast	77.8	77.3	77.8
Midwest	79.7	74.0**	78.4
South	74.8	65.3**	73.0
West	78.3	74.2**	77.9
Total	**77.2**	**70.7****	**76.2**

Source: authors using CPS data.
Note: Difference between metro/nonmetro (** -- significant at 0.01).

Figure 2. Source authors using Bureau of the Census information

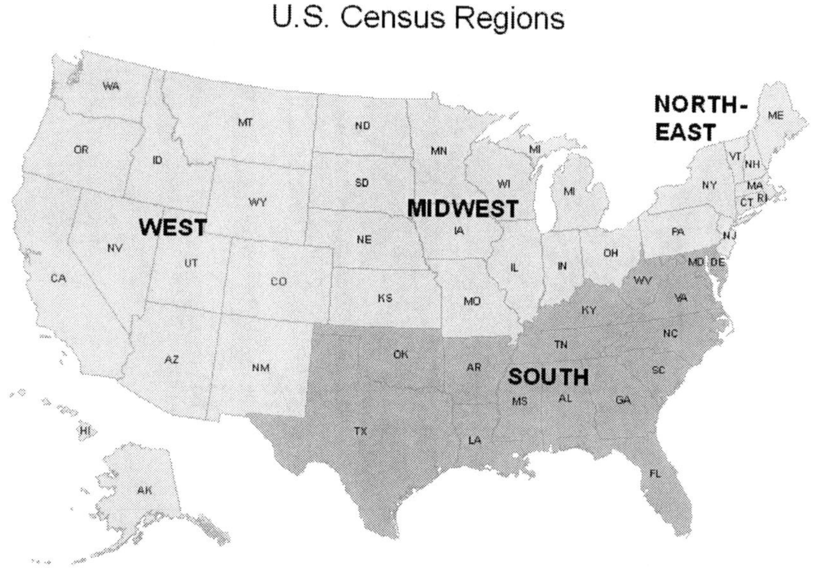

For example, distance education programs, such as on-line MBA programs, are often designed for groups of specific individuals. Subsidies of telecommunication rates, for another example, are directed toward the household.

When it comes to the Internet if one member of the household uses the Internet, the rest are likely to use it too. The consequences can be seen by comparing Table 3 and Table 4; the percentage of at least one person going on-line in a household is less than percentage of all individuals. The gap between urban and rural households is wider than for the gap between all urban and rural persons. The gap between rural and urban households in the Northeast, however, is still narrow.

Income differences have often been offered as a key explanation for the disparity in Internet use by individuals or households (Choudrie and Dwivedi, 2006; Dwivedi and Irani, 2009; Flamm and Chaudhuri, 2007; Stenberg and Morehart, 2007). The reason why can be seen in Figure 3. Clearly low income households are less apt to exploit the Internet than high income households. Income, of course, is not the end of the story as income is highly correlated to, or determined by, education, age, and other factors, but household

Table 4. Households with at least one person going on-line at home or elsewhere, 2007

	Metropolitan (percent)	Nonmetropolitan (percent)	Total (percent)
Northeast	71.0	69.5	70.9
Midwest	74.0	65.7**	72.1
South	70.7	58.3**	68.3
West	75.5	68.6**	74.9
Total	**72.6**	**63.3****	**71.1**

Source: authors using CPS data.

Note: Difference between metro/nonmetro (** -- significant at 0.01).

Figure 3. Percent of households accessing Internet anywhere, 2007 (Source authors using CPS data)

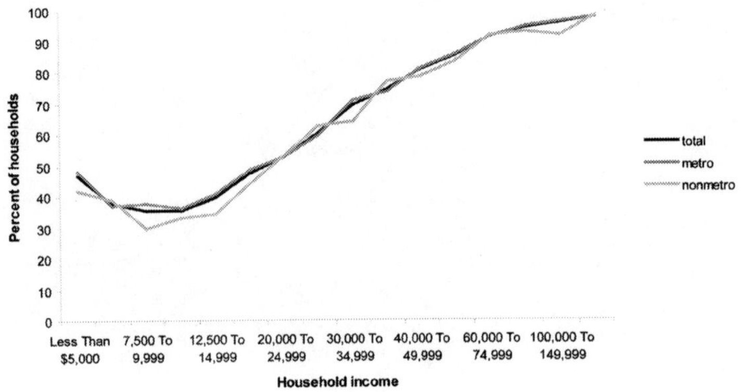

income by itself does raise one circular argument: does non-use of the Internet lead to higher household income or does less household income lead to less Internet use anywhere in the household members' lives.

People access the Internet from many different locations, including their homes, schools, libraries, and at friends homes. Over 80 percent of households with incomes greater than $40K had household members make use of the Internet in the home or elsewhere. While there was an overall gap between rural and urban households in accessing the Internet anywhere, there was no significant gap between rural and urban households of the same income (Figure 3).

While home Internet access is the most common location for individuals to go on-line, fewer access the Internet at home than all locations combined (67 percent versus 71 percent). This was true across all regions of the country with home Internet most common in the Northeast (Table 5). The drop-off of home access rates between urban and rural residents is larger than for residents accessing the Internet anywhere, suggesting the greater dependence rural residents may have for going on-line outside of the home.

The rural South had significantly less home Internet access than the rest of the country (Table 5). The South also had the greatest drop-off between urban and rural residents in home Internet access. The Northeast again had little difference between urban and rural residents. Sixty-two percent of all households had access to the Internet in the home (Table 6) as compared to 71

Table 5. Percent of persons going on-line at home, 2007

	Metropolitan (percent)	Nonmetropolitan (percent)	Total (percent)
Northeast	71.4	68.7*	71.1
Midwest	69.2	61.4**	67.5
South	65.6	51.9**	63.1
West	69.8	60.6**	69.0
Total	**68.5**	**58.1****	**66.9**

Source: authors using CPS data.
Note: Difference between metro/nonmetro (** -- significant at 0.01 * -- significant at 0.05).

Table 6. Percent of households with at least one person going on-line at home, 2007

	Metropolitan (percent)	Nonmetropolitan (percent)	Total (percent)
Northeast	64.1	61.0*	63.7
Midwest	63.1	53.7**	60.9
South	61.7	46.4**	58.7
West	67.0	56.9**	66.1
Total	**63.7**	**51.9****	**61.8**

Source: authors using CPS data.
Note: Difference between metro/nonmetro (** -- significant at 0.01 * -- significant at 0.05).

percent of households that had at least one member access the Internet in any of a number of locations. The West had the highest percent of households with in-home access partially reflecting the more urbanized nature of its population distribution vis-à-vis other regions of the country.

The drop-off of in-home Internet access rates between urban and rural households was great, though again the Northeast rural-urban dynamic showed little of this pattern. A majority of households in the rural South did not have access to the Internet in-home.

Income is a major factor in whether a household has in-home Internet access (Figure 4). Over 70 percent of all households with incomes greater than $40K had in-home Internet access. Differences between rural and urban households' in-home access rates are largely non-existent between households of the same income-level. The steeper slope between in-home access rates and going on-line anywhere rates suggest that price may be a factor in in-home Internet access adoption.

Taking into account only households that have at least one household member accessing the Internet changes the home Internet access dynamic; once someone is using the Internet anywhere the more likely there will be home Internet access. This higher rate is reflected in the income distribution shown in Figure 5. The rate differences between rural and urban residents are small.

Most persons who have in-home Internet access have broadband connections (Table 7). This rate varies little across urban regions across the country. The same can not said for rural residents.

Figure 4. Home internet access by income, 2007 (Source authors using CPS data)

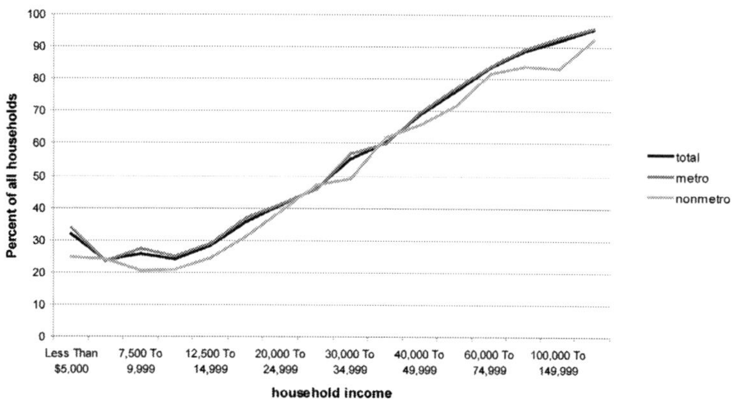

Figure 5. Internet users with household connection by income 2007 (Source authors using CPS data)

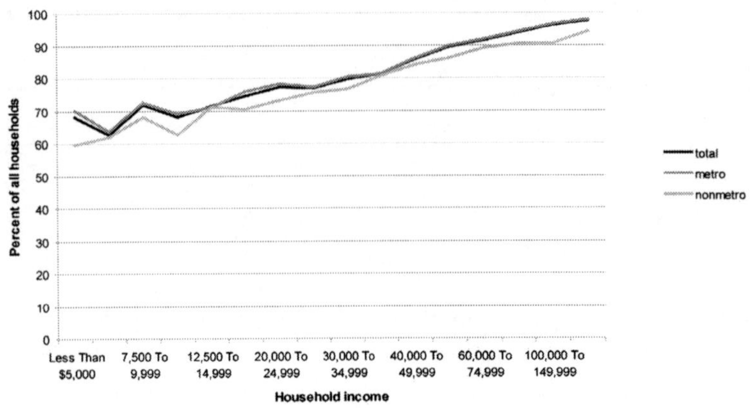

A marked drop-off in broadband access exists between urban and rural residents, even in the Northeast. Only 70 percent of rural residents with in-home Internet access have broadband access as compared to 85 percent of urban residents. The data suggest that broadband availability is an issue for rural areas across the country.

Broadband access rates do not diverge much when taking into consideration persons or households (Table 8). This lack of variation between household and individual broadband access rates suggests that once the household dynamic helps determine whether the household will have in-home access, the household dynamic no longer matters greatly in deciding in favor of, or against, taking the next step, having broadband access.

The broadband rural-urban dichotomy is more readily apparent when household income is taken into account (Figure 6). The relationship shown in Figure 6 suggests that income is not much of a factor in opting for broadband over dial-up for an in-house Internet connection. The gap between rural and urban household use of broadband connections thus would suggest that broadband availability is more of a challenge for rural than urban households unless there is some systemic differences between rural and urban households that otherwise could explain the gap. The systemic household dissimilarity, if it exists, would have to explain why rural households are as likely as urban households using the Internet but do not

Table 7. Percent of on-line persons with home broadband access, 2007

	Metropolitan (percent)	Nonmetropolitan (percent)	Total (percent)
Northeast	88.1	68.5**	86.1
Midwest	83.8	72.3**	81.5
South	83.8	66.7**	81.0
West	85.8	76.9**	85.1
Total	**85.2**	**70.2****	**83.1**

Source: authors using CPS data.
Note: Difference between metro/nonmetro (** -- significant at 0.01).

Figure 6. Households with broadband access by income, 2007 (Source: authors using CPS data)

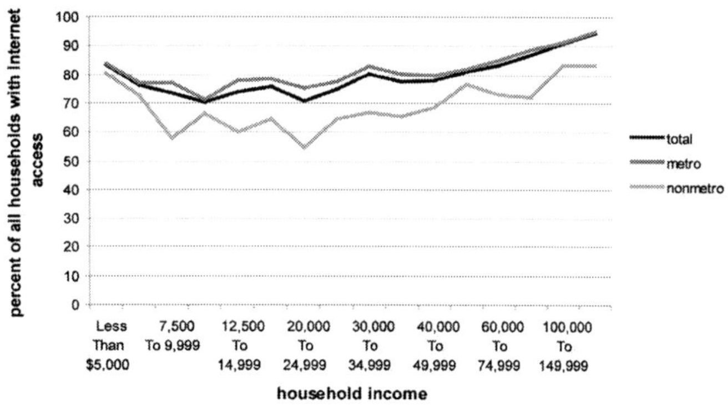

Table 8. Percent of on-line households with broadband access, 2007

	Metropolitan (percent)	Nonmetropolitan (percent)	Total (percent)
Northeast	87.3	68.8	85.4
Midwest	82.9	70.6	80.4
South	83.0	67.3	80.5
West	85.3	75.2	84.4
Total	**84.4**	**69.7**	**82.3**

Source: authors using CPS data.
Note: Difference between metro/nonmetro (** -- significant at 0.01).

opt for broadband if they already use the Internet at home.

Distance education has been considered a critical element in the delivery of education (Poley). At the same time the presence of children in the household has been a determinant, or at least deemed a likely contributing factor, in a household having in-home Internet access (Stenberg and Morehart, 2006; Choudrie and Dwivedi, 2006). As a result in-home Internet access may improve household well-being through educational programs; education programs may thus drive household demand for in-home Internet access. Households with children do have higher rates of in-home Internet access (Table 9). Households with teenage children are more likely to

Table 9. All types of Internet access at home by household composition, 2007

	Metro	Nonmetro	Total
Not a parent	65.0	53.4**	63.3
No children under 18 years of age	69.7	59.4**	67.7
Only children less than 6	70.8	55.1**	68.7
At least one child 6-13 and none older than 13	72.8	65.4**	71.7
At least one child older than 13	81.4	76.5**	80.6
Total	**68.5**	**58.1**	**66.8**

Source: authors using Bureau of the Census CPS data.
Note: Difference between metro/nonmetro (*--significant at 0.05, ** -- significant at 0.01).

Table 10. Broadband in homes with Internet access by household composition, 2007

	Metro	Nonmetro	Total
Not a parent	85.7	71.4**	83.9
No children under 18 years of age	82.3	66.9**	79.7
Only children less than 6	90.5	78.4**	89.1
At least one child 6-13 and none older than 13	87.3	71.8**	85.2
At least one child older than 13	85.4	71.8**	83.3
Total	**85.2**	**70.3****	**83.1**

Source: authors using Bureau of the Census CPS data.
Note: Difference between metro/nonmetro (*--significant at 0.05, ** -- significant at 0.01).

have in-home Internet access than any other child/non-child-household type (Table 9).

Rural households, however, have uniformly less access than urban households across all child/non-child-household types. In as much as distance education becomes beneficial to future household economic well-being, continuation of this rural-urban dichotomy would put rural households at a disadvantage vis-à-vis urban households.

Once a household has in-home Internet access the step toward broadband is not readily affected by child/non-child-household type (Table 10). The rural-urban dichotomy, however, remains and further differentiates rural and urban households. In as much as broadband becomes necessary or beneficial in the distance education system, rural households would be at a disadvantage vis-à-vis urban households.

In the current (slow growth) national economic environment a number of Internet activities may become more critical for countless households including: (1) job searches and (2) home businesses. The 2007 CPS data gives us some information on both. Unemployed adults, while showing some drop-off in Internet use as compared to employed persons, still had high on-line Internet access rates when considering getting on-line anywhere (Table 11). The big drop-off in on-line activity was for people who were not in the labor force due to retirement or disability. Unemployed individuals looking for work were more likely to have had access than other people not employed. Rural people in the labor force had a lower access rate than urban people.

The picture starts to change for home Internet access of all on-line individuals (Table 12). Home

Table 11. Online activity using any access technology by labor force status, 2007

	Metro	Nonmetro Percent	Total
Employed-At work	83.7	78.6**	82.9
Employed-Absent	85.9	79.4**	84.8
Unemployed-On layoff	72.8	62.3**	70.6
Unemployed-Looking	77.7	72.8*	76.9
Retired-Not in labor force	52.6	43.7**	50.9
Disabled-Not in labor force	47.1	43.2*	46.1
Total	77.2	70.6**	76.1

Source: authors using Bureau of the Census CPS data.
Note: Difference between metro/nonmetro (*--significant at 0.05, ** -- significant at 0.01).

Table 12. Home Internet access using any technology by labor force status, 2007

	Metro	Nonmetro Percent	Total
Employed-At work	74.8	65.0**	73.3
Employed-Absent	75.0	65.2**	73.5
Unemployed-On layoff	60.3	49.6**	58.1
Unemployed-Looking	63.8	51.4**	62.0
Retired-Not in labor force	48.8	41.3*	47.4
Disabled-Not in labor force	39.6	34.9*	38.4
Total	68.5	58.1**	66.9

Source: authors using Bureau of the Census CPS data.
Note: Difference between metro/nonmetro (*--significant at 0.05, ** -- significant at 0.01).

Internet access affordability likely becomes an issue as the rate for those who are unemployed or no longer in the labor force falls below the national average (Table 12). This appears to be true for both rural and urban residents as well. As these individuals likely no longer have access to the Internet at work, going on-line at such locations as libraries and schools would be expected.

A broadband connection is again the choice for most homes with in-home Internet access across all labor force categories (Table 13). No matter whether fully employed, retired or of some other employment classification, whether the household is in an urban or rural location, if a household has in-home Internet access, the household will most likely have a broadband connection. The biggest difference between rural and urban residents falls within the unemployed or not-in-labor force groups.

Many households have home businesses covering a wide range of professions such as farmers, doctors, and artisans (Table 14). These home business households may become more commonplace in the current economic environment as households often during economic hard times try to compensate for loss of jobs or reduced work hours by starting home businesses. Home businesses are even more commonplace in rural areas than urban areas (Table 14). In-home Internet access is much more common in households with home businesses (81 percent) than all households in the aggregate (62 percent). This is also true for both urban and rural home businesses (83 percent versus 64 and 70 versus 52, respectively).

Table 13. Broadband in homes with Internet access by labor force status, 2007

	Metro	Nonmetro Percent	Total
Employed-At work	86.6	72.1**	84.6
Employed-Absent	87.3	75.5**	85.6
Unemployed-On layoff	75.7	54.2**	72.0
Unemployed-Looking	83.7	72.0**	82.1
Retired-Not in labor force	75.0	59.0**	72.3
Disabled-Not in labor force	75.0	63.2**	72.4
Total	85.2	70.2**	83.1

Source: authors using Bureau of the Census CPS data.
Note: Difference between metro/nonmetro (*--significant at 0.05, ** -- significant at 0.01).

Table 14. Internet users by home business status, 2007

	Metro	Nonmetro Percent	Total
Households with home businesses	11.7	14.8	12.2
Home businesses with all-kinds of home Internet access	83.3	70.1**	80.7
Proportion of home Internet access with broadband	87.6	71.3**	84.6

Source: authors using Bureau of the Census CPS data.

Note: Difference between metro/nonmetro (*--significant at 0.05, ** -- significant at 0.01).

The broadband connection divergence between all households and households with home businesses, like in the prior comparisons, is small; once a household has in-home Internet connections they are likely to have a broadband connection (84 percent for on-line households with home businesses versus 82 percent for all on-line households in the case of home businesses).

In summary, a broadband connection is the choice for the greater majority of on-line households. The evidence suggests that more rural households would have broadband connections if these connections were as readily available as in urban areas. Some of the CPS data does suggest there is lost economic opportunity for rural households because of the challenge of their getting broadband access.

The data, however, also suggests that some of the shortfall in rural Internet activity may result from other factors that precede the decision to get a broadband connection. These factors include the lower average income for rural households, higher average age of rural residents, and lesser educational attainment of rural residents as compared to their urban counterparts.

FUTURE RESEARCH DIRECTIONS

More research is needed in many areas to better comprehend broadband and rural economies. Better information on broadband availability, use, cost, and technical characteristics is needed to better grasp broadband Internet's potential effect on rural economies. Broadband availability data are only now being collected below the ZIP Code area; the more geographically granular the data the better our understanding will become on unserved and underserved areas. Price data are largely unavailable, hindering economic analysis of supply and demand of the regional broadband market.

Detailed broadband Internet use data also has not been collected since 2003. Our understanding is limited on what uses people have for broadband. Perhaps the greatest need in understanding Internet processes falls in the area of education. We know certain Internet-based education activities are taking place, but the extent to which Internet practices are taking place in rural education systems (in either the school or at-home educational activities) is still unclear.

CONCLUSION

A number of facts are clear about the Internet. Internet use has grown rapidly and so has its integration into the rest of the economy. Connecting to the Internet via DSL lines, cable, satellite, and wireless networks increases bandwidth dramatically, making the Web much more useful. Increased Internet access speed has ignited an explosion of electronic commerce, video on demand, telecommuting, collaborative scientific projects, videoconferencing, and virtual environments. High speed platforms known as broadband Internet are viewed by many as necessary to fully utilize the potential from the Internet.

Rural communities have not been left out of the ever changing digital economy though there has been an issue of equal access across the rural-urban milieu. Rural households are almost as likely as urban households to use the Internet. Broadband Internet access in rural areas, however, has been less prevalent than in much more densely populated areas of the country.

Broadband users do use the Internet more intensively than dial-up users. Broadband users now outnumber dial-up users and the continued evolution of data dense technologies will bring to an end the use of dial-up technologies. The high adoption rate of broadband technologies by urban Internet users indicates people value what the Internet has to offer. Rural Internet users do have less in-home broadband access and the circumstantial evidence suggests that the difference may lie in the higher cost or less availability of broadband Internet access in rural areas.

Government policies to encourage deployment of broadband services in rural areas have increased availability, and in some cases encouraged, competitive pricing. Unfortunately, there is little, if any, data available that links Internet access choice with the cost of service. As a result, it is difficult to distinguish between financial and other motives when examining broadband adoption. In migration studies people leaving one location to go to another has often been cited as "voting with one's feet." Here users have been "voting with one's hands" by using the Internet more intensively, on more activities, using newer (and presumably better) technologies, and bringing the Internet into the home, into the schools, and into their work.

More activities are moving and adapting to the Internet. Some of these activities have great potential value for the rural economy. Education programs -- primary secondary, higher education, and continuing education have increasingly been shifting to, or at least the offerings have been becoming richer on, the Internet. Telework has become a more reasonable option for workers and businesses and has shown some growth in activity. Some medical services lend themselves readily to the Internet environment and show potential in either reducing costs for rural communities or offering in-situ services not otherwise readily available. Rural businesses, such as retail businesses, have been adopting more e-commerce and Internet practices, offering some improvements in business economic vitality. Individuals are using the Internet to get involved with their communities.

Access to the Internet through broadband technologies may no longer be perceived as a luxury. With the increasing data intensiveness of household on-line activities, broadband technologies may have become a necessity to fully utilize and gain benefit from the Internet further buttressing the argument for government policy intervention.

REFERENCES

Choudrie, J., & Dwivedi, Y. (2005). The demographics of broadband residential consumers in a British local community: The London Borough of Hillingdon. *Journal of Computer Information Systems*, *45*(4), 93–101.

Choudrie, J., & Dwivedi, Y. (2006). Investigating factors influencing adoption of broadband in the household. *Journal of Computer Information Systems*, *46*(4), 25–34.

Dwivedi, Y., & Irani, Z. (2009). Understanding the adopters and non-adopters of broadband. *Communications of the ACM*, *52*(1), 122–125. doi:10.1145/1435417.1435445

Flamm, K., & Chaudhuri, A. (2007). An analysis of the determinants of broadband access. *Telecommunications Policy*, *31*, 312–326. doi:10.1016/j.telpol.2007.05.006

Greenstein, S., & Prince, J. (2006). *The diffusion of the Internet and the geography of the digital divide in the United States*. Working Paper 123182, National Bureau of Economic Research, USA.

Horrigan, J. (2008). *Broadband adoption, rural America, and economic decisions*. Paper presented at the Economic Research Service's Broadband in the Rural Economy workshop, Washington, DC.

Leamer, E. E., & Storper, M. (2001). *The economic geography of the Internet age*. Working Paper 8450, National Bureau of Economic Research, USA.

Malecki, E. (2008). *The rural digital economy: competition and the geography of telecommunications*. Paper presented at the Economic Research Service's Broadband in the Rural Economy workshop, Washington, DC.

NTIA. (2000). *Falling through the Net: Toward digital inclusion*. Washington, DC: U.S. Department of Commerce.

NTIA. (2002). *A nation online: How Americans are expanding their use of the Internet*. Washington, DC: U.S. Department of Commerce.

NTIA (National Telecommunications and Information Administration & Economics and Statistics Administration). (2004). *A nation online: entering the broadband age*. Washington, DC: U.S. Department of Commerce.

Oden, M., & Strover, S. (2002). *Links to the future: The role of information and telecommunications technology in Appalachian economic development*. Washington, DC: Appalachian Regional Commission.

Parker, E. (2000). Closing the digital divide in rural America. *Telecommunications Policy, 24*(4), 281–290. doi:10.1016/S0308-5961(00)00018-5

Parker, E., & Hudson, H. (1992). *Electronic byways: state policies for rural development through telecommunications*. Boulder: West View Press.

PEW Internet & American Life Project. (n.d.).Retrieved from http://www.pewinternet.org/index.asp.

Poley, J. (2008). *Rural distance education*. Paper presented at the Economic Research Service's Broadband in the Rural Economy workshop, Washington, DC.

Stanton, L. (2004). Factors influencing the adoption of residential broadband connections to the Internet. In *Proceedings of the 37th Hawaii International Conference on System Sciences*.

Stenberg, P. L. (2006). Investment and household adoption of communication and information services across the United States. In Johansson, B., Karlsson, C., & Stough, R. (Eds.), *The emerging digital economy: Entrepreneurship, clusters and policy* (pp. 263–276). New York: Springer.

Stenberg, P. L. (2008). *Rural broadband internet use and the rural economy: An quasi-experimental design analysis of the broadband effect on rural economies*. Paper presented at the Economic Research Service's Broadband in the Rural Economy workshop, Washington, DC.

Stenberg, P. L., & Morehart, M. (2006). *Determinants of internet use for rural and farm economic sectors*. Paper presented at the Telecommunication Policy Research Conference, Arlington, VA.

Stenberg, P. L., & Morehart, M. (2007). The internet in U.S. farm and rural businesses. In Johansson, I. (Ed.), *Entrepreneurship and Development-Local Processes and Global Patterns*. Sweden: University West.

U.S. General Accounting Office. (2001). *Characteristics and choices of internet users*. Washington, DC (GAO-01-345).

ENDNOTE

1. The views expressed are those of the authors and do not necessarily reflect the views of the Economic Research Service or the U.S. Department of Agriculture.

Chapter 11
Information Communication Technology (ICT) for Rural Women's Life in Bangladeshi Villages

Md. Mahfuz Ashraf
University of Dhaka, Bangladesh

Helena Grunfeld
Victoria University, Australia

Syeda Rownak Afza
BRAC University, Bangladesh

Bushra Tahseen Malik
Brainstorm Bangladesh, Bangladesh

ABSTRACT

Information Communication Technology (ICT) has the potential to contribute to development, especially in rural areas of developing countries. But the mechanisms through which ICT can be combined with development agendas and an understanding of the actual development process and impacts of ICT are less well understood or properly defined in the academic literature. This research is an attempt to contribute to understanding this process by analysing the impact of two ICT initiatives in Bangladesh, aimed at improving the lives of rural women. An interpretive approach in the qualitative research tradition was adopted to identify emergent themes in this study. Our findings indicate that these ICT projects have significantly improved the socio-economic opportunities of many women. This paper will be useful for those academics, practitioners and policymakers who wish to enhance their understanding of ICT projects in rural areas of developing countries.

DOI: 10.4018/978-1-60960-011-2.ch011

INTRODUCTION

ICT and its role in gender equality and empowerment in rural areas of developing countries have been a concern of international donor agencies, development partners, non-government organizations, and academicians. Despite a growing understanding of the potential role of ICTs in promoting social inclusion and gender equity, there is nevertheless a recognized knowledge gap relating to the impact of ICT initiatives, particularly with respect to impact on women. This research is an attempt to contribute to filling this gap by studying two different ICT projects in rural Bangladesh; the Grameen Phone Community Information Center (GPCIC), which provides mobile broadband connectivity in rural community centers and the Village Phone (VP), which facilitates small scale women enterprises through enabling women to sell mobile phone services in rural communities. The VP initiative is one way of contributing to the achievement of the Milennium Development Goals (MDGs), particularly Target 11, which calls for an increase in the 'ratio of women to men in wage employment in the non-agricultural sector'. This initiative has opened new doors for women to become entrepreneurs, by providing them with opportunities to change their socio-economic condition, thereby overcoming oppression, social control, and other humiliating features associated with being a woman in a male-dominated society.

The aim of this research is to illustrate the impact of these two ICT projects on the lives of rural women at an individual level. In its simplest form, an impact assessment is a systematic process of identifying and analyzing the impact (positive and negative) of an intervention. A key function of impact assessments is to enable researchers, users, and policy makers to compare impacts with objectives, taking into account the surrounding social and economic environments and human contexts, and to identify any unintended impacts.

LITERATURE REVIEW

Noting that there had been insufficient attention to understanding the impacts of ICT on women and that few projects initiated by the International Telecommunication Union (ITU) had taken into account the needs and perspectives of women, the 1998 World Telecommunication Development Conference (WTDC) resolved to establish a task force on gender issues (ITU, 1998). To emphasize the importance of gender issues, this task force was converted into a Standing Working Group on Gender Issues at the WTDC in 2002 (ITU, 2002). This brought renewed emphasis on training and education for women in ICT, a theme that recurred at the World Summit on the Information Society (WSIS), which emphasized the importance of training and education of women to motivate and promote "participation and active involvement of girls and women in the decision-making process of building the Information Society" (WSIS, 2005).

Sometimes ICT is an integral part of an organization's activities, as was the case with the Self Employed Women's Association (SEWA), India, which has used many facets of ICT, including training poor women in the use of video cameras and other audiovisual equipment, and educating them via satellite communication about the importance of insurance (Roger & Rajesh, 2006). In an endeavour to attract women to participate in its village knowledge centre (VKC) activities, the M. S. Swaminathan Research Foundation (MSSRF) requires that at least one woman is engaged in the management of each VKC, resulting in many of the operators and volunteers in the VKCs being women. This involvement has contributed to increasing women's influence and status, not only from working with the centers, but also from small enterprise activities associated with this activity, such as production of paper and biopesticides (Colle & Roman, 2001; Ofir & Kriel, 2004). Another example from India is the 'Putting ICTs in the Hands of Women of Kanpur and "Chikan" Embroidery Workers of

Lucknow', which aimed at exploring whether ICTs can expand the capacity of women engaged in the informal sector to improve their livelihoods (Batchelor & Norrish, 2004). In a survey of 17 infoDev funded ICT4D projects, Batchelor et al. (2003) included a number of projects designed to empower women in different ways, including the Brazilian 'Strengthening Women's Leadership in Community Development through Internet Radio' and the African 'Abantu for Development' project. Abantu, an NGO established by African women in Europe in 1991, which entered into the ICT4D arena in 1999, has trained women's groups in Kenya in basic software and Internet skills and brought together different sectors of the business community to produce gender-sensitive ICT policy recommendations. Other ICT4D initiatives in Africa include WOUGNET, a non-profit non-governmental organization established in 2000 by several women's organizations in Uganda to promote and support the use of ICTs and the African Gender Research in Africa into ICTs for Empowerment (GRACE) project. The latter project has now been expanded to North Africa and Asia (WSIS, 2008). Also in Uganda, Prahalad & Hammond (2004) reported that centers run by the Women's Information Resource Electronic Service provided female entrepreneurs with information on a range of issues, from market prices to details on credit and trade support services. Preferring money over information, many women did not initially value information as a resource, but were then taught how to use information to improve their businesses (Bridges,2003). Africa is also home to the Digital Diaspora Initiative, supported by the United Nations Development Fund for Women and African ICT entrepreneurs in the Diaspora. It aims at helping women in the country of origin of its members, use new ICTs and create enabling environments for African women to influence all aspects of ICT use and policy.

There is a debate about the value of ICTs for development in general and for women's empowerment in particular. Critics of ICTs point to the more pressing needs of women in developing countries for safe water, adequate food, improved health, and better education, rather than the provision of access to ICTs. The counter argument is that health, water, food, education and ICT are not in opposition to each other, but rather that ICTs can be a tool to provide information on health and food, as well as a carrier of education for women and girls.

Researching ICT in terms of impact on equality in the gender dimension, especially as it relates to. community empowerment has become a challenge for academicians, practitioners and international donor agencies, particularly with increasing focus on gender equality and empowerment by donor countries (Tarasiewicz, 2003; Walsham & Sahay, 2006; Manning, 2008). While the potential of ICT for socio-economic development and empowerment for different disadvantaged and marginalised groups, such as women, the poor, disabled, and indigenous groups is recognized and has been addressed in many programmes, only a few research publications available in the public domain focus on how members of these groups have used ICT for their empowerment. For example, the United Nations Development Program (UNDP) has provided assessment guidelines for ICT and gender initiatives in 25 UNDP country offices (UNDP, 2004), but these are based on a questionnaire at the organizational, rather than the individual level. Point 114 of the WSIS Tunis Agenda (WSIS, 2005) noted the launch, in June 2004, of the *Partnership on Measuring ICT for Development*, and its efforts to develop a common set of ICT indicators, including point 114 (d) specific gender-disaggregated indicators to measure the digital divide in its various dimensions. The gender evaluation methodology (GEM[1]), developed by the Women's Networking Support Programme of the Association for Progressive Communications as a tool to facilitate the process of learning about using ICTs for gender equality, is another initiative focusing on assessing the impact of ICT programs and policies on women's empowerment from a program, rather than an individual perspective. Also designed for the organizational and national levels are indicators suggested by the ITU's (2006) for measuring inputs, outputs and outcomes associated with achievement of the MDGs. But its indicators for MDG 3, "promote gender equality and empower women" do not refer directly to empowerment. The closest is the outcome at the national level, which is defined as "positive changes in women's status and employment based on ICTs" (p.29). Orbicom, an international network of UNESCO chairs, has attempted to measure the impact of ICT on gender equality from a social development perspective (Orbicom, 2005). A subsequent Orbicom report by Huyer & Hafkin (2005) aimed to provide a framework for data analysis relating to measuring contributions by women to the knowledge society.

Notwithstanding the importance of the many initiatives designed to measure the role of women in ICT and the various indices developed by different agencies to monitor the impact of ICT4D, we consider anecdotal evidence and profiling of success stories to be useful complements to indexing efforts, by giving deeper insights on issues of importance to women. It is particularly useful as precursors to developing quantitative indicators. To illustrate impacts of ICT on gender empowerment, we now present two ICT projects in rural Bangladesh that forms the basis of our empirical study.

Grameen Phone Community Information Center (GPCIC) in Jamalpur Village

Grameen Phone is the leading telecommunications service provider in Bangladesh with more than 23.26 million subscribers as of December 2009 (BTRC 2009). Grameen Phone is a joint venture initiative between Norway's Telenor ASA (62%) and Bangladesh's Grameen Telecom (38%). Grameen Phone initiated a new program to provide a comprehensive ICT solution through its new program, the Grameen Phone Community Information Centers (GPCIC). The GPCIC are set up on shared premises in select rural areas around the country. The centers provide computers, printers, scanners, and web cams for the rural population. They are operated independently as small business by local entrepreneurs. GPCIC provides mobile Internet connections in every center for access to the Internet and has enabled some 20 million people in rural areas to access the Internet and use e-mail for the first time. Following a successful pilot phase, conducted with the GSM Association, GPCIC is teaming up with local entrepreneurs to set up at least one center in each sub-district throughout the country. At the GPCIC centers, users can access the government websites, e.g. to obtain passport forms, birth and death certificate forms and other related information. Market prices of agricultural products and other agricultural information can easily be obtained from GPCIC. The GPCIC's are also of help to students and professionals for gathering reports and news. Information related to local and foreign job search sites is available at the centers. Many services that would otherwise have been beyond the reach of an individual living in a rural area are now available at the GPCIC. Health and medical information will also be made available through the GPCIC soon.

The GPCIC in this study located in the village of Jamalpur, 38 km northwest of Dhaka, the capital city of Bangladesh. The village is in the District (Administrative Unit) of Manikgonj, which has a long history as a trading center. The total area of the Jamalpur village is about 5-6 sq. km. This village is bounded in the south by a small river called the Dholeshori and in the west by a highway connecting to a main road to Dhaka. There are approximately 780 households in this village; with most villagers living in tin-shed houses. There are only a few brick houses in the village. In Bangladesh the commercial activities of a typical village revolve around the "*hat*" (weekly market) and *bazaar* (daily market). In Jamalpur village, Saturdays and Tuesdays are the days for *hat*. Villagers from different villages come to this *hat* with vegetables, fish, chicken, and cattle for trading and the village is in festive mood. The total population of the village is 2,530 of which 60 percent are male and 40 percent female (CIA, 2008).

Village Phone (VP) Project in Goma-Krisnokathi Village

VP (which is actually a typical cellular mobile service), was initiated as a self-employed income generating program, that provides modern telecommunication services in rural Bangladesh, where the VP operators, mainly women, resell network services.(Richardson, Ramirez, & Haq, 2000). These grassroots women entrepreneurs, operate their businesses in rural villages where no telecommunications services previously existed; they offer the use of their phones to community members on a per-call basis. Offering services at affordable rates to customers, they earn enough to repay their loans they had to take to finance the purchase of their phones and earn incomes that allow them to make investments in their children's health, nutrition and education, and in other business ventures. More importantly, individuals living in rural communities gain access to affordable telecommunication services, linking them to their friends, family, and business contacts within Bangladesh and around the globe. Expanding access to the rural poor by providing rural telecommunications infrastructure through a sustainable business model, this programme has been praised as a unique and solid model that promotes development and poverty alleviation through the use of ICTs.

Figure 1. Map of Bangladesh and respective divisions for field study

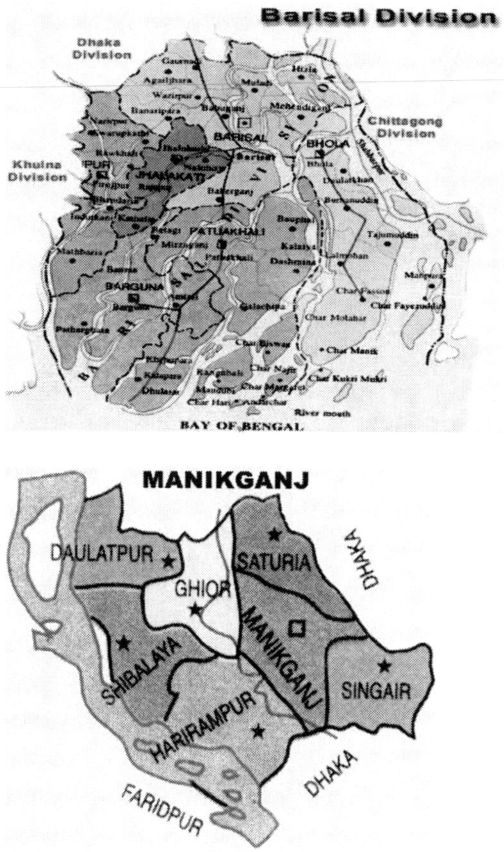

Despite facing interconnection problems with the incumbent Bangladesh Telegraph and Telephone Board during the early years of its operations (Bhuiyan, 2004) which started in 1997, the program had grown to more than 260,000 VP operators in over 50,000 villages in 439 Upazilas (sub-districts) of the country by mid-2008. This initiative is acknowledged as a sustainable development tool by governments and development agencies such as the World Bank, the United Nations, the International Finance Corporation and USAID.

The VP project in this study is located in the Goma- Krisnokathi village, Barisal district, in a remote area of Bangladesh, some 420 km south of Dhaka The area of the village is approximately 16 sq. km. and it is surrounded by 3 rivers, and 3-4 canals. The village has only one land phone line. The population of approximately 6,000 live in some 550 households, 60% of which are Muslims and the rest are Hindus. With the skills of most residents limited to basic agricultural skills, the major livelihood for males is in the agricultural sector (80%) and fishing (10%), with the remaining 10% occupied in the service sector. Many of the villagers work outside the village. Most of the women are housewives, however some are involved in poultry, the mobile telecommunication business, sewing and making handicraft products. Almost 90% of the villagers are landless and live under the poverty line, depending on landowners for their living. The average income for men is approximately 3,000 taka (USD 43) per month, corresponding to an annual income of USD 516, which represents 40% of the estimated per capita GDP in Bangladesh in 2007 (CIA, 2008). Their income varies with the seasons. In the harvesting season, they earn much more compared to the rainy season.

Research Design

The methodology adopted in this research is based on an interpretive approach in the qualitative research tradition. The interpretive approach advocates the development of understanding of human behaviour by considering surrounding contexts (Walsham, 2002). It is therefore a suitable framework when there is a need for explorative research to understand complex human behaviour (Evert, 2003). Previous studies have not adequately addressed these aspects with respect to impacts of ICT in the social context of rural women in Bangladesh. The use of an interpretative methodology in the absence of primary data when theorising a problem is supported by Menou & Potvin (2007): statement, 'One cannot overemphasize that the kind of data, especially those which are needed for building indicators and the indicators themselves have to be selected from the beneficiaries perspective and with their participation, e.g. through a grounded theory approach'(p. 17).

Primary data were collected from six (6) female operators of the VP project and three (3) female users of the GPCIC project through focus group discussions and personal interviews. Participants were selected on the basis of their willingness to take part in interviews and the availability of the field investigator's time during a one month field visit in the two villages. The most common characteristics of the six female operators are as follows:

1. they are married and are in the 30 year age range,
2. their average monthly income is less than USD 100, and
3. they are directly or indirectly involved in agriculture or business.

The research findings are presented in a story telling fashion to identify emergent issues from interview transcripts. Listening to their stories gives us an insight into how they interpret ICD4D or ICT-led development from their perspective.

FIELD RESULTS

VP Project

A. Economic Progress

VP is an opportunity for rural women to initiate their own business in their localities. Where required, the female operator first gets a loan to purchase a handset and to finance infrastructure development within her area, if necessary. The VP operators can set up the business adjacent to their homes.

Recently I engaged in a business concerning cellular phone. I obtained license from Grameen Phone company. My income increased from Tk.1,000 to Tk.6,000 per month.

The improvement in rural women's economic or financial condition through the VP project is not straightforward. We found some challenges — a conservative social structure, in which conservative elements use social pressure in their attempts to hinder economic progress by women., This is demonstrated by the experiences of one of the operators:

People came to my shop to call their relatives and for business purposes. My business was running well, but after some days some people objected to my family, mentioning that a single woman cannot operate business in the middle of the Bazaar (market). I had a fear like this because our society is quite conservative for women.

However, this negative perception of the role of women was challenged by other women, as illustrated by another operator:

I motivate others to become more financially independent, and it makes them feel happy and proud. Look at me! I have proved that a woman can also earn income like man. A female can do it as well, or sometime more successfully.

B. Decision Making Capacity

Since female operators have been contributing financially to their families, their importance, decision-making capacity, and power in family affairs have increased, as stated in the following vignette, in which one operator talks about the impact on VP operators in general:

It has significantly changed her economic condition and given her complete satisfaction and confidence, as she can now do something for her family and herself. Her decisions now get priority from her husband on every aspect of life, like children's education and birth control, as she has proved herself to be successful in her business.

The operators are also becoming important persons at the village level, as stated by one operator, describing how she perceives the situation for VP operators:

She is being treated as a successful entrepreneur in the society and thus she plays an important role in its social activities.

C. Social Progress

In general, although not well educated themselves, the female operators have become aware of the importance of education, hence they are giving high priority to sending their children to school. Further, they have become aware of different social issues that are conducive to their development. The following vignette indicates the social progress of an operator:

My level of knowledge, skill and awareness of the wider environment is drastically changed as I have to manage different people, have to go different places for business purposes and thus my bargaining and negotiating power has increased.

In summary, the VP business has improved the confidence of the female operators and has enabled them to be active citizens. This confidence has prepared them to speak out on a number of issues, including unethical conduct they may encounter and to join civil society and work with government agencies.

GPCIC PROJECT

In this section, we present a number of stories told by the GPCIC users from their perspectives of ICT led development. According to Dart and Davies (2003), there are five advantages to using stories for evaluation type of research: engagement, insight, sense-making, memory and discussing hard messages. We used stories because they give greater insight than hard data in understanding the benefits that have emerged from the GPCIC project.

Story 01: A Successful Female Social Entrepreneur

Jamalpur village (at the center of GPCIC) is known by farmers and local buyers. Some farmers buy goods from the *Bazaar*(local market) and stock them in their backyard, waiting until the market price is at a satisfactory level for them to sell their products to anyone willing to pay right away. They have been doing this for generations. The local buyers do not face as many challenges as a farmer does. The farmer has to do much more hard work in order to bring fresh and quality goods in the market (Bazaar). Mrs. Toslim (34 years of old) was born in this village area. She has the equivalent of an intermediate certificate; but obtained other skills and knowledge by volunteering in different businesses, learning about photography, computers, web browsing, e-mail, fax, letter writing etc.

After graduation one of her relatives informed her about GPCIC. She then went to Dhaka and met with Grammen phone and GPCIC officials. They agreed to franchise Mrs. Toslim to run GPCIC in her village. She discussed with local villagers, including students, teachers, and farmers the benefits they can derive from the GPCIC. They were very excited or hopeful about this project and have become frequent users. Many students come on the daily basis to use computers, access the Internet, take digital photos, download music and various application forms, e.g. for admission to academic institutions and passports, and to check national secondary and higher-secondary results. The teachers come to GPCIC to prepare lectures. Through the center, Mrs. Toslim earned respect from many community members and has become successful in this community.

Story 02: Economic Progress through Information

Born in the village of Jamalpur, the 34 year-old Mrs Aman is a farmer, living with her husband and 5 children. Her husband was also a farmer, but due to an accident, he cannot walk properly. So, Mrs. Aman has to look after the agricultural and all other responsibilities related to their business. A few years ago Mrs. Aman thought about planting *Boroi* (trees that grow some kind of berries) and improve her income by selling Boroi berries at the market. No one in Baira Union had Boroi trees. She was the first one to think about it, so she was very excited about doing this quickly and properly. She told the chairman of her union about her plans and he considered it an excellent idea and was eager to see the Boroi Trees in Mrs. Aman's field. The Chairman and the villagers helped her in realizing her plans.

The Boroi trees seemed to be growing nicely but in 2008 the trees started to die, as a result of insects destroying the leaves. "Most of the trees were infected and dying quickly and they were not in good shape at all" said Mrs. Aman. She was looking for a solution all over her village, asking anyone who might know anything about Boroi trees. The Chairman was also unable to provide any information that could solve the problem.

Mrs. Aman then came to GPCIC to look for information about the problem and find a possible solution. Mrs. Toslim, manager of GPCIC, visited Mrs. Aman's *Boroi* plants and took notes and pictures of some of the infected trees and leaves. Back at the center, she checked the website www.ruralinfo.bd.com, designed by GPCIC, for a solution. From the web, the manager found that Voxosofer and Admier spray can kill the insects, thereby helping the *Boroi* trees to grow. Having been advised by Mrs Toslim about this remedy, Mrs. Aman acted accordingly. A few weeks later Mrs. Aman visited the GPCIC and told the manager that her trees were in good shape and healthy because of timely action. Mrs. Aman said "this year I will sell lots of *Boroi* and make good profit" and has improved her livelihood significantly from selling the Boroi berries.

Story 3: Changes of Female's Socio-Economic Life

Married at the age of 17, Kulsum, is a 33 year old divorced woman, with two daughters, both born within five years of her marriage, Her husband left her two years after the birth of their second daughter, as a result of her not giving birth to a son. After getting divorced she stayed with her brother's family, obviously with much shame, as it is very humiliating for women to get divorced in the rural Bangladeshi context. Her brother earned less than $80 a month and also had to feed his own family of five. Kulsum learned tailoring skills under the local Youth Ministry, Government of Bangladesh's project *jubounnoun odihidopotor* (youth development program). She also took loan from a local NGO and bought a sewing machine to start earning her own income. Her business was moderate.

One day she heard about the GPCIC and the Internet. She went to the center and asked the manger how Internet can help her to improve her business. Mrs. Toslim helped her by searching for images of different types of *kurtas* (traditional dresses for women in Bangladesh) on Google and showed her. She got some ideas about patterns and colours and experimented with making some *kurtas* on her own. Her collection was very much appreciated by her regular customers and through word-of-mouth she got many customers for her dresses. While she gained confidence and improved her economic situation she repaid the loan to the local NGO. Now she has a reputation that she has some unique designs in that area and she has also extended her line to include children's wear. Though she works in a limited scale, she is thinking about employing an assistant to enable her to meet orders in a timely manner. She visits the GPCIC regularly to search the Internet for new ideas for her dresses. She has a dream that someday she will have her own collection displayed on the Internet and people will order from there. She has also achieved much social recognition and economic independence. Now she can give her children education without the help of others and can meet her own requirements. She feels more empowered when she sees herself as an independent individual.

CONCLUSION AND FUTURE RESEARCH DIRECTIONS

This research is explorative in nature; hence an interpretive research methodology was suitable. But while the limited scale and scope of the study gave insights into some women's perceptions about their livelihood improvements, it did not manage to provide a wider understanding of the socio-economic impact of the two projects. In order to do this, the study has to move from the exploratory stage into a larger scale study. Using the GPCIC and VP projects as examples, the aim of this research was to highlight some impacts of ICT on rural women's lives at an individual level. With the focus on women as operators of VP franchises and users of GPCIC's services, we have illustrated how socio-economic progress is defined and perceived by the women, whose stories we included in this chapter. The benefits identified by these women are consistent with the work of other researchers in this area, who have also found that ICT has potential to contribute to socio-economic progress in developing countries through addressing various developmental challenges, such as poverty alleviation (Arunachalam, 2002;Anwer 2006; Ashraf, Hanisch, Swatman 2007). The open-ended interviews enabled the women to express the benefits experienced by them in their own words, thereby complementing existing knowledge based on qualitative research on the gender dimension of ICT led development in Bangladesh. We found, through a systematic analysis of some women's voices, that ICT in the form of the GPCIC and VP projects can act as a medium for economic and social progress in rural areas, thereby playing an important role in poverty reduction for the women, their children and families.

REFERENCES

Adeye, C. (2002). *ICT and Poverty: a literature review.* Retrieved from http://web.idrc.ca/uploads/user-S/ 10541291550ICTPovertyBiblio.doc

Aminuzzaman, S., Baldersheim, H., & Jamil, I. (2003). Talking back: Empowerment and mobile phones in rural Bangladesh: A study of the village pay phone of Grameen Bank. *Contemporary South Asia, 12*(3), 327–348. doi:10.1080/0958493032000175879

Anwer, M. (2006) *ICT for Better Society*. Paper presented at 9th International Conference on Computer and Information Technology. Dhaka, Bangladesh.

Arunachalam, S. (2002). *Reaching the unreached: How can we use ICTs to empower the rural poor in the developing world through enhanced access to relevant information?* (p. 11). India: M.S.S.R. Foundation. M S Swaminathan Research Foundation.

Ashraf, M. M., Hanisch, J., & Swatman, P. (2007). *ICT Intervention Towards Development In The Third World Countries: some perspectives of IS research.* Paper presented at 9th International Conference on Computer and Information Technology (ICCIT). Dhaka, Bangladesh.

Batchelor, S., & Norrish, P. (2004). *Framework for the assessment of ICT pilot projects: beyond monitoring and evaluation to applied research. InfoDev*. Washington, DC: World Bank.

Bayes, A., Vn Braun, J., & Akhteret, R. (1999). *Village payphones and poverty reduction: insights from a Grameen Bank initiative in Bangladesh. ZEF – Discussion Papers on Development Policy No. 8, Center for Development Research.* Bonn: ZEF.

Bhuiyan, A. J. M. S. A. (2004). Universal access in developing countries: a particular focus on Bangladesh. *The Information Society, 20*(4), 269–278. doi:10.1080/01972240490480983

Bridges.Org. (2003). *Case Study Series on ICT-Enabled Development: The Women's Information Resource Electronic Service.* Retrieved from *http://*www.bridges.org/case_studies/135

BTRC. (2009). Retrieved February 12, 2010 from http://www.btrc.gov.bd

Cecchini, S.S., C. (2003). Can information and communications technology applications contribute to poverty reduction? Lessons from rural India. *Information Technology for Development, 10*(2), 73–84. doi:10.1002/itdj.1590100203

Chapman, R., Slaymaker, T., & Young, J. (2003). *Livelihoods approaches to information and communication in support of rural poverty elimination and food security*. London: Overseas Development Institute.

CIA. (2008). Retrieved from http://www.isilo.com/download /CIAWFB08.htm

Colle, R.D.R., R. (2001). *Modules for training telecentre staffs: An Interim Report with sample modules.* Cornell University.

Dart, J., & Davies, R. (2003). A Dialogical. Story-Based Evaluation Tool: the most significant change technique. *The American Journal of Evaluation, 24*(3).

Evert, G. (2003). All research is interpretive! *Journal of Business and Industrial Marketing, 18*(6/7), 482–492. doi:10.1108/08858620310492365

Gurumurthy, A. (2003) *Bridging the digital gender divide: Issues and insights on ICT for women's economic empowerment.* U.N.D.F.f. Women.: Impression communications India.

Gurumurthy, A. (2006). Promoting gender equality? Some development-related uses of ICTs by women. *Development in Practice, 16*(6), 611–616. doi:10.1080/09614520600958298

Harris, R., & Bhatarai, M. (2003). *Fact-Finding Review of ICT in Development in a Rural-Urban Setting*, in *Assessing the feasibility of ICT as a development instrument for rural-urban linkages in Nepal.* UN-HABITAT and Roger Harris Associates. Rural-Urban Partnerships Programme (RUPP) under Support Services Policy and Programme Development (SPPD).

Harris, R. R. R. (2006). *Empowering the poor. Information and communications technology for governance and poverty reduction: a study of rural development projects in India.* Bangkok: UNDP-APDIP.

Heeks, R., Arun, S., & Morgan, S. (2005). *Researching Women's ICT-Based Enterprise for Development: Methods, Tools and Lessons from Fieldwork.* UK: IDPM, University of Manchester.

Huyer, S. Hafkin, N., Ertl, H, & Dryburgh, H. (2005). From the Digital Divide to Digital Opportunities: Measuring Infostates for Development. In Sciadas (Ed.), *Women in the Information Society.* Montreal: Orbicom.

ITU. (1998). *Resolution 7. Gender and Telecommunication Policy in Developing Countries.* World Telecommunication Development Conference, Valletta.

ITU. (2002). *Gender Mainstreaming.* Retrieved September 28, 2009 from http://www.itu.int/ITU-D/pdf/5196-011-en.doc

ITU. *(2006).* World telecommunication/ ICT development report: measuring ICT for social and economic development. *Geneva.*

Izugbara, C. O. (2004). Gendered micro-lending schemes and sustainable women's empowerment in Nigeria. *Community Development Journal, 39*(1), 72–84. doi:10.1093/cdj/39.1.72

Jerinabi, U., & Arthi, J. (2005). *Impact of Information Technology and Globalisation on Women's career.* International Conference on Information and Communication Technology of Management. Melaka, Malaysia.

Keogh, D., & Wood, T. (2005). *Village Phone Replication Manual: Creating Sustainable Access to Affordable Telecommunications for the Rural Poor.* New York: The United Nations Information and Communication Technologies Task Force.

Manning, R. (2008). *Development Co-operation Report (2008): Efforts and policies of the members of the Development Assistance Committee.* OECD Journal of Development.

Marker, P., McNamara, K., & Wallace, L. (2002). *The Significance of Information and Communication Technologies for Reducing Poverty.* London: Department for International Development.

Mbarika, V., Okoli, C., Byrd, T., & Datta, P. (2005). The Neglected Continent of IS Research: A Research Agenda for Sub-Saharan Africa. *Journal of the Association for Information Systems*, *6*(5), 130–170.

Menou, M.J., & Potvin, J. (2007). *Toward a conceptual framework for learning about ICT's and knowledge in the process of development- A GK LEAP background document.*

Miles, M. B., & Huberman, A. M. (1994). *Qualitative data analysis: an expanded sourcebook* (2nd ed.). Thousand Oaks, CA: Sage Publications.

Nevile, A. (2007). Amartya K. Sen and social exclusion. *Development in Practice*, *17*(2), 249–255. doi:10.1080/09614520701197200

Ofir, Z., & Kriel, L. (2004). *Evaluating policy influence of ICTs for rural areas: the MSSRF information villages research project*. Johannesburg: Evalnet.

Orbicom (2005). *From the digital divide to digital opportunities: measuring infostates for development* (G. Sciadas. Ed.). Montreal, Canada: NRC press-presses du CNRC.

Patel, R., & Parmentier, M. J. C. (2006). The persistence of traditional gender roles in the information technology sector: a study of female engineers in India. *Information Technologies and International Development*, *2*(30), 29–46.

Pillarisetti, J.R.M., M. (1998). Human development and gender empowerment: methodological and measurement issues. *Development Policy Review*, *16*, 197–203. doi:10.1111/1467-7679.00059

Prahalad, C. K., & Hammond, A. (2002). Serving the World's Poor, Profitably. *Harvard Business Review*, *80*(9).

Richardson, D., Ramirez, R., & Haq, M. (2000). *Grameen Phone (GP) village phone project: final report*. Canada: C.I.D.A. Guelph Ontario.

Roger, H., & Rajesh, R. (2006). *Empowering the Poor: Information and Communications Technology for Governance and Poverty Reduction.* UNDP, Asia Pacific Development information Program, Bangkok, Thailand.

Sen, A. (2000). Social exclusion: concept, application and scrutiny. In *Social Development Paper No. 1*. Manila: Asian Development Bank.

Tarasiewicz, M. (2003). Taking Gender Into Account. *UN Chronicle*, *40*(4), 37.

Torero, M. B. J. (2006). *Information and communication technologies for development and poverty reduction: the potential of telecommunications.* Baltimore: Johns Hopkins University Press.

UNDP (2004). *Bridging the gender digital divide*. UNDP/UNIFEM. UNDP regional centre for Europe and the CIS and UNIFEM Central and Eastern Europe: Solvak Republic.

UNDP. (2005). *Promoting ICT for human development in Asia: Realizing the Millennium Development Goals*. Regional Human Development Report. UNDP.

UNIFEM. (2005). *Support to NEPAD*. Retrieved August 18, 2009 from http://www.un.org/africa/osaa/reports/UN%20folder2006-UNIFEM.pdf

Walsham, G. (2002). Interpretive case studies in IS research: Nature and method. In Avison, D. E., & Myers, M. D. (Eds.), *Qualitative research in information systems: a reader* (pp. 101–113). London: SAGE.

Walsham, G. (2005). Development, global futures and IS research: a polemic. *The Journal of Strategic Information Systems*, *14*, 5–15. doi:10.1016/j.jsis.2004.11.009

Walsham, G., & Sahay, S. (2006). Research on information systems in developing countries: current landscape and future prospects. *Information Technology for Development*, *12*(1), 7–24. doi:10.1002/itdj.20020

WSIS. (2005)*Tunis Agenda: for The Information Society*. Retrieved from http://www.itu.int/wsis/d ocs2/tunis/off/6rev1.html Accessed date: 18.09.2009

WSIS. (2008) *WSIS follow-up Report 2008*. Retrieved from http://www.unctad.org/e n/docs/none20081_en.pdf Accessed date: 18.09.2009

ADDITIONAL READING

Initiative, C. (2006). *Gender Indicators*. G.E.M. Team. Canada: The communication initiative.

Kanungo, S. (2004). On the emancipatory role of rural information systems. *Information Technology & People*, *17*(4), 407–422. doi:10.1108/09593840410570267

Knight-John, M. (2008). Making a Business out of a Village Phone. In Zainudeen, R. S. A. (Ed.), *ICT infrastructure in emerging Asia: policy and regulatory roadblocks* (pp. 116–130). New Delhi: Sage Publications.

Nanavaty, R. (2000). Satcom for Bareoot Women Managers. In Schware, S. B. R. (Ed.), *Information and communication technology in development: cases from India* (pp. 163–167). New Delhi: Sage Publications.

UNCTAD. (2006). *Information Economy Report*. New York, Geneva: UNCTAD.

Warnock, K. W. R. (Ed.). (2005). *Information and Communication Technologies and Large-Scale Poverty Reduction: Lessons from Asia, Africa, Latin America and the Caribbean*. London: PANOS.

WOUGNET. (2003). *Women of Uganda Network meeting: information sharing on the application of ICTs*. Kampala.

KEY TERMS AND DEFINITIONS

ICT: Information and communication technologies (ICT) is an umbrella term that covers all technical means for processing and communicating information. The term has gained popularity partially due to the convergence of information technology (IT) and telecommunication technologies. ICT defines a broad range of technologies, including methods for communication (communication protocols, transmission techniques, communications equipment, media (communication)), as well as techniques for storing and processing information (computing, data storage, etc.)

Socio-Economic Development: Socio-economic development is the process of social and economic development in a society. Socio-economic development is often measured with indicators, such as GDP, life expectancy, literacy and levels of employment. Changes in less-tangible factors are also considered, such as personal dignity, freedom of association, personal safety and freedom from fear of physical harm, and the extent of participation in civil society. Causes of socio-economic impacts are, for example, new technologies, changes in laws, changes in the physical environment and ecological changes.

ENDNOTE

1. This methodology is different from the 'gender empowerment measure', also known as GEM, which is used in the UNDP Human Development Reports to reflect gender inequalities and rank countries according to the level of female participation in political decision-making, access to professional opportunities and earning power (Pillarisetti & McGillivray, 1998).

Chapter 12
The Application of Social Networking as a Marketing Platform to Young Adults:
An Emerging Market Perspective

Justin Henley Beneke
University of Cape Town, South Africa

ABSTRACT

Social networking is often touted as being a prominent application responsible for driving the adoption of residential broadband services. The growth of social networks is phenomenal – in many cases more than doubling in size on an annual basis. This study considers how social networking may be utilized for commercial purposes to spread word-of-mouth communication. The chapter therefore considers the characteristics of young adult social network users, how they behave and interact with other users on such platforms, as well as the manner in which marketers can make the most of this platform without experiencing a consumer backlash. The research suggests that if a symbiotic relationship does need exist between broadband proliferation and the adoption of social networking, both have a vested interest in each other's continued success.

BACKGROUND

Word-of-mouth (WOM) communication has received extensive attention from academics and marketing practitioners for many years (De Bruyn & Lilien, 2004). In fact, one of the most commonly accepted concepts in consumer behaviour is that WOM communication plays a fundamental role in shaping consumers' attitudes and behaviours (Brown & Reingen, 1987). With the increasing development of communication and information technologies, the consumers' interactive environment has changed to reflect an amorphous network of communication possibilities, with electronic interaction (particularly the Internet) leading the social reform. As a result, peer-to-peer communication has become increasingly saturated with information, enhancing the effect of WOM through a process commonly described as social networking. Social networking refers to the building of communities that bring people together on the Internet, and has manifested itself in many ways

DOI: 10.4018/978-1-60960-011-2.ch012

Figure 1. Trendline of unique visitors to Facebook versus MySpace (Source McCarthy (2009))

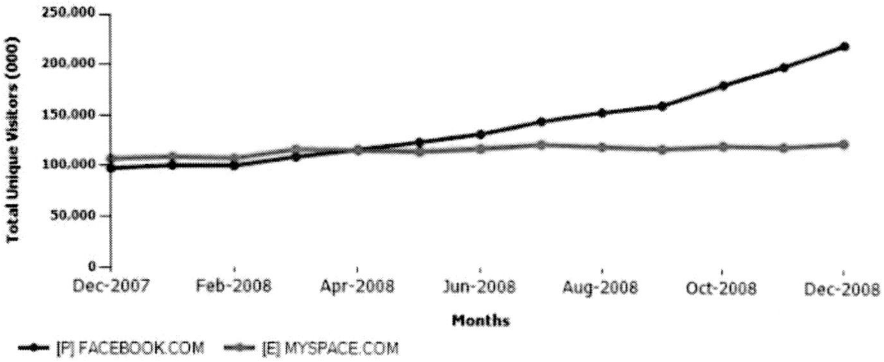

(Raskin, 2006). The most predominant social networking sites include MySpace, Facebook, Friendster, YouTube, Blog-it, Mig33 and Xanga. Mxit, a South African initiative, features amongst the most popular mobile-driven social networking platforms. These sites are, in different forms, social utilities that connect people through the ability to share information such as messages, photographs and personal characteristics and preferences. As of mid 2009, Facebook alone had over 300 million users worldwide, a four-fold increase within a single calendar year (Wikipedia, 2009). MySpace.com, now owned by News Corporation, launched as a website to help unknown musicians reach greater audiences. It soon progressed into a virtual community of over 84 million users who post profile pages detailing their life in words, pictures and music (Wolfe, 2006). It is presently rumoured to have around 110 million users and is the dominant player in the United States market (McCarthy, 2009). Both Facebook and MySpace allow members to interact socially, building a network of friends as well as friends of friends with whom they share their thoughts, perceptions and general life experiences (Raskin, 2006).

The graph below visually depicts traffic to the two social networking stalwarts between December 2007 and 2008, revealing that Facebook appears to be winning in the share of total unique visitors.

Marketers are increasingly faced with the challenge of breaking through commercial clutter in an attempt to capture the attention of over-stimulated consumers, who are becoming more and more adept at tuning out traditional forms of advertising (Wolfe, 2006). With the extensive adoption of online communication, many experts believe social networking websites to have vast potential as viral marketing platforms (i.e. tools for spreading product information through WOM communication). However, although evidence of previous viral marketing success exists, little is known about the attitudes, motivations and behaviours of those consumers who engage in social networking, as well as which online promotion and communication techniques best stimulate viral marketing amongst social network users (Phelps *et al*, 2004). The lack of knowledge in this area may lead to the implementation of ineffective marketing, and requires further investigation in order to fully harness the potential of this platform.

RESEARCH DESIGN

Both qualitative and quantitative research was employed for the empirical component of this study.

In terms of the former, experience interviews were conducted with four individuals knowledgeable on the subject of online marketing and social networks.

These individuals are as follows:

- Ian Calvert, Chief Executive Officer - Instant Grass, South Africa
- Alex Van Tonder, Trend advisor - Instant Grass, South Africa
- Angela Banks, Brand Manager - Unilever, South Africa
- Sarah Manners, Public Relations Manager - Quirk E-Marketing, South Africa

In terms of the quantitative aspect of the study, an internet survey was utilized in order to capture further data about social network users. The research instrument (questionnaire) was promoted on Facebook in order to ensure that all respondents were social network users. However, the actual survey was hosted on a standalone site. There were many advantages as well as disadvantages to using online survey distribution. The main advantages were the speed and cost effectiveness of distributing the questionnaire in this manner, as well as the ability to capture data in real-time, thereby increasing efficiency and accuracy of data collection. This method also allowed for complete respondent anonymity, therefore allowing respondents to give more honest and personal feedback. The disadvantage was that it is difficult to obtain a representative sample as under a tenth of the South African population has Internet access (Goldstuck, 2009). However, as the research is only pertinent to those individuals who are Internet users and actively utilize online social networks, it was only necessary for that particular market segment to be probed. Another potential disadvantage was the fact that response rates to Internet based surveys are generally low. This was combated by proactively seeking respondents on Facebook and, through the snowball effect, eventually acquiring a database of 202 respondents.

Table 1 on the following outlines the demographic profile of the sample.

USAGE OF SOCIAL NETWORKING

The degree to which online social networkers have access to interaction with other users plays a fundamental role in the effectiveness of any given WOM campaign. Figure 2 illustrates the distribution of online contacts ('friends') with whom social networkers interact.

The research shows that the vast majority (90%) of respondents have more than 100 contacts on their choice of social network. This skewed distribution is congruent with research conducted by Vilpponen (2006), who indicates that online social network users tend to communicate with

Figure 2. Number of online contacts

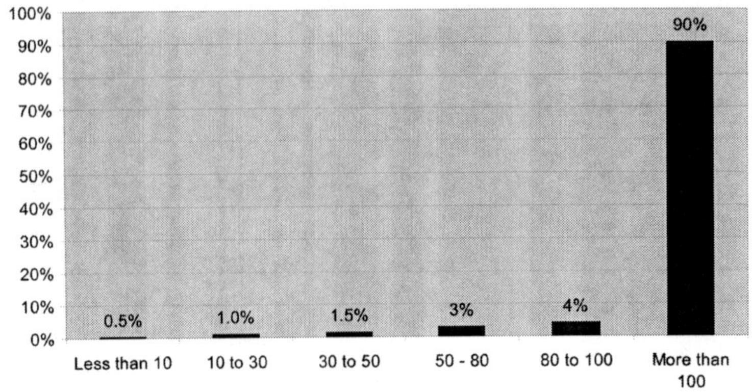

Table 1. Profile of respondents

Variable	Categories	Percentage
Age	16 - 24	92.0
	25 - 35	8.0
Gender	Male	50.5
	Female	49.5
Occupation	Professional/Tech	6.5
	Admin/Managerial	8.0
	Clerical Sales	3.5
	Service Delivery	3.0
	Artisans	4.0
	Production/Mining	0.5
	Student	70.0
	Scholar	3.0
	Athlete	1.0
	Armed Forces	0.5
Home Language	English	97.0
	Afrikaans	2.5
	Zulu	0.5
Marital Status	Single, no children	96.0
	Married, no children	0.5
	Married, with children	2.0
	Divorced, no children	0.5
	Divorced, with children	1.0
Monthly Income	R100 – R499 (US$ 14-70)	7.0
	R500 – R999 (US$ 70-140)	10.5
	R1000 – R1499 (US$ 140-210)	17.0
	R1500 – R1999 (US$210-280)	17.0
	R2000 - R2499 (US$ 280-350)	8.5
	R2500 – R2999 (US$350-420)	7.5
	R3000 – R3999 (US$420-560)	3.0
	R4000 – R4999 (US$560-700)	5.5
	R5000 – R6999 (US$700-840)	8.0
	R7000 – R15999 (US$ 840-2 240)	10.5
	R16000+ (US$ 2 240 +)	6.0
Highest level of education	High School	8.5
	Matric Pass	14.5
	Tertiary Institution	77.0

large-scale audiences. This high-level network structure indicates that social networks are likely to offer a suitable medium through which to initiate and deliver WOM communications.

Heinz *et al* (2005) observes that online social network users display a low sensitivity to privacy, and exhibit a propensity to sharing a large amount of information about themselves to both friends and even strangers. Complimentary to this observation, 70% of respondents admitted to revealing their personal information (such as age, telephone number and their relationship status) on their profile.

Additionally, a large percentage (41%) of respondents admitted to spending between two and six hours on a social network per week. Figure 3 further reveals that 21% of respondents spent between six and ten hours and 16% spent more than ten hours interacting on a social network each week.

When questioned as to whether they discuss brands or products with each other, 80% of respondents specified that they do. This suggests that online environments are indeed an appropriate (and potentially credible) medium through which to stimulate WOM.

The survey also asked respondents which social networks they were aware of, as well as which networks they actually subscribed to.

Figure 4 reveals that two thirds of the respondents were aware of Facebook, MySpace, YouTube and Mxit, listed in descending order of awareness. Blog0it, Mig33, Friendster and Xanga were considerably less well known.

As per Figure 5, Facebook was revealed to be the strong preference of social networking site. Nearly eighty percent of respondents were registered Facebook users. MySpace, YouTube and Mxit had minimal usage (none over thirty percent) with the remainder having experiencing very little usage. This confirms the popularity of Facebook, over and above the competition, in South Africa.

Characteristics and Motivations of Social Networkers

Motivation occurs when a consumer strives to satisfy a need that has arisen. The term 'need' essentially refers to something that is either physical or psychological that the individual must have in order to function and develop (Cant *et al*, 2006). Needs can be either innate or acquired. Innate needs are physiological and help to sustain biological life, while acquired needs are those learned in response to an individual's culture and environment. In a marketing context, needs on their own are not sufficient: the consumer must express the

Figure 3. Number of hours spent social networking per week

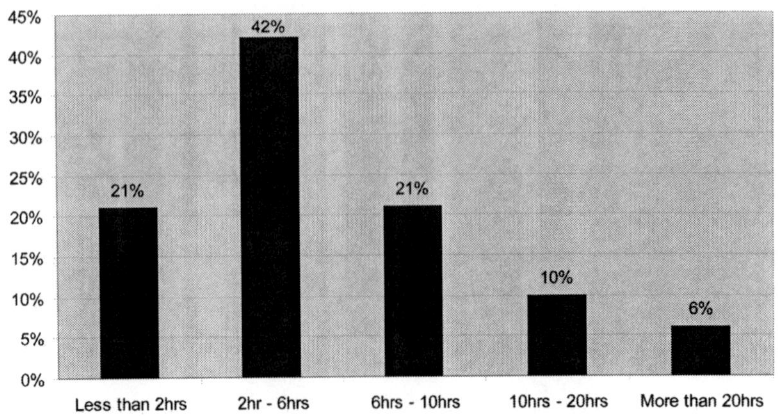

Figure 4. Awareness levels of various social networks

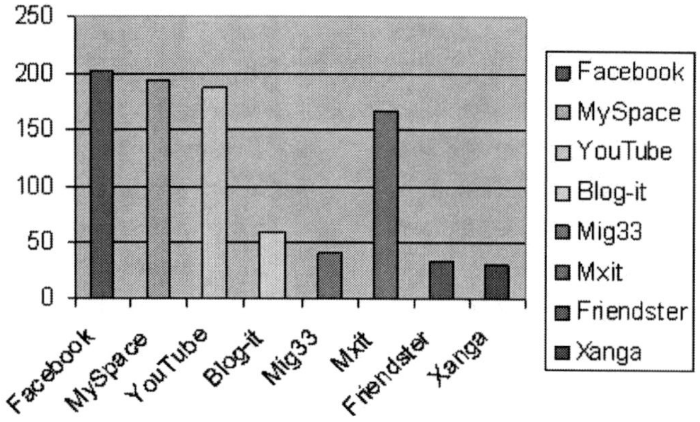

desire to satisfy his or her needs. This desire, or internal driving force to satisfy a given need, is known as 'motivation'. Motivation is therefore produced by a state of tension that exists as the result of an unfulfilled need (ibid).

The arousal of a need stems from the presence of a stimulus causing recognition of that need. Environmental arousal occurs when specific cues in the environment activate a consumer's need. Advertisements, for example, often generate a psychological imbalance in the viewer's mind, arousing the need to counter that imbalance (Solomon, 2004). Once aroused, the need produces a drive state in which the consumer experiences emotions and physiological arousal, leading to goal-directed behaviour in which the consumer seeks to act in a way that will relieve the need state. Most common examples of relieving the need state result in consumers actively searching for information and/or discussing products with other consumers (Cant *et al*, 2006).

Figure 6 details Maslow's Hierarchy of Needs with associated examples, as well as the transitivity of those examples to an online environment.

Figure 6 illustrates how physiological needs must first be satisfied before the need for secu-

Figure 5. Subscription to social networks

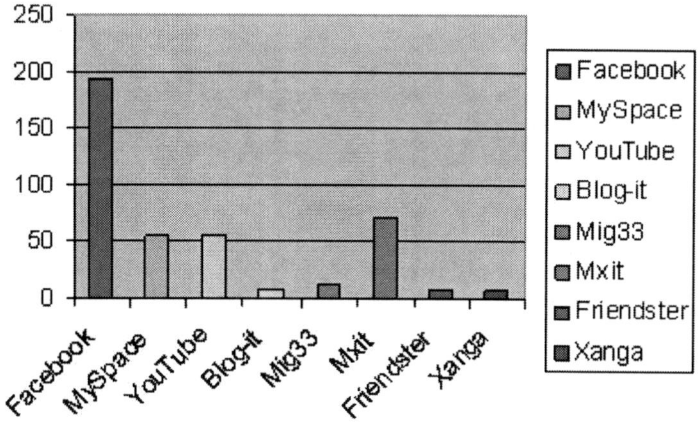

rity and safety is activated. In an online community, physiological needs incorporate issues such as basic system access, while safety needs include issues such as protection from hackers and viruses, and maintaining online privacy. Consumers who engage in online social networking are necessarily those who have already satisfied their physiological and security/safety needs, as they have already attained online access and clearly signal their feeling of security by engaging in social networking behaviour. Therefore, it seems that the needs most likely to stimulate online WOM communication behaviour are those of social, self-esteem and self-actualisation (Kim, 2006).

In an effort to uncover more specific motives for online WOM communication, a multiple frequency analysis was conducted on a 12-item multiple response table. Here, respondents were requested to indicate which factors best described their motivation for interacting and communicating with others on a social network. Table 2 includes the items which were included for selection, as well as the response rates for each.

Of all respondents, 98% indicated that they utilise social networks to keep in contact with friends, while 78% said they use online networks to find long lost friends. Half of the respondents stated that they engage in online social networking because their friends do, whilst 30% of respondents admitted the desire to be part of a group or community. Additionally, it was found that 22% of respondents use online social networks to meet new people, whilst 18% noted that they engage in online networks to share information about themselves. This empirical evidence partially verifies research by Kim (2006), who argues that the needs most likely to stimulate online WOM are those of social, self-enhancement (or self-esteem) and self-actualisation. While the research leans more heavily towards the need for social interaction, limited is given to the need for self-actualisation with a handful of respondents indicating the use of social networks to establish or enhance their personal identity.

The concept of self-enhancement is supported by Engel *et al* (1993) as well as Sundaram *et al* (1998), who identify the tendency of consumers to initiate WOM in order to share product infor-

Figure 6. A comparative study of the application of Maslow's Hierarchy of Needs to both offline and online environments (Source Kim (2006))

Need	Offline (Maslow)	Online Communities
Physiological	Food, clothing, shelter, health	System access; the ability to own and maintain one's identity while participating in a web community
Security & Safety	Protection from crimes and war; the sense of living in a fair and just society.	Protection from hacking and personal attacks; the sense of having a "level playing field"; ability to maintain varying levels of privacy
Social	The ability to give and receive love; the feeling of belonging to a group.	Belonging to the community as a whole, and to subgroups within the community.
Self-Esteem	Self-respect; the ability to earn the respect of others and contribute to society.	The ability to contribute to the community, and be recognized for those contributions.
Self-Actualization	The ability to develop skills and fulfill one's potential.	The ability to take on a community role that develop skills and opens up new opportunities.

Advancement in Hierarchy

mation in the attempt to enhance their self image among other consumers. The empirical evidence shows that 13% of respondents participate in social networks because they like the attention, whilst 9% indicated the need to influence the way they are seen by others. Considering the above data, the concept of self enhancement appears to have partial significance as a motive to communicating in online social networks.

The motivational concepts of social-interaction and self-enhancement are further substantiated by Alex Van Tonder of Instant Grass. Van Tonder argues that in order to establish WOM information transfer amongst users of social networks, it is imperative for companies to involve the consumer. Van Tonder further claims that the most effective media for stimulating WOM is that which resonates with, or is focused on, the consumer herself. The reason for this is that consumers are increasingly interested in enhancing their sense of self, a desire which they partially achieve through enhanced social-interaction.

In addition to the concepts of social-interaction and self-enhancement, it was found that 60% of respondents interact and communicate on a social network for the purpose of entertainment. Entertainment as a motivational factor is, likewise, supported by Engel *et al* (1993), who argue that a motivational cue exists when consumers find entertainment in discussing certain promotions/advertisements. This is further emphasised by Phelps *et al* (2004), who note that consumers are more likely to pass on information that is entertaining, or involves some aspect of enjoyment. According to Calvert of Instant Grass, a WOM campaign stands or falls on the basis of how amusing or entertaining it is.

Interestingly, analysis shows little evidence of altruism and dissonance reduction as motivational factors. Only 5% of respondents indicated that they communicate on a social network to help others, while less than 2% admitted to using social networks to find out more about a specific product. This may be due to a mindset of consumers firstly engaging with a brand through the appropriate manufacturer's or service provider's web site, possibly due to time pressure and the likelihood of receiving official information.

Table 2. Reasons why social networkers choose to interact with each other

a)	To meet new people	22%
b)	To be part of a group/community	30%
c)	Because my friends do	50%
d)	To keep in contact with friends	98%
e)	To find long lost friends	78%
f)	To share information about myself	18%
g)	I like the attention	13%
h)	To influence the way other people see me	9%
i)	To find out more about a specific product	2%
j)	For entertainment	60%
k)	To help others	5%
l)	To establish my personal identity	7%

GOING VIRAL

The term 'Viral Marketing' was first used in 1996 by a venture capital firm, Draper Fischer Jurvetson, while describing the marketing strategy incorporated by the free e-mail service Hotmail, which involved the tactic of appending promotional messages to outgoing e-mails in order to encourage new subscriptions (Porter & Golan, 2006). Since then, the term viral marketing has become vitally important and an integral part in the ever evolving sphere of twenty first century marketing.

With viral marketing being such a fresh and overused term, disagreement exists about its precise definition. Some (e.g. Pastore, 2000) view it as WOM advertising in which consumers tell other consumers about the product or service, whilst others (e.g. Modzelewski, 2000) argue that "true" viral marketing differs from WOM in that the value of the 'virus' to the promoter is directly related to the number of individuals it attracts. Shirky (2000) suggests that soon viral marketing will mean WOM advertising to most people. But, more importantly, he adds that the concept describes a way of acquiring new customers by encouraging honest communication among consumers and thereby lending credibility to the transfer of the message.

Irrespective of the definition, viral marketing is instrumental in diffusing information among participants – "members have the power of millions and can spread links faster than any television network ever could" (Brier, 2005). This allows for dynamic interaction between marketers and consumers. Viral marketing is extremely popular amongst marketers because of the ease associated with executing the campaign, relatively low cost (even when compared with direct mail) and the high and rapid response rates (Sarah Manners, Quirk E-Marketing). This is particularly welcoming for companies that have modest marketing budgets, yet still need to engage effectively with a large audience.

It may be argued that the difference between traditional and electronic WOM is due to externalities. These network externalities occur when "the buyer of the last unit of a good has a higher benefit than the buyer of the first because the sale of the earlier units has created some benefits in a related dimension" (Economides, 1991). For example, when a social network begins it does not seem as attractive to potential adopters due to the small number of users. But as the number of users grows, the utility a potential adopter can attain by joining the network increases, until it reaches a critical mass point which causes the majority of potential adopters to become users. This can have a significant effect on rates of adoption and therefore diffusion of information and purchase of goods (Vilpponen *et al*, 2006). As a result, viral marketing within a network structure can be highly successful in increasing awareness and adoption of innovations, but the type of product or brand being marketed plays a fundamental role in the ease and efficiency of these rates (ibid).

Van Tonder points out that technical products and online products are best suited for online network marketing. More generally, Banks (Unilever) believes that visible brands, especially contemporary youth brands such as the deodorant brand Axe, are most suited to social network marketing. This is because some product categories are more "colourful" and interesting, making them more conducive to viral marketing. Alcoholic beverages, for example, are almost synonymous with viral marketing due to the easy association of the product with sexual, social and humorous cues.

WOM is acknowledged to be one of the strongest drivers of market-place influence but suffers the misfortune of remaining an enigma for some marketers as the theories that drive WOM do not connect with research methodologies which allow for accountable planning, implementation and evaluation of strategies (Carson, 2005). Furthermore, in a world where consumers feel that they are being bombarded with information, commercial-oriented communication may not always be welcomed in social networking circles. Research by Carroll *et al* (2006) underlines the tendency of consumers to want control over the amount of information received. Calvert (Instant Grass) supports this notion, expressing that consumers, in an online environment particularly, do not want their space invaded and have a large degree of control as to what information they are exposed to. To this end, Sicilia *et al* (2006) emphasise that the manner in which the message is received is beyond the control of the sender. According to Calvert, one of the most predominant trends in marketing over the last couple of years is that marketers have increasingly less control over the message delivered.

In such a scenario, fraught with risk, the promoter of the message needs to ensure that the correct media approach is utilised, the target audience is properly identified and that it applies a user-friendly viral approach to each online marketing campaigns it wishes to undertake.

Characteristics of Success in Viral Marketing

In this new technological era it seems the marketplace has become a conversation. In particular, a conversation controlled by the consumer. The structured world of the past, where consumers were viewed as being passive and attentive, no longer exists. Consumers are now more knowledgeable as the communication environment in which they exist has been changed and enriched (Vilpponen *et al*, 2006). Consumers can post reviews about a product or brand in minutes and theses views can be accessed by millions all over the world with ease. This has created a shift in balance of power over brand perceptions from the respective companies to the consumer. Brands will be built by the WOM created by consumers as they collect information from a myriad of sources. This has been labelled as the world of "mosaic marketing' in which consumers are calling the shots (Runnacles, 2006).

Due to the clutter of marketing messages in the media world, marketers now need to develop new ways of interacting with consumers and engaging them in order to break through. But how do you obtain the edge in an over-saturated market that is being bombarded by mass media?

Firstly, in-depth research needs to be conducted to gain a full understanding of the target market (Smith, 2006). Once a greater understanding has been realised, then messages relating to "basic human motivations" (Porter & Golan, 2006) should be directed towards the consumer market while still incorporating a personal approach appropriate with the information gathered in the earlier research. One of the major promotion and communication techniques that drive online viral marketing is the use of incentives such as the opportunity to win a prize (Angela Banks, Unilever). There may be other major drawcards which, without research, would otherwise have remained hidden.

Secondly, keeping it light hearted and upbeat is important. Branded content, or "advertainment" as it has been named, is one of the ways in which brands can aim to utilize social networks to create an interactive medium with which to communicate and intermingle with the target market (Kirby, 2004). This branded content includes entertainment media such as video trailers, logos and rap songs (Beeching, 2006). Branded content involves creating entertainment that inspires interaction and involvement to encourage social networkers to pass them on and post them, rather than a company merely creating an online banner or television advertisement (Saunders, 2006). According to the respondents of the survey, almost all (87%) felt entertainment was crucial, while 60% of respondents say that the element of surprise (shock value) is also important in inducing them to talk about a product.

Similarly, a study by Phelps *et al* (2004) stated that marketers should focus on the desires for fun, entertainment and social connections to attract the attention of such users. It is these attributes that are seen to be the most appropriate when it comes to attracting the attention of consumers as well as leading them towards viral marketing, in the sense that they inevitably feel the need to pass on the message to fellow peers and hence transfer the communication. Another key consideration centres on the form of this communication – it would appear that consumers do not relate this kind of 'advertising' to common mass media advertising (Rodgers & Thorson, 2000). This aspect allows consumers to participate in the adverts without the feeling that they are being manipulated or coaxed into a purchasing decision, which is far more effective than 'forced' traditional advertising.

Whilst the "wow" factor is a necessary component of advertainment, brand consistency should be upheld. In this respect, companies should release online viral material that retains brand and campaign themes thus providing an ever increasing return on investment to the company (Kirby, 2004).

Thirdly, using social networking to break company hierarchies and truly connect with customers seems to be smart business practice. Here, embracing the blogosphere may allow the company an opportunity to effectively utilize the online environment to create interactive relationships with its customers. The two fundamental points that differentiate blogging from mainstream media are that it is firstly a consumer generated form of communication and secondly, readers of blogs are seeking and expect to find objective conversations (Corry & Mundell, 2006). For such reasons, most consumers appear to trust the information found on blogs as the opinions and information given are generally not influenced by a bias towards promoting a specific product or point of view. An increasing trend has been shown towards consumers utilising blogs to supplement product information found on corporate websites (ibid).

Somewhat strangely, companies have now taken the next step and developed their own corporate blogs. Here, executives record their thoughts, opinions and observations on general business issues as well as matters relating to their specific company (Corry & Mundell, 2006). However, in the creation of a company blog, it is crucial that the creators retain a degree of freedom in what they are producing. Here, it is recommended that a company position the content of the blog in a direction that captures readers' interest, such as providing an appropriate blend of entertainment and controversy (ibid).

Perpetuating the Infectiousness

The quickest and most effective way to create new product awareness is to create "buzz" about the product amongst the target audience (Angela Banks, Unilever). Buzz, a term typically used by industry professionals, describes the excitement accompanied with the release of new brands or products as well as price discounts. Buzz is thought to account for heightened sales as well as providing the catalyst for positive WOM about the product/brand (Sarah Manners, Quirk E-Marketing).

In the attempt to promote consumer communication of a product, marketers need to distinguish between which factors stimulate product memorability and which do not. Marketers refer to this as the "stickiness" effect. In the viral marketing context, this is what spurs consumers to pass along viral information (Welker, 2002). Another aspect of successful viral marketing concentrates not on incubating a message, but rather on spreading it. In other words, the viral message has to be contagious (Brier, 2005) and in turn should be used to gauge the effectiveness of a viral marketing campaign.

There are four '*Questions of Contagiousness*' which are essential in determining the level of contagiousness pre-implementation:

Did you heed the connectors? There are millions of consumers out there writing blogs, telling marketers exactly what they like and don't like and sharing their personal opinions with the world. Other consumers listen to these people who love to be at the forefront of any opinionated discussion. Therefore it is up to the marketer to find these consumers, listen to them, engage them and let them help you by becoming co-marketers (Brier, 2005).

Did you ask people you know? It is vital that marketers test their approach on people that they know, ask them what they think and most importantly see if they would be willing to pass on the message. As previously discussed, the more of a connector they are the more valuable their opinion. This demonstrates the strength of viral marketing – potential carriers (co-marketers) are not hard to find, they are all around us (Brier, 2005).

Is it honest? Nothing spreads faster and is more damaging to a marketing campaign than bad publicity. Therefore this question enforces the focus on being honest and up-front with consumers. This is because if you are pretending to be something you are not and you get caught, that negative publicity can be just as contagious as any positive viral campaign… possibly even more so. It is important for companies to remember that they are engaging with the consumers on a personal level and any dishonesty will certainly prove to be crippling for the campaign and possibly the company involved as well (Brier, 2005).

Did you have fun? One of the major factors regarding viral adoption is that the consumers must be engaged in the message. For this to be truly achieved it is advised that the marketers make the experience of creating the campaign as enjoyable as possible to allow the consumers to feed off their enjoyment and be attracted to the campaign, and thus distribute the message through a viral format. This aspect of viral marketing falls under the premise that if the marketer didn't enjoy creating the campaign then there is little chance that the consumer will enjoy experiencing the campaign and, hence, not distribute the message (Brier, 2005).

CONCLUSION

In terms of managerial implications for WOM strategy development, the research suggests that online promotion and communication techniques should ideally:

- be creative, interactive, emotive (e.g. humorous or shocking) and distinguishable
- be deemed compatible with the product's appropriateness in an online context
- maintain a predominant factor of entertainment
- refrain from being intrusive in any manner
- use incentives to attract consumers to pass along the message
- promote friendship generation or group membership
- encourage self-enhancement through assisting online social networkers in generating recognition from other users
- result in the formation of brand specific communities and company blogs which exhibit the qualities of honesty and transparency

ACKNOWLEDGMENT

Inspiration and some material for this chapter was drawn from a UCT Marketing Honours (2007) project by M. Coppin, N. Howarth and J. McKillop. The research programme was convened by the author of this article

REFERENCES

Beeching, P. (2006). How to win friends and influence people. *Market Leader* [online], 34.

Brier, N. (2005). Viral Marketing: Making the Right Connection. *Admap Magazine*, 465, October.

Brown, J. & Reingen, P. (1987). Social ties and word of mouth referral behaviour. *Journal of Consumer Research* [online]. [Accessed 12 March 2009].

Cant, M., Brink, A., & Brijball, S. (2006). *Consumer Behaviour* (1st ed.). Cape Town: Juta & Co.

Carroll, A., Barnes, S., Scornavacca, E. & Fletcher, K. (2005). Consumer perceptions and attitudes towards SMS advertising: recent evidence from New Zealand. *Journal of Consumer Psychology* [online] [Accessed 13 March 2009].

Carson, J. (2005) Word of mouth marketing: A New Mandate. *Admap Magazine*, 465, October

Corry, W., & Mundell, J. (2006) Demystifying blogs – embracing objective communication. Market Research Society Annual Conference [accessed on April 27, 2009]

Engel, J., Blackwell, R., & Miniard, P. (1993). *Consumer Behavior* (8th ed.). Fort Worth: Dryden Press.

Goldstuck, A. (2009). *Internet Access in South Africa 2008*. Johannesburg: World Wide Worx.

Gremler, D., Thorsten, H., Gwinner, K., & Walsh, G. (2004). Electronic word of mouth via consumer-opinion platforms: What motivates consumers to articulate themselves on the Internet? *Journal of Interactive Marketing, 18*(1).

Heinz, H., Gross, R., & Acquisti, A. (2005). *Proceedings of the 2005 ACM workshop on social network privacy* [online]. Retrieved March 11, 2009 from http://www.heinz.cmu.edu/~ acquisti/papers/privacy-facebook-acquisti-slides.ppt#584,61, Conclusions.htm

Kim, A. (2006). Community Building on the Web. We Media [online]. Retrieved May 3, 2009 from http://www.hypergene.net/wemed ia/weblog.php?id=P40

Kirby, J. (2004). Using "Word of Mouse" in Brand Marketing. *Admap Magazine*, 454.

Maslow, A. (1970). *Motivation and Personality* (2nd ed.). New Jersey: Pearson Education.

Mayzlin, D. (2006). Promotional Chat on the Internet. *INFORMS. Marketing Science, 25*(2). doi:10.1287/mksc.1050.0137

McCarthy, C. (2009). Facebook now twice as big as MySpace? Oh boy! [online]. Retrieved January 23, 2009 from http://news.cnet.com/8301-13577_3-10148855-36.html?part=rss&subj=news&tag=2547-1023_3-0-5

Modzelewski, F. (2000, September 11), Finding a Cure for Viral Marketing. *Direct Marketing News* [online][accessed on May 1, 2007]

Ozcan, K. (2004). *Consumer-To-Consumer Interactions In A Networked Society: Word of mouth theory, consumer experiences, and network dynamics* [online]. PhD dissertation. University of Michigan. Retrieved May 4, 2009 from http://www.insna.org/IN SNA/power_inf.html

Pastore, M. (2000). *The value of word of mouth* [online]. [Accessed on April 24, 2009]

Phelps, J., Lewis, R., Mobilio, L., Perry, D., & Raman, N. (2004). Viral marketing or electronic word of mouth advertising: Examining consumer responses and motivations to pass along e-mail. *Journal of Advertising Research, 44*(4).

Porter, L., & Golan, G. (2006). From subservient chickens to brawny men: A comparison of viral advertising to television advertising. *Journal of Interactive Advertising, 6*(2).

Raskin, R. (2006). Facebook faces its future. *Young Consumers* [online] [Accessed 09 March 2009].

Rodgers, S., & Thorson, E. (2000). The interactive advertising model: How users perceive and process online ads. *Journal of Interactive Advertising, 1*(1).

Runnacles, M. (2006). Branding in the internet bazaar- struggling for control in an uncontrollable world. *Market Leader*, 33.

Saunders, J. (2006). Growing up digitally: change drivers in marketing. *Market Leader*, 33.

Shirky, C. (2000, July 25). The Toughest Virus of All. *Business 2.0* [online] [accessed on May 3, 2009]

Sicilia, M., Ruiz, S., & Reynolds, N. (2006). Attitude formation online. *Journal of Consumer Psychology* [online]. [Accessed 13 March 2009].

Smith, J. (2006). Becoming a part of the community. *Admap Magazine*, October.

Solomon, M. (2004). *Consumer Behavior: buying, having and being* (6th ed.). New Jersey: Pearson Prentice Hall.

Steyer, A., Garcia-Bardidia, R., & Quester, P. (2006). Online discussion groups as social networks: An empirical investigation of word of mouth on the internet. *Journal of Interactive Advertising*, *6*(2).

Sundaram, D., Mitra, K., & Webster, C. (1998). Word of mouth communications: A motivational analysis. *Advances in Consumer Research* [online]. [Accessed 01 May 2009].

Vilpponen, A., Winter, S. & Sundqvist, S. (2006). Electronic WOM in Online Environments: Exploring Referral Network Structure and Adoption Behaviour. *Journal of Interactive Advertising* [online]. [Accessed 14 March 2009].

Welker, C. (2002). The paradigm of viral communication. *Information Services and Use* [online] [Accessed 11 March 2009]

Wikipedia. (2009). Facebook [online]. Retrieved October 2, 2009 from http://en.wikipedia.org/wiki/Facebook

Wolfe, J. (2006). Space race. The Advertiser, August, 36-39 [online]. Retrieved March 11, 2009 from http://www.ana.net

Chapter 13

Consumer Usage of Broadband Internet Services:
An Analysis of the Case of Portugal

Janice A. Hauge
University of North Texas, USA

Mark A. Jamison
University of Florida, USA

Mircea I. Marcu
University of Florida, USA

ABSTRACT

We analyze the intensity and patterns of use of fixed and mobile broadband consumers in Portugal. If usage across types of consumers is similar after controlling for individual characteristics identified to be important drivers of adoption, then it is more likely that consumers view mobile and fixed broadband as somewhat substitutable. Such a result is important for studies of broadband impacts; specifically, for discerning whether mobile broadband service will have a similar level of impact upon social and economic development as fixed broadband services have had. Results indicate that broadband uses are similar across fixed and mobile users, suggesting that the technologies are somewhat substitutable from customers' perspectives and raising the possibility of limited differential effects on innovation and other social goals. Results of interest include the characteristics of Internet users by technology, and differences of usage patterns reflected by individual characteristics.

INTRODUCTION

The issues of substitutability across broadband technologies and the context in which broadband penetration occurs are important for a variety of reasons. First, if the frequency and intensity of Internet use across types of consumers is similar after controlling for individual characteristics identified to be important drivers of adoption, then it is more likely that consumers view mobile and fixed broadband Internet access as somewhat substitutable. This is important for discerning whether mobile broadband service will have a similar level of impact upon social and economic development as fixed broadband services have had.

DOI: 10.4018/978-1-60960-011-2.ch013

Several empirical studies have linked the use and adoption of broadband technology to various measurements of both economic and job growth; however, such analyses have not been undertaken specifically with regard to mobile broadband diffusion. In part, this is due to data limitations that exist due to the relatively recent emergence of mobile broadband technology.

Additionally, as the goal of ubiquitous broadband adoption increasingly is proposed, policymakers will find it necessary to determine which demographic or socio-economic groups are least likely to be included in the achievement of this goal. To that end, a clear understanding of characteristics of individuals who access the Internet through various means, and such individuals' patterns of use can help to determine appropriate policies for reaching such individuals and for providing information on benefits of Internet applications likely to be important to them.

To add to studies focusing primarily on fixed broadband adoption, we include analyses of mobile and nomadic broadband adoption and consumer usage patterns. Specifically, we consider characteristics of broadband users by access type and relevant socio-economic and demographic variables, and we analyze usage patterns by access type, also including relevant explanatory variables.

BACKGROUND

The economic importance of broadband is well accepted, but there is much that we do not know about how various technologies of broadband delivery differ in their commercial viability, effectiveness, and value. In some countries, such as the United States, customers often can choose between fixed technologies (such as DSL, fiber to the home (FTTH), and cable), and can access wireless broadband through WiFi and third generation mobile (3G). In other countries, where cable television is less well developed, customers generally do not have the option of choosing cable for broadband access. Japan is emphasizing FTTH in its broadband policies, and also relies on DSL. There also are countries, such as Portugal, where wireless broadband is expanding rapidly. Whether customers view these various technologies as providing equivalent broadband access is important for public policy reasons: a country that is predominant in one broadband technology may be so because regulatory policies include technology biases. Such a country could be at a competitive disadvantage if its populace would find a different mix of technologies to be more productive economically and socially. On the other hand, if alternative broadband technologies are close substitutes, then a country could waste resources promoting a change in technology mix.

Another important consideration for understanding the roles of various broadband technologies is the context within which broadband penetration occurs. According to Schwab and Porter (2007), the most competitive economies in the future will be those that are innovation-driven. Broadband is instrumental in creating opportunities for innovation in a modern economy. A study by Van Ark and Inklaar (2005) supports this assertion, finding that the economies that have experienced the greatest economic impacts from information technologies are those that have leveraged those technologies to create entirely new products and ways of doing business. Still, broadband alone does not promote innovation; the Global Competitiveness Index includes numerous economic and legal features of a country that should be present if broadband is to reach its potential impact. [1]

Studies of fixed broadband adoption and deployment are numerous. Holt and Jamison (2008) provide an overview of various economic impact studies, but such studies of broadband's effects suffer from the problem of endogeneity (i.e., information that comes from the model cannot be used to explain the model). For example, if it is observed that economic development and broadband adoption are positively correlated, how does one know whether economic development results from broadband adoption, leads to broadband adoption, or both? The accepted wisdom is that broadband is both a cause and an effect of economic development.

In order to understand the benefits of broadband both for individual consumers and for achievement of social goals, it is critical to understand consumers' usage of broadband. This includes the degree to which consumers make choices based on technology preferences, and consumers' usage patterns and intensity. In other words, determining consumers' broadband applications and the amount of time spent using broadband enables policymakers to adopt the most appropriate methods of achieving social and economic goals.

Demand studies for broadband generally have found that demand is positively correlated with income, education, and greater use of other information technologies. For example, Crandall *et al.* (2002), Kridel *et al.* (2001), Garcia-Murillo (2005), and Prieger and Hu (2008) find that lower-income groups are less likely to subscribe to broadband than higher-income groups. Goldfarb and Prince (2008) concur in this finding and add that more highly educated consumers are more likely than less-educated consumers to purchase broadband.

With respect to race and ethnicity, Fairlie (2004) uses US Current Population Survey data to show that blacks and Hispanics are less likely to have a computer in the home than are members of other racial or ethnic groups; this of course means these households do not have broadband access in their homes. Other studies address race and ethnicity more directly. Prieger (2003) and Hu and Prieger (2008) find that race has no impact on suppliers' willingness to deploy DSL, once variations in income and other economic factors are considered. Leigh (2003) finds similar results. However, Flamm and Chaudhuri (2007), GAO (2006), and Prieger and Hu (2008) find that race impacts broadband adoption, perhaps because of differences in computer skills (Krueger, 2003) or network effects (Goolsbee and Klenow, 2002). These race impacts may result from factors that are correlated with race, but that are unobserved in the researchers' data or not fully captured in the statistical analysis even if the data are there. For example, Prieger and Hu (2008) discuss whether some blacks and Hispanics lack spare time to be online or find online content less valuable than do other racial groups.

In contrast to other studies, Goldfarb and Prince (2008) survey 18,439 Americans and find that, conditional on adoption, low-income, less-educated consumers spend more time online than their higher income, more educated counterparts, a result that is best explained by differences in the opportunity cost of leisure time according to the study.[2]

With respect to the impact of broadband, Holt and Jamison (2008) provide an overview of various studies, but such studies of broadband's effects suffer from the problem of endogeneity (i.e., information that comes from the model cannot be used to explain the model). For example, if it is observed that economic development and broadband adoption are positively correlated, it is impossible to know whether economic development results from broadband adoption, leads to broadband adoption, or both. The accepted wisdom is that broadband is both a cause and an effect of economic development. Clearly research has proven that advanced communications technologies have a significant economic impact across countries and increasingly so, as such advanced technologies are more rapidly deployed. Unfortunately, rarely does this research include usage or adoption patterns, and deployment clearly is an imperfect proxy for actual broadband use. For this reason, we focus on consumer usage patterns and intensity of broadband use in an effort to expand the existing literature to include choices based on broadband access technology.

BROADBAND USAGE IN PORTUGAL

Broadband access to the Internet was offered in Portugal through cable modem technology beginning in 1999. In 2000, the telecommunications industry was fully liberalized, and local loop unbundling was mandated in 2001. Subsequently, the telecommunications incumbent Portugal Telecom (PT) began offering broadband Internet access through ADSL, and currently offers broadband Internet access both through DSL and cable.[3]

Among EU Member States, Portugal has the highest ratio of fixed broadband subscribers using a provider other than the incumbent. However, Portugal also has one of the lowest growth rates in fixed broadband. In fact, by the broadband performance index developed, Portugal ranks poorly - in the fourth of five clusters. The 2008 ITIF Broadband Rankings listed Portugal as 18th, with a composite score only slightly higher than average (10.15 compared to an average of 10.00).[4] Currently Portugal residents increasingly use mobile broadband. Portugal makes intensive use of mobile communications services with 58 percent of voice traffic originating from mobile networks. The mobile market also is credited with exhibiting low churn and high customer loyalty.

Data[5]

For our analyses we draw upon data largely obtained from three surveys conducted in Portugal in 2006 and 2008.[6] The surveys were stratified by geographic regions of the country that correspond generally with the Nomenclature of Territorial Units for Statistics (NUTS II) geographical coding used by the European Union to indicate divisions of countries for statistical purposes. These regions include: Açores, Algarve, Centro Litoral, Grande Lisboa, Grande Porto, Interior, Madeira, and Norte Litoral.

Education levels vary across the regions, with the highest education levels being in the most urban and highest income areas (Grande Lisboa and Grande Porto), and the lowest education levels (i.e., illiterate) being in some of the lowest income regions (for example Madeira and Alentejo). The vast majority owns a television; however few have more than one car. On average, in 2006 about 40 percent of Portugal households accessed the Internet from their homes. Of those with home Internet access, 61 percent used dial-up Internet access but that proportion is shrinking. Norte Litoral, Grande Lisboa and Centro Litoral had greater home Internet access via cable than by phone.

A primary focus is to understand usage patterns and intensity of use among fixed, mobile, and nomadic consumers.[7] As prior literature has found education to be a significant determinant of broadband usage, we illustrate fixed, mobile, and nomadic usage by education level from 2008 survey data. This and other such delineations serve to frame the empirical models used in the analyses.

Analyses and Results[8]

Characteristics of Broadband Consumers

Prior research focuses on characteristics primarily of fixed broadband users. Including characteristics of fixed, mobile, and nomadic broadband users in Portugal provides evidence that across technologies, there is some variation in characteristics among users. We begin by describing the empirical models used in the analysis.

Empirical models for analyses of user characteristics employ utility models in which the utility of individuals from alternative modes of Internet access depends on characteristics of the individual or household, and on characteristics of the respective mode of Internet access.[9] Specifically, the multinomial logit model,[10] which is an extension of the logit regression model used to represent the choice between mutually exclusive options, is appropriate to use.[11]

Figure 1. Fixed, mobile, and monadic usage by education, 2008

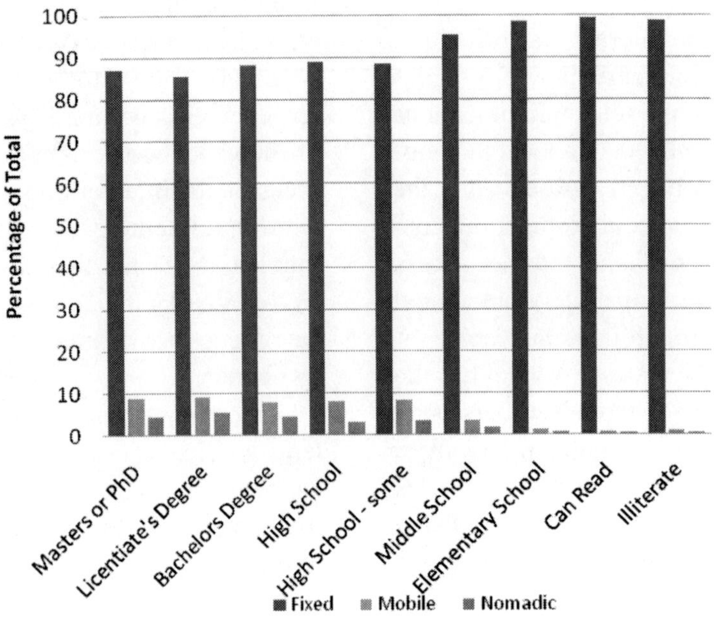

The dependent variable for the model focusing on characteristics of broadband users by technology is a binary (dummy) variable representing households' broadband choices. Explanatory variables include the age, education level, and employment status of the main income contributor, the number of people in the household, the number of children in the household, the occupation of the primary respondent, wealth, region, and habitat. Results of the multinomial logit estimations used to compare individuals' characteristics by the manner in which they access the Internet for the years 2006 and 2008 are reported in relative risk ratios (rather than coefficients); these represent the probability that respondents choose an alternative, such as mobile broadband, relative to the omitted (reference) alternative, which in this case is fixed broadband.

The results indicate that older persons and persons with lower education levels, lower incomes, and smaller households are the least likely to have Internet access of any form. Madeira is the region where people are least likely to have no Internet access. The relative probability of a wealthier respondent not having Internet access in the home is 0.36, meaning most people of higher income do have Internet access in their home.

The choice of narrowband versus fixed broadband follows a similar pattern, with older persons being more likely to have narrowband than fixed broadband. High school graduates were 2.7 times more likely to have no Internet access than to have fixed broadband compared to university graduates; compared to respondents with a university degree, those who graduated from high school were approximately half as likely to have mobile instead of fixed broadband (relative probability 0.55). Households with at least two cars were 1.73 times more likely to get mobile broadband instead of fixed broadband than households with a single car or without a car.

In general, having at least one child appears to make it more likely that a household will not have home Internet access. Similarly, larger households are less likely to have Internet access at home, but household size does not affect significantly the choice between different types of Internet access.[12] Initially the probability of not having Internet at home relative to that of having fixed broadband decreases with family size, but that probability increases for families with at least five individuals;[13] the minimum probability of not having home Internet access instead of having fixed broadband Internet access is for a household size between four and five members[14].

The results also indicate that income (using a proxy by a dummy variable for two or more cars and owning a dishwasher) and education are the primary factors related to the choice between fixed and mobile broadband.

Finding differences across end-users who choose multiple types of access is a valuable goal to help determine the degree to which such access types are substitutes. In this respect there exists limited support for the notion that multiple access types are indeed substitutes: of approximately 8,600 respondents in the 2006 BCS survey, only 151 (1.76 percent) report having more than one type of home Internet access, and 80 percent of those had only two types. Still, it is useful to attempt to ascertain reasons for multiple modes of access to better understand substitutability across such modes, as these respondents indicate that various access types are not perfect substitutes. It is possible that fixed broadband might be deemed more valuable for those with a fixed location who want higher bandwidth, while mobile may be more useful for those who also are more mobile.

Nomadic users often are also fixed broadband subscribers or mobile broadband subscribers. This results from the growth of home WiFi networks and the converged device market, which in Europe has grown from about 11 million units sold in 2005 to more than 15 million in 2006. Furthermore, the converged device customer profile has been changing in Europe. As recently as 2005, most high-end smartphones were used primarily for business. Now consumers are using these convergence devices, such as the BlackBerry Pearl or Palm Treo 750, for web browsing, checking personal e-mail, managing family schedules, and the like. This has happened in part because of a decrease in prices: T-Mobile UK has given the BlackBerry Pearl away for free with some wireless calling plans and, in 2008, T-Mobile Germany launched the iPhone in packages priced for the residential user.[15]

Figure 2 illustrates the education and income level (by proxy) of those respondents reporting two or more types of Internet access in their homes in 2006. Those reporting multiple access types generally have a higher level of education, and also greater income. While income is expected to be a significant predictor of multiple modes of access, the link between multiple modes of access and education is not as predictable. Results indicate that age falls steadily with an increase in the number of types of access (from an average age of 36.23 years to 31.29).[16] Regional variations exist as well; noticeably, Madeira has about the same number of homes with two, three, or more types of access and Algarve has only one type. Being island regions, the Açores and Madeira might be expected to be similar; however, they do not appear to have the same options available. It remains to be determined why individuals need more than one access type; in other words, for a small subset of our sample database it is clear that types of access are not perfect substitutes; there must be value in different forms of access.

Analysis of Usage Patterns

The value of understanding usage patterns lies in finding potential differences among consumers for various broadband technologies. Usage is an indirect indicator of the value consumers place on their Internet access; that value is expected to vary across types of consumers (i.e., across ages and education levels) as well as across types of applications (i.e., filing taxes versus downloading music). In aggregate, hours of use are approximately 18 hours per week for fixed broadband and 21 hours per week for mobile broadband. The most frequently reported number of hours of use for fixed users was 10 hours per week (18

Figure 2. Characteristics of respondents with multiple access types, 2006

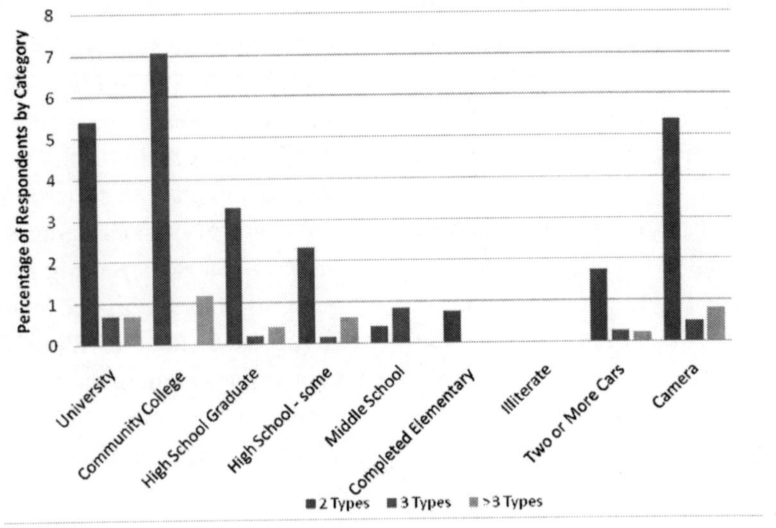

percent of respondents) followed by 20 hours per week (17 percent of respondents). 20 hours per week was the most frequently reported response of mobile users (22 percent) followed by 5 hours and 10 hours per week (18 percent and 16 percent, respectively).

There are different ways of measuring usage. We focus primarily on hours and types of usage; however, access bandwidth also is important. For example, respondents were asked which package offered by their provider they had chosen. In some instances the package information obtained in the survey included bandwidth (e.g., Sapo offers a 2MB, 4MB, and 8MB package, among others; TvTel offers 256 1GB, 256 3GB, 256 SL and 512 SL; Via Networks offers 512K, 768K, and 1024K, among others), but in other instances only the name of the plan was given, not the features.

Surveys show that hours of use and types of usage are fairly consistent across age groups and regions in Portugal, even though variations are statistically significant in econometric models (details of which are below). Similarly, types of usage are similar across access types. Mobile access is slightly greater for financial applications than fixed access is, and fixed acces is slightly greater for entertainmet, news, and downloading of music, games, and movies. This might occur because mobile broadband is associated with higher-income users than is fixed broadband. Figure 3 illustrates these results.

Empirical models analyzing the difference in usage between narrowband and broadband users consider both the impact of fixed versus mobile access on hours of use and the purpose of use (making purchases, bank transfers, filing taxes), after controlling for individual/household characteristics. This suggests the extent to which narrowband and broadband are substitutes, and similarly the extent to which fixed and mobile broadband are used for similar purposes.

The empirical results indicate that older respondents tend to have lower hours of use; those age 25 to 44 have the highest usage among various age groups.[17] Wealthier respondents have higher hours of use. More densely populated and urban areas have higher usage as well. The more densely populated areas of Portugal also have higher per capita GDP and the marginal cost of providing broadband should be lower in more densely populated areas than in less densely populated

Figure 3. Patterns of us by access type, 2006

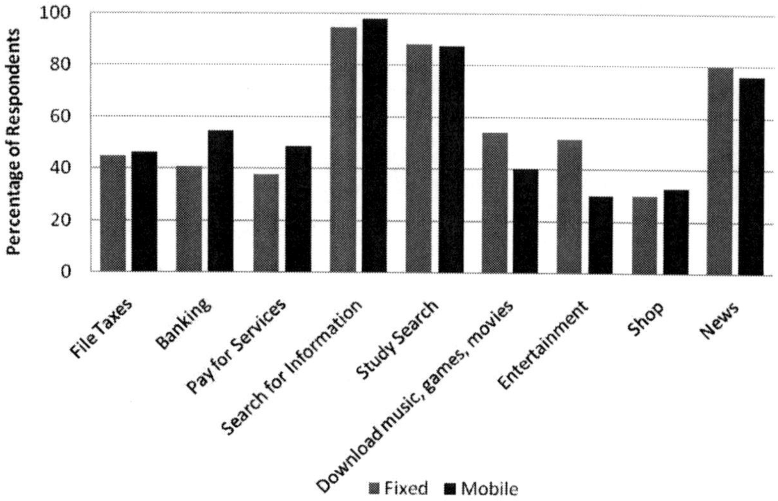

areas. Finally, there exists a negative relationship between satisfaction with speed and hours of use.

Logistic regressions suggest that mobile subscribers are not different from other broadband subscribers with respect to financial applications; only narrowband proved to be significant in terms of access mode. Education remains a primary indicator of Internet usage. There are variations across regions once other socio-economic characteristics are controlled for, for example, almost all regions, Madeira being the exception, have lower propensity to use Internet financial services than Grande Lisboa (the reference region).

With respect to non-financial uses, in general a few trends stand out. First, age is significant for entertainment uses (those reporting downloading music, games, and videos, as well as entertainment); younger users consistently access the Internet for non-financial uses at a greater level than older users. Also, education is significant for news-related uses; users with more years of education access the Internet for news at a greater level than less-educated users. As might be expected, the length of time the respondent has had home Internet access is significantly correlated with online shopping, perhaps indicating that those more familiar with Internet use become more trusting of Internet shopping opportunities. Mobile broadband users are less likely than DSL broadband users to use broadband for most of the non-financial uses.

Surveys indicated that fixed service customers are more likely to be frequent broadband users than other access customers, but nomadic customers may be different than fixed or mobile customers. Both fixed and mobile show similar usage profiles: between 40 and 50 percent of the customers of each access type use broadband several times each day, a little more than 25 percent use it at least once each day, and less than 10 percent use it less than once per week. The nomadic customers have a flatter distribution: less than 40 percent use it several times a day, more than 30 percent use it once per day, and over 20 percent use it twice per week. This is consistent with what we might expect of nomadic users: at least some of them would have to physically go to a hotspot to access the Internet and this inconvenience would lower the number of times they use broadband relative to users that have it available at home (i.e., fixed and mobile subscribers) or with them as they move about (i.e., mobile subscribers).[18]

Portions of our analyses considered differences between users of fixed, mobile, and nomadic broadband. While information on nomadic broadband users is limited, there appear to be few differences between fixed, mobile, and nomadic broadband customers in terms of hours of use, implying that at least fixed and mobile broadband are viewed as substitutes by many customers. The differences shown using both 2006 BCS and 2008 ICSCE datasets include the following statistically significant findings:

- Mobile broadband users are slightly less satisfied than fixed and nomadic with their service, but fixed and nomadic broadband customers have about the same degree of satisfaction.
- Customers of the three types of access – fixed, mobile, and nomadic – are similar in their satisfaction with their service speeds.
- Most customers do not switch providers.
- Mobile broadband customers are heavier users in terms of hours of use than are customers of fixed broadband.
- Mobile broadband subscription is higher relative to fixed in older age groups.
- Most regions are similar in their use of fixed and mobile broadband, but Açores and Madeira, both island regions, show relatively more mobile broadband usage than the other regions.
- Most respondents who are accessing the Internet from home multiple times per day are doing so via fixed communications.

FUTURE RESEARCH DIRECTIONS

The surveys in Portugal serve to provide important information on customers and usage. Additional data on prices for narrowband and broadband (both fixed and mobile), as well as nomadic, would allow the determination of the elasticity between technologies. Speed information and volume of traffic both would improve the strength of the models employed. Actual expenses per month for service bundles including broadband, television, fixed telephony, mobile telephony, as well as actual usage of each service for fixed and mobile (number of hours, not just number of occurrences) are needed. It is difficult to analyze bundles without detailed knowledge about the package, but since this is how customers are purchasing the services, these details are important to gather.

Several important research questions could drive future directions, including those relating to nomadic use, market competition, and resolving cause and effect questions. Regarding nomadic usage, the analyses indicate that nomadic subscriptions in Portugal follow a different pattern than does the development of hotspots. This implies that hotspots are a broader market than nomadic Portuguese customers. The location choices and technology impacts of hotspots should be explored, as should the types of customers who find nomadic use to be a substitute for fixed or mobile broadband descriptions. Such analyses could inform important public policy questions about regulatory oversight (if any) that might be appropriate for nomadic subscriptions and hotspots. This could be particularly important in countries that are growing rapidly and have large geographic space.

Market performance is an important question. Further work on price elasticities and cross elasticities would be valuable in the study of a country's markets and the development of appropriate regulatory policies.

As a matter of important developments for research techniques, the choice of broadband plan and usage are endogenous, and the hours of use of different services are likely correlated. Future work should examine carefully the cause and effect relationships, paying particular attention to the interdependencies.

Finally, it is understood that broadband has economic impacts. Less well understood are the societal impacts, more specifically, the impacts of citizen participation in government, educational achievement, and other societal engagements. For example, in Brazil children in low-income areas are given computer training, after which the training facility sets them up with their first job. As another example, in some African countries, mobile communications are used to monitor elections.[19]

CONCLUSION

In the literature to date there remains a great deal that is unknown with respect to how various technologies of broadband delivery differ in their commercial viability, effectiveness, and value. Currently available data have enabled us to provide a wealth of information not previously available on the characteristics of customers and their associated usage patterns and the degree to which customers view different types of Internet access to be substitutes.

The analysis of usage objectives and patterns of use determined that fixed and mobile customers are remarkably similar. Results also are in agreement with other studies that have found that more educated, wealthier, and younger customers tend to have greater usage hours. There also exists evidence that customers use the Internet (whether it is via fixed, mobile, or nomadic access) in similar ways. Slight differences were found in usage patterns by customer characteristics and also across geographical regions; however, these differences are as expected. For example, in the autonomous regions of Açores and Madeira, one might expect mobile Internet access to be more prevalent relative to other regions due simply to the terrain and cost of providing fixed versus mobile broadband.

While this analysis is specific to Portugal, there is no reason to assume that findings for Portugal are not illustrative of a general trend across other similarly developed countries. This makes the conclusions of these analyses of the deployment, adoption and use of fixed and mobile broadband access useful in a much broader context.

REFERENCES

ANACOM (2009). Consumo de Banda Larga survey (BCS, 2006).

ANACOM (2009). Inquérito ao Consumo das Comunicações Electrónicas (ICSCE, 2008).

ANACOM (2009). Índice Europeu da Satisfação de Consumidor (ECSI, 2008).

Crandall, R., Lehr, W., & Litan, R. (2007). The Effects of Broadband Deployment on Output and Employment: A Cross-Sectional Analysis of U.S. Data. *Issues in Economic Policy*, *6*, 1–34.

Crandall, R., Sidak, G., & Singer, H. (2002). The empirical case against asymmetric regulation of broadband Internet access. *Berkeley Law and Technology Journal*, *17*(1), 953–987.

Cummings, R., Harrison, G., & Rutström, E. (1995). Homegrown Values and Hypothetical Surveys: Is the Dichotomous Choice Approach Incentive Compatible? *The American Economic Review*, *85*(1), 260–266.

Fairlie, R. W. (2004). Race and the digital divide. *Contributions to Economic Analysis & Policy*, *3*(1).

Flamm, K., & Chaudhuri, A. (2007). An analysis of the determinants of broadband access. *Telecommunications Policy*, *31*, 312–326. doi:10.1016/j.telpol.2007.05.006

Garcia-Murillo, M. (2005). International Broadband Deployment: The Impact of Unbundling. *Communications & Strategies*, *57*, 83–108.

General Accountability Office (2006). *Broadband Deployment Is Extensive throughout the United States, but It Is Difficult to Assess the Extent of Deployment Gaps in Rural Areas*, GAO-06-426, May.

Goldfarb, A., & Prince, J. (2008). Internet adoption and usage patterns are different: Implications for the digital divide. *Information Economics and Policy, 20*, 2–15. doi:10.1016/j.infoecopol.2007.05.001

Goolsbee, A., & Klenow, P. J. (2002). Evidence on learning and network externalities in the diffusion of home computers. *The Journal of Law & Economics, 45*, 317–343. doi:10.1086/344399

Holt, L., & Jamison, M. A. (2008). Broadband and contributions to economic growth: Lessons from the U.S. Experience. PURC Working Paper, Department of Economics, University of Florida.

Hu, W., & Prieger, J. (2008). The Timing of Broadband Provision: The Role of Competition and Demographics. In Dwivedi, Y. (Eds.), *Handbook of Research on Global Diffusion of Broadband Data Transmission* (pp. 241–259). Hershey, PA: IGI Global.

Kridel, D., Rappoport, P., & Taylor, L. (2001). The demand for high-speed access to the Internet: the case of cable modems. In Loomis, D., & Taylor, L. (Eds.), *Forecasting the Internet: Understanding the Explosive Growth of Data Communications* (pp. 11–22). Boston, MA: Kluwer.

Krueger, A. (2003). The digital divide in educating African-American students and Workers. In Conrad, C. (Ed.), *Building Skills for Black Workers* (pp. 51–76). New York: University Press of America.

Pereira, P., & Ribeiro, T. (2006). The Impact on broadband access to the Internet of the dual ownership of telephone and cable networks. NET Institute, working paper #06-10.

Prieger, J. (2003). The supply side of the digital divide: Is there equal availability in the broadband internet access market? *Economic Inquiry, 41*(2), 346–363. doi:10.1093/ei/cbg013

Prieger, J., & Hu, W. (2008). The Broadband Digital Divide and the Nexus of Race, Competition, and Quality. *Information Economics and Policy, 20*(2), 150–167. doi:10.1016/j.infoecopol.2008.01.001

Schwab, K., & Porter, M. (2007). *The Global Competitiveness Report 2007-2008*. Geneva, Switzerland: World Economic Forum.

Standard and Poor's. (2009). *Telecommunications: Europe*. New York: Standard and Poor's Equity Research Services.

Van Ark, B., & Inklaar, R. (2005). *Catching Up or Getting Stuck? Europe's Troubles to Exploit ICT's Productivity Potential*. Groningen Growth and Development Centre, University of Groningen.

World Economic Forum. (2008). The Global Competitiveness Report 2008-2009. Geneva, Switzerland.

ADDITIONAL READING

Aron, D., & Burnstein, D. (2003). Broadband Adoption in the United States: An Empirical Analysis. In Shampine, A. L. (Ed.), *Down to the wire: Studies in the diffusion and regulation of telecommunications technologies* (pp. 119–138). Haupaugge, NY: Nova Science Publishers.

Bauer, J., Kim, J., & Wildman, S. (2003). *Broadband uptake in OECD countries: Policy lessons and unexplained patterns.* Paper prepared for presentation at the European Regional Conference of the International Telecommunications Society, Helsinki, Finland.

Broadband Stakeholders Group (2004). The Impact of Broadband-Enabled ICT, Content, Applications and Services on the UK Economy and Society to 2010. BSG Briefing Paper 27.

Brown, S., & Sibley, D. (1986). *The Theory of Public Utility Pricing*. Cambridge University Press.

Cardona, M., Schwarz, A., Yurtoglu, B., & Zulehner, C. (2007). Demand estimation and market definition for broadband Internet services. Retrieved February 16, 2010, from http://ssrn.com/abstract=1081261

A Case Study in the Shippagan Area, a Rural Zone in Atlantic Canada. *Journal of Information, Information Technology, and Organization, 2*, 79-94.

Cava-Ferreruela, I., & Alabau-Munoz, A. (2006). Broadband Policy Assessment: A Cross-national Empirical Analysis. *Telecommunications Policy, 30*(8), 445–463. doi:10.1016/j.telpol.2005.12.002

Correa, L. (2006). The Economic Impact of Telecommunications Diffusion on UK Productivity Growth. *Information Economics and Policy, 18*(4), 385–404. doi:10.1016/j.infoecopol.2006.04.001

Costa, C. (2009). Fixed Broadband Adoption Factors and Implications Towards the Promotion of Information Society. Working paper available from the author.

Denni, M., & Gruber, H. (2005). *The Diffusion of Broadband Telecommunications: The Role of Competition*. Paper presented at International Telecommunication Conference, Pontevedra, Spain.

Distaso, W., Lupi, P., & Manenti, F. (2006). Platform competition and broadband uptake: theory and empirical evidence from the European Union. *Information Economics and Policy, 18*, 87–106. doi:10.1016/j.infoecopol.2005.07.002

econrsss.anu.edu.au/~aleigh/pdf/ Digital%20 divide%20update.pdf.

Edwards, G., & Waverman, L. (2006). The Effects of Public Ownership and Regulatory Independence on Regulatory Outcomes. *Journal of Regulatory Economics, 29*(1), 23–67. doi:10.1007/s11149-005-5125-x

Entner, R. (2008). *The Increasingly Important Impact of Wireless Broadband Technology and Services on the U.S. Economy: A Follow up to the 2005 Ovum Report on the Impact of the US Wireless Telecom Industry on the US Economy*. Ovum.

Ford, G., & Koutsky, T. (2005). Broadband and Economic Development: A Municipal Case Study from Florida. *Applied Economic Studies*, 1-15. See also Ford, G. & Koutsky, T. (2005). Broadband and Economic Development: A Municipal Case Study from Florida. *Review of Urban and Regional Development Studies, 17*(3), 216–229. doi:10.1111/j.1467-940X.2005.00107.x

Ford, G., Koutsky, T. & Spiwak, L. (2008). The Broadband Efficiency Index: What Really Drives Broadband Adoption across the OECD? Phoenix Center Policy Paper Number 33.

Gillett, S., Lehr, W., Osorio, C., & Sirbu, M. (2006). *Measuring Broadband's Economic Impact. Final report, Prepared for the U.* S. Department of Commerce, Economic Development Administration.

Gutiérrez, L. H. (2003). The Effect of Endogenous Regulation on Telecommunications Expansion and Efficiency in Latin America. *Journal of Regulatory Economics, 23*(3), 257–286. doi:10.1023/A:1023412226826

Ida, M., & Kuroda, Y. (2006). Discrete choice analysis of demand for broadband in Japan. *Journal of Regulatory Economics, 29*(1), 5–22. doi:10.1007/s11149-005-5124-y

Insurance, Publishing, and Health Care. *IEEE Journal on Selected Areas in Communications, 10*(9), 1369–1381.

Jamison, M. A. (2004). Effects of Prices for Local Network Interconnection on Market Structure in the U.S. In Bohlin, E., Levin, S., Sung, N., & Yoon, C. (Eds.), *Global Economy and Digital Society* (pp. 301–320). Amsterdam: Elsevier Science.

Jamison, M. A. (2007). The 1st Academic Conference on the Auspicious Occasion of His Majesty the King's 80th Birthday Anniversary: The Importance of Telecommunications Development. In *NTC Annual Review 2007* (Vol. 1, pp. 58-87). Bangkok: National Telecommunications Commission of Thailand.

Jamison, M. A. (2008). Methods for Increasing Competition in Telecommunications Markets. Working Paper, Public Utility Research Center, University of Florida.

Jorgenson, D. (2001). Information Technology and the U.S. Economy. *The American Economic Review*, *91*(1), 1–32. doi:10.1257/aer.91.1.1

Jorgenson, D., & Vu, K. (2007). Information Technology and the World Growth Resurgence. *German Economic Review*, *8*(2), 125–145. doi:10.1111/j.1468-0475.2007.00401.x

Lee, S. & Marcu, M. (2008). An Empirical Analysis of Fixed and Mobile Broadband Diffusion. *Under review and available from the authors at mircea@ufl.edu.*

Leigh, A. M. (2003). Digital divide and broadband divide – some multiple regression results. Unpublished manuscript. Retrieved February 16, 2010 from

OECD Information and Communication Technologies, OECD Communication Outlook 2007. (2007).

Pereira, P., & Ribeiro, T. (2006). The Impact on broadband access to the Internet of the dual ownership of telephone and cable networks. NET Institute, working paper #06-10.

Rappoport, P., Kridel, D., & Taylor, L. (2002). The Demand for Broadband: Access, Content, and the Value of Time. In Crandall, R. W., & Alleman, J. H. (Eds.), *Broadband: Should We Regulate High-Speed Internet Access?* Washington, DC: The Brookings Institution.

Rappoport, P., Kridel, D., Taylor, L., & Alleman, J. (2001). Residential Demand for Access to the Internet. In Madden, G. (Ed.), *Emerging Telecommunications Networks: The International Handbook of Telecommunications Economics* (*Vol. 2*, pp. 1–20). Cheltenham, UK: Edward Elgar Publishers.

Röller, L., & Waverman, L. (2001). Telecommunications Infrastructure and Economic Development: A Simultaneous Approach. *The American Economic Review*, *91*(4), 909–923. doi:10.1257/aer.91.4.909

Selouani, S. & Hamam, H. (2007). Social Impact of Broadband Internet:

Shideler, D., Badasyan, N., & Taylor, L. (2007). The Economic Impact of Broadband Deployment in Kentucky. Federal Reserve Bank of St. Louis. *Regional Economic Development, 3*(2), 88- 118.

Stanton, L. J. (2004, January). *Factors Influencing the Adoption of Residential Broadband Connections to the Internet*. Paper presented at the 37th Hawaii International Conference on System Sciences, Manoa, Hawaii.

Underwood, J., Ault, A., Banyard, P., Bird, K., Dillon, G., Hayes, M., et al. (2005). *The Impact of Broadband in Schools*. Final project report for Becta, Coventry.

Underwood, J., Ault, A., Banyard, P., Dillon, G., Durbin, C., Golland, D., et al. (2004). *Connecting with Broadband: Evidence from the Field*. Final project report for Becta, Coventry.

Varian, H. (2002). The Demand for Bandwidth: Evidence from the INDEX Project. In Crandall, R. W., & Alleman, J. H. (Eds.), *Broadband: Should We Regulate High-Speed Internet Access?* (pp. 39–56). Washington, DC: The Brookings Institution.

Wakefield, T., McNally, D., Bowler, D., & Mayne, A. (2007). *Introduction to Mobile Communications: Technology, Services, Markets*. Boca Raton, Florida: Auerbach Publications.

Wallsten, S. (2001). An Econometric Analysis of Telecom Competition, Privatization, and Regulation in Africa and Latin America. *The Journal of Industrial Economics*, *49*(1), 1–19. doi:10.1111/1467-6451.00135

Wallsten, S. (2008). *Understanding International Broadband Comparisons*. Technology Policy Institute.

Ward, M. (1995). *Measurements of Market Power in Long Distance Telecommunications*. Federal Trade Commission, Bureau of Economics Staff Report.

Waverman, L., Meschi, M., & Fuss, M. (2005). The Impact of Telecoms on Economic Growth in Developing Countries. *The Vodafone Policy Paper Series*, *3*, 10–23.

Wright, D. (1992). Strategic Impact of Broadband Telecommunications in

KEY TERMS AND DEFINITIONS

Fixed Broadband: Method of accessing the Internet via modem or ISDN, ADSL or other XDSL access, or cable.

Internet Access Substitutability: the degree to which consumers are willing to trade one form of Internet access for another form, i.e., fixed broadband Internet for mobile broadband Internet.

Internet Usage Patterns: Internet consumers' typical uses of various applications, including financial uses, entertainment uses, and research, among others.

Mobile Broadband: Method of accessing the Internet via mobile phone or PDA with broadband Internet connection, or data transmission card.

Multinomial Logit Model: Empirical model that requires the distribution of the random error terms to be independent and identical over the alternatives.

Nomadic Broadband: Wireless access whereby the end-user can move among different access locations, such as hot spots, using a method other than one classified as mobile broadband.

Relative Risk Ratio: Empirical result that is reported rather than a standard coefficient. The ratio represents the probability that respondents choose an alternative relative to the omitted (reference) alternative, i.e., choosing mobile broadband relative to fixed broadband.

ENDNOTES

1 The Global Competitiveness Report is published annually by the World Economic Forum. The rankings are calculated from publicly available data and from an annual survey conducted by the World Economic Forum. The number of countries ranked varies based on available survey data within countries. The report is available at http://www.weforum.org/en/initiatives/gcp/Global%20Competitiveness%20Report/index.htm. According to the most recent report, the main purpose of the ranking and report is to improve the understanding of the key factors that determine economic growth, and to explain "why some countries are much more successful than others in raising income levels and opportunities for their respective populations, offering policymakers and business leaders an important tool in the formulation of improved economic policies and institutional reforms."(page xi).

2 Our findings indicate that broadband usage is increasing in income and education. This contrasts with the Goldfarb and Prince study, but does not contradict it. Their finding relates to all Internet usage and shows that

time is less valuable for lower income and less educated households. This time preference could lead lower income, less educated customers to use more dialup Internet than broadband, leaving open the possibility that once a customer has adopted broadband, the customer's usage might be positively correlated with income and education.

3 Pereira and Ribeiro (2006).

4 ITIF 2008.

5 For detailed information on data used, analyses, and results, please see the full paper, available at www.purc.ufl.edu.

6 Specifically, data are from the following surveys: 2006 Consumo de Banda Larga survey (broadband consumption survey, abbreviated as BCS), 2008 Inquérito ao Consumo das Comunicações Electrónicas survey (survey of consumption of electronic communication services, abbreviated as ICSCE), and 2008 Índice Europeu da Satisfação de Consumidor (European consumer satisfaction index, abbreviated as ECSI). The surveys differ and therefore cannot be combined for empirical analysis (i.e., data is cross-sectional). Information on the manner in which the surveys were given and the households selected is available from the authors upon request.

7 Nomadic users are characterized based on their responses to 2008 ICSCE survey question Q.93, "What type of Internet connection do you use at home?" Responses were as follows: modem or ISDN (1), ADSL or other XDSL access (2), cable (3), mobile phone or PDA with broadband Internet connection (4), phone connected to Internet through narrowband (5), broadband wireless connections other than mobile phone and PDA (6), data transmission cards (7), and other or do not know, which also were coded responses. By these categories, respondents answering (6) were considered nomadic users (those answering 4 and 7 were considered mobile). The frequency of responses is available upon request.

8 We note at the outset of our analysis that some of the data represents subjective opinions and impressions of the respondents, including statements about what they intend to do. For example, in the next section we examine respondents' statements about their intent to change service providers. Survey respondents' answers to such questions can be imprecise. Cummings et al. (1995) find that when survey respondents are asked whether a product is worth a particular price, more respondents will say "yes" if they are told they have no obligation to buy the product at the stated price than if there is an obligation to purchase at the price. In the case of the survey data we use, respondents' statements concerning their satisfaction, usage and uses, and intent to switch providers might be imprecise. If the errors in their answers are random, then the effect on our research is to decrease the confidence we can place in our statistical results leading us to understate the validity of our findings. If the errors are systematic, in other words, if respondents consistently understate usage, then the effect on our research is to bias our results either up or down, depending on the direction of error. Since we cannot know whether respondents made errors in their answers and, if they did, the direction of those errors, we cannot do better than report our results with this caveat.

9 Utility also depends on a random disturbance that has extreme value distribution of type I, which gives rise to logit models of choice between alternatives.

10 A multinomial logit is appropriate for any binary dependent variable y_i and a continuous independent variable x_i, $Pr(y_i=1)=F(x_i' \text{ b})$ where, as before, b is a vector of parameters to be estimated and F is the logistic cumulative distribution function.

11 The multinomial logit requires the distribution of the random error terms to be independent and identical over the alternatives. This can produce biased estimates if the cross-elasticities between all pairs of alternatives are identical. The major weakness of the multinomial logit is that the choice between alternatives (i.e., DSL versus cable) depends solely on the characteristics of those alternatives being compared, excluding the characteristics of any other alternatives possible.

12 It is unclear from the data why the presence of children and larger household size decreases the probability of having Internet access given the common understanding that Internet access is useful for education and entertainment. We note that while "study search", which would appear to be an educational use, is a prominent use of the Internet in Portugal, general information searching and obtaining news are comparable and might not be related to children or household size. This issue would appear to require further study to find an answer in which we could place more confidence.

13 The actual coefficients are -0.7594721*dc_3+0.0799416*dc_3^2.

14 The minimum is 0.7594721/(2*0.0799416)=4.75 household members.

15 IDC, as cited by Standard and Poor's (2009).

16 Result is not shown in the figure.

17 Results are provided in the Appendix, section 1b. In the 2008 ICSCE survey we do not have a question asking hours of use; Q92 asks "How often do you usually use Internet in your home?" with possible responses of: several times a day, once a day, two to three times a week, and less than once a week. The 2008 estimation is therefore an ordered logit instead of a negative binomial model as used for the 2006 analysis.

18 Respondents are asked what type of home Internet access they use, and those responding "nomadic" may therefore interpret this question as asking what type of Internet access they use either most frequently for personal reasons, or access other than at and for work.

19 For more on these and other examples of societal impacts, see Jamison (2007).

Section 4

Technological Advances and its Impact on Emerging Applications

Chapter 14
Evolution in Broadband Technology and Future of Wireless Broadband

Banani Nandi
AT&T Shannon Laboratories[1], USA

Ganesh K. Subramaniam
AT&T Shannon Laboratories, USA

ABSTRACT

In the last two decades, the availability of worldwide high speed Internet networks together with progressive deployment and adoption of broadband connections to Internet has significantly enhanced the ability to transmit video, audio and voice through the same channel with high performance. In addition to that, since the beginning of this millennium, there has been a rapid growth in diffusion of wireless broadband. In this respect, mobile broadband is gaining popularity lately in both developing as well as developed countries. The main reason for this growth in wireless broadband lies in its ability to serve remote areas at relatively lower cost than landline and its potential to transmit information seamlessly from anywhere in the world. The purpose of this chapter is to examine this changing diffusion behavior of fixed versus mobile broadband technology and highlight the future possible path of adoption for these technologies by users around the world.

INTRODUCTION

Explosive innovation in information technology and rapid expansion of mobile communications are changing the way we access and transmit information, bringing significant efficiency gain in our socio-economic activities. In the last two decades, availability of worldwide high speed Internet network together with progressive deployment and penetration of broadband (BB) connection to Internet has significantly enhanced the capacities of transmitting video, audio and voice through the same channel with high performance. However, in the early part of last decade, BB connection was limited only to fixed access technology. Since then, the rapid growth of mobile devices for both business and personal use has significantly enhanced the demand for wireless multimedia services such as data, voice and video.

DOI: 10.4018/978-1-60960-011-2.ch014

This has driven the need for development of new wireless standard and deployment of various types of wireless BB connections. Thus, in the new era, with the increasing speed and stability of mobile broadband connections, another revolutionary change is coming in the communications process which will facilitate access to and transmit of information from anywhere in the world by using a variety of exciting new mobile devices[2].

In last two decades, the use of wireless voice telephone communications has significantly increased in developed as well as in developing countries. Also, in recent years, with the increased capacity and speeds of wireless transmission and introduction of many new mobile devices with data access capability, people are increasing their use of this technology for their data communication needs. This change in behavior has created the demand for wireless access connections with higher bandwidth. Moreover, due to its cost advantage of deployment, this technology is not only becoming the communications platform for provisioning broadband access to homes in the urban areas but also a viable solution for providing services for homes in both remote and underserved areas. The purpose of this chapter is to study this evolving change in broadband technology and the role the wireless broadband access technology plays in this evolution[3]. More specifically, the object is to study the current trends in wireless broadband landscape and assess where the technology is heading as a whole.

Finally the aim of this chapter is to analyze the observed trend in the application of various broadband technologies by using historical data from the US and the rest of the world. For the BB adoption data, we use functional data analysis (Ramsay and Silverman, 1997) to estimate the velocity and acceleration associated with BB penetration in various countries around the world. In addition to that, this study formulates a simple time series model to provide a directional forecast for the future trend in the deployment and adoption of fixed as well as mobile broadband technology.

The layout of the chapter is as follows. The second section introduces the characteristics of various broadband access technologies and describes the benefits associated with their adoption. A brief description of the data is given in section III. In the subsequent section (IV), country specific differences in diffusion of fixed as well as mobile broadband technology are analyzed while taking into account the underlying dynamics in rate of growth of adoption. In section V, we study the trend in growth of fixed and mobile broadband subscribers and compare that with observed subscription behavior in landline versus wireless voice telephone service. Based on historical data for broadband adoption, we develop a simple model to predict the future trend in broadband adoption behavior for selected countries and the results are presented in section VI. The summary and conclusions are provided in the final section..

BROADBAND ACCESS AND ITS BENEFITS

Definition and Characteristics of Broadband Access

In operational terms, BB refers to a set of electronic communications technology solutions. The primary features that distinguishes this technology from other competitive transmission technologies include its relatively high bandwidth, always on functionality [4] and capability for high speed information transmission in both directions: downstream from Internet to the user's computer and upstream from user's computer to the Internet. An integrated view of these features, however, does not suggest a specific speed or a specific service. In fact there is no universally accepted standard regarding the speed of transmission that characterizes BB. Instead a number of speed transmission standards have been proposed by various organizations around the world. The Recommendation 1.113 of the ITU Standardization Sector (ITU-T) defines BB as a transmission capacity that is faster than primary rate of ISDN at greater than 1.5 or 2.0 Mbit/s. However, this definition is not strictly followed (ITU 2006). The Organization of Economic Cooperation (OECD) considers BB to correspond to transmission speed equal to or greater than 256 Kbit/s. The U.S. Federal Communication Commission (FCC) has defined BB as starting at 200 Kbit/s and ITU currently use the definition of BB as a combined capacity (upstream and downstream) totaling 256 Kbit/s or more. Finally, National Research Council (NRC, 2002) proposes keeping its specification altogether open allowing it to evolve with the technological dynamics of transmission mechanism and changing consumers' needs of transmission speeds.

Empirical observation on BB services around the world indicates that these services can be delivered through a number of different wired and wireless network technologies. These access technologies differ among themselves by their corresponding cost of deployment and functionality. The commonly known access technologies deployed are Digital Subscriber line (DSL) and Cable modem. These technologies are primarily for providing fixed BB access service and can be referred to as evolved stages of existing network infrastructures[5] that require less cost for implementation in comparison to building a completely new wired infrastructure. DSL technology allows to transfer data faster over traditional copper telephone lines whereas cable modem perform the same function using the coaxial cables that deliver pictures and sounds to TV sets. Power line and Rail line communications[6] are other options for fixed BB which can leverage existing power line infrastructure and railway infrastructure respectively. Delivering BB over Power line (BPL) is an emerging technology currently available in very limited areas. However, BPL is a potential BB platform for future and IEEE are working on defining its new high speed standard. On the other hand, Fiber optic cable is a newer technology available to provide fixed BB and can transmit data at speed far exceeding current DSL and Cable modem speeds, typically by tens or even hundreds of Mbit/s. The actual speed the user experiences depends on a variety of factors such as how close to the user's computer the service providers are bringing the fiber and how the service providers configure the services including the bandwidth to be used. However, deployment cost for this technology is much higher than providing service through DSL or Cable modem and thus requires for relatively significant amount of investment before serving as an effective channel for transmission of information. With recent demand for high bandwidth, reduced price for fiber optic cables and increased competition from cable providers for VOIP service, telecommunication companies in many countries have deployed a significant amount of fiber optic network of different forms such as Hybrid Fiber Copper (HFC), Fiber-to-the-Curb (FTTC), Fiber-to-the-Home (FTTH) etc during the years 2007 through 2009 and are continuing that momentum into the year 2010[7].

In addition to the wired BB technology, satellite and terrestrial wireless technologies are also capable of providing high speed BB Internet access to users. Satellite technology is being used to deliver Internet access to the very remote regions of the world. These technologies provide viable options to rural and remote areas of the world which has no other BB access options.

Theoretically, wireless technology provides high speed BB access service to users by connecting homes and businesses to the Internet using a radio link between the user's location and the service provider's facility and can support users both at stationary or mobile state. However, actual adoption of such service depends on demand and supply side factors. In recent years, we have been observing an increase in the deployment of broadband access networks around the world. However, successful deployment and adoption of wireless BB access in any country depends on availability of well developed backhaul infrastructure.

Internationally the ITU encompasses mobile or fixed technologies that provide connections at speed higher than the primary rate (e.g., 2 Mbit/s). However, the definition differs across countries based on their available technologies and stages of economic development. There is also a three-part definition of Broadband Wireless Access (BWA) in accordance with ITU-R Recommendations as follows: (ITU, 2006)

Wireless access systems are broadband radio systems that may be deployed either indoors or outdoors. These systems include:

1) Fixed wireless access which may be defined as wireless access application(s) in which the location of the end users terminations and the network access point to be connected to the end users are fixed.
2) Mobile wireless access which may be defined as wireless access application(s) in which the location of end user termination is mobile
3) Nomadic wireless access which may be defined as wireless access application(s) in which the location of the end-user termination maybe in different place but it must be stationary while in use.

A wireless local area network (WLAN) is defined as a local area network in which one segment uses electromagnetic waves to transmit and receive data over short distances in place of wired network access. Of all the WLAN technologies, the most popular and widely known one is IEEE 802.11b, currently redefined as Wireless Fidelity (Wi-Fi). Wi-Fi coverage is limited to a maximum of 300 feet radius. On the other hand, World Wide Interoperability to Microwave Access (WiMAX) is IEEE 802.16 standard which uses point-to-multipoint architecture and is capable of delivering last mile BB connections in a larger radius than Wi-Fi. WiMAX is seen as a high-speed wireless backbone or middle-mile technology, linking different ISPs to the Internet whereas service like Wi-Fi can be used for connections over the last leg of the network to users. Thus, WiMAX can be used as a complementary service to WLAN (ITU, 2006, Odinma, Oborkhale, Eng and Kah, 2007, Niyato and Hossain, 2007). Technical aspects of current and upcoming wireless broadband applications are discussed in many existing research works[8].

In the mobile BB technology space, 3G (for third generation) and LTE marketed as 4G (for fourth generation) are two well known technologies in the current generation of mobile networks. The 3G systems expect to provide data at a minimum of 144 Kbit/s for all radio environments and 2 mbit/s to low-mobility and indoor environment, though these high speeds may not be available for all the time in service offerings. The 3G standard as defined by ITU is International Mobile Telecommunications or IMT-2000 and the standards for 4G are currently being defined.

Wi-Fi, WiMAX and 3G cellular are very different from one another but they complement each other as well as compete with each other under some situations. WiMAX is expected to provide fixed and nomadic wireless connections. The future 4G infrastructure will consists of a set of various networks using IP (Internet Protocol) as a common protocol. Ultimately 4G will integrate fixed, mobile and other BB technologies in order to provide robust, seamless communications with high speed data transfer[9].

Regarding the spectrum band for BWA, generally, 450-500 MHz and 600-1000 MHz bands could be used for broadband mobile access services given the significant enhanced propagation characteristics of the bands. Many of the bands between the ranges 1.5 – 2.5 GHz are currently identified as IMT-2000 bands which could be used for delivering both fixed and mobile broadband access services. Sometimes such technological application is referred as "IMT-2000 and beyond" (that which is 3.5G or 4G technology). These bands are identified for hybrid satellite and terrestrial wireless systems that permit seamless continuation of roaming for BWA operation. 3.4 – 3.7 GHz bands are allocated over all regions for licensed BWA whereas 5.1 – 5.7 GHz bands are allocated across all regions for unlicensed BWA users (ITU, 2006).

BWA networks with their mobility and portability at the core, offer a variety of useful hybrid architectures that will provide users with rich, multimedia experience virtually at any time or any place. In addition to that, recent technological advances have increased the spectrum capabilities to allow licensees to do more with the same level of spectrum. However, the potential ability to improve the lives of people ultimately depends on the spectrum regulators making it available to service providers.

Benefits of Broadband Access

Existing literature on endogenous growth theory [10]suggests that efficient transfer of knowledge and information, because of its ability to produce spillover effects, holds the key to higher economic growth across world. However, efficient transmission of information is only feasible if a country has well developed information and communications technology infrastructure (ICT). A number of studies in the literature (Wildman, 1992; Loveman, 1994; Roller and Waverman, 1996; Gera, Wuling & Lee, 1998; Nadiri & Nandi, 2001; Stiroh, 2002) documented a close link between information and communication technology (ICT) infrastructure with productivity growth at the industry level as well as at the aggregate economy level. Many of these studies claim that deployment of BB networks can enhance ICT contribution towards growth by improving the quality of existing ICT infrastructure and thus generating an outward shift of the aggregate production function. Many researchers study the benefits of BB diffusion (Crandall & Jackson, 2001; Firth & Mellor, 2005; Crandall, Lehr & Litan, 2007). Crandall, Lehr and Litan (2007) in their recent study showed that employment in both manufacturing and service industries is positively related to BB penetration. Regarding output growth, they also found positive association with BB penetration but pointed that because of noise in the data, estimates did not appear as statistically significant. In addition to that, various other studies argue that both fixed and mobile BB access capacities can increase the ICT contribution by widening the choice of services through applications in voice, audio, video, e-commerce, e-education, e-health, e-government, online interaction and information sharing. Social impacts of many such applications are highlighted by another group of researchers (Greenstein & Spiller, 1996, Sawyer, Allen and Lee, 2003). In addition to the landline BB networks, the increasing deployment of wireless broadband access networks, because of its cost advantage and special characteristics, is bringing significant benefits all over the world. As ITU mentioned in its report (2006), wireless broadband technologies are stimulating the global economy by enabling users to (1) freely access the internet from anywhere (2) connect to the internet seamlessly using one single device and (3) by offering access to resources and opportunities from all parts of the world. Since the wireless broadband access encompasses both fixed and mobile technologies, the mobility feature has become more appealing to users in recent years. It allows the user more flexibility to access service from anywhere and irrespective of stationary or mobile state of user. In addition to fixed BB, mobile BB technology using larger range directional equipment can provide BB service to remote or sparsely populated areas where fixed BB service such as DSL and Cable model would be costly to deploy. This technology is more useful for many developing countries where fixed communications networks are not well developed. Therefore, access to this technology is allowing them to Leap-Frog the modern BB access technology even in the absence of developed landline infrastructure.

Sharing of knowledge can be enhanced by obtaining equitable access to the Internet which is considered a source of information regarding scientific, economic, social, political and cultural activities. BB technology facilitates such knowledge transfer across the globe and generates spillover effects[11]. Combined view of all these findings reflects the potential role of BB technology as a catalyst for higher growth and development in the information age.

DATA

This study primarily uses the data reported in the ITU the Telecommunications Indicator database, released in June of 2009 (ITU 2009). This database contains time series data (up to 2008) for various telecommunications indicators which includes data for fixed as well as mobile BB subscribers and covers data for 217 countries around the world. Time series of subscription data for various BB services for the period of 1998 to 2008 are used in this study. In addition, time series data for traditional mainline subscriptions and wireless subscriptions are used for the period of 1980-2008. However, due to missing data for many countries, appropriate sections of the available time series data are used for analyzing various topics addressed in this chapter. Definitions (ITU 2009) of various telecommunications indicators used in this analysis are summarized in Appendix-A. We also obtained additional data from the official websites of ITU, OECD and FCC to complement the data available in ITU reported database (ITU 2009).

In the following two sections, we analyze these data to study the trend in the growth for various BB technologies across different countries.

BROADBAND DIFFUSION

Together with innovations of mobile phone and Internet connection, the broadband (BB) technology has offered a major step forward towards world wide access to information and its associated spillover benefits. Based on the prospective benefits from BB technology, countries around the world are trying to increase the BB diffusion rate. In recent years, we have observed a significant growth in the fixed BB penetration for several countries. As a result of, the accelerated growth in the use of mobile phone devices in recent years and the introduction of new wireless devices that allow users to access Internet, there is increased opportunity for using Broadband Wireless Access (BWA. This BWA will allow end users to access Internet easily from anywhere and at anytime. Realizing this emerging need and potential benefits, several countries are already putting efforts in the deployment of BWA networks. In the next section, we will focus primarily on analyzing the growth and technological composition of fixed BB access network in various countries and in the following section, we will focus on analyzing the current and future trends in use of mobile BB technology.

Summary Statistics for Fixed Broadband

Use of Internet services have been gaining momentum since the early part of 1990s. Based on ITU's data (2009); the number of Internet users around the world reached a total of 1.40 billion by the end of 2007, about 20.92% of the world population. Increased demand for Internet use together with expected benefits from high speed Internet connections have induced the rapid deployment of BB access network around the world. As reported by ITU, the total number of fixed line BB subscribers reached 349.2 million by the end of 2007 accounting for 5.22% of world population which is about 73.56% of the world Internet subscribers. As per ITU's definition, a fixed BB Internet subscriber refers to a subscriber who pays for high-speed access to the public Internet (a TCP/IP connection), at speeds equal to, or greater than, 256 kbit/s, in one or both directions. Based on this definition, the estimate for global fixed line BB subscribers in 2008 is expected to be at 406 million[12]. The

Table 1. Distribution of Fixed Broadband Subscribers by Region, 2007

Region	% of World Subscribers
South East Asia and Pacific	36.41%
Europe and Central Asia	33.34%
Latin America and Caribbean	5.73%
Middle East and North Africa	1.33%
North America	23.01%
Sub-Saharan Africa	0.18%
TOTAL	100.00

Source: ITU World Telecommunications Indicators Database, June, 2009

distribution of total BB subscribers of 2007 [13]by various regions is reported in Table-1.

It is apparent from Table 1 that more than 36% of the world subscribers belong to Asia-Pacific regions. On the other hand, Europe together with central part of Asia and North America, with their share of world subscribers at about 33% and 23% respectively, are also playing a significant role towards BB adoption.

Regarding the choice of network technology for delivery of BB services, a majority of the subscribers in the world are still obtaining their BB service using DSL followed by cable modem. However, the percentages of both DSL and Cable modem users have declined over the last few years whereas the number of subscribers in other technologies such as Fiber cable, satellite, LAN and Broadband over Power Line has increased. The distribution of BB subscribers by various technologies for OECD countries for 2008 is summarized in Table 2 below. Similar distribution pattern is expected for the entire world with little higher share in DSL., However, as the ITU database (2009) has missing data for several countries for 2008, we summarized OECD reported data in Table 2[14].

If we rank the countries based on their total number of BB subscribers, then the scenario for 2007 is as follows:

Table 2. Distribution of Subscribers of OECD countries by Technology, 2008

Broadband Platform	% of Subscribers
Cable Modem	29%
DSL	59%
Other Technologies (FTTH, Satellite, Ethernet, LANs, BB over Power line etc.)	12%

Source: OCED Broadband Statistics, December, 2008, OECD website

This ranking changes when we analyze the BB penetration rates for various countries where penetration rate is defined by number of subscribers by 100 inhabitants (or per 100 populations).

Table 4 shows the list of top 10 countries by their BB penetration rate for last 7 years. Numbers within parenthesis are showing the penetration rate per 100 inhabitants for each country

Table 4 shows that over the last few years, there is a significant growth in number of subscribers per 100 inhabitants in all countries[15]. However, the observed growth rates across various countries are very different. Due to this reason, the ranking of countries by BB penetration rate has changed significantly over the last few years.

Table 3. Top 10 Countries by Broadband Subscribers Numbers, 2007

Country Name	Number of BB Subscribes
U.S.A.	70,187,400
China	66,414,000
Japan	28,287,000
Germany	19,600,000
U.K.	15,600,000
France	15,550,000
Korea	14,709,998
Italy	10,122,126
Canada	9,070,254
Spain	8,070,254

Source: ITU World Telecommunications Indicators Database, June, 2009

Table 4. Top 10 countries by Broadband Penetration Rate per 100 inhabitants

Year/Rank	2002	2003	2004	2005	2006	2007	2008
1	Korea (Rep.) (21.3)	Korea, Rep. (23.3)	Korea, Rep. (24.8)	Iceland (26.5)	Denmark (31.86)	Denmark (36.3)	Sweden (37.3)
2	Hong Kong, (14.6)	Hong Kong (18.0)	Hong Kong (21.9)	Korea, Rep (25.2)	Netherland (31.79)	Iceland (34.8)	Denmark (36.8)
3	Canada (11.5)	Canada (14.8)	Netherland (19.8)	Netherland (25.2)	Monaco (29.5)	Netherland (33.5)	Netherland (35.0)
4	Taiwan (9.4)	Iceland (14.5)	Denmark (18.9)	Denmark (24.9)	Iceland (29.53)	Finland (33.3)	Norway (34.0)
5	Iceland (8.6)	Taiwan (13.4)	Iceland (18.4)	Liechtenstein (24.8)	Korea (29.2)	Switzerland (32.1)	Switzerland (33.0)
6	Denmark (8.6)	Denmark (13.1)	Canada (17.1)	Switzerland (23.8)	Liechtenstein (28.4)	Korea (30.6)	Iceland (32.9)
7	Belgium (8.4)	Belgium (12.4)	Taiwan (16.5)	Hong King (23.6)	Finland (27.14)	Norway (29.0)	Korea (32.0)
8	Sweden (7.7)	Japan (11.7)	Switzerland (16.4)	Finland (22.4)	Norway (26.95)	Hong Kong (26.1)	Finland (30.6)
9	Austria (6.6)	Netherland (11.6)	Belgium (15.6)	Norway (21.4)	Switzerland (26.48)	Sweden (25.9)	Luxembourg (30.3)
10	Netherland (6.5)	Switzerland (11.4)	Finland (15.3)	Canada (20.8)	Sweden (25.87)	U.K. (25.6)	Canada (29.0)

Source: ITU World Telecommunications Indicators Database, June, 2009

Note: Number within parenthesis shows broadband subscribers per 100 inhabitants

Figure 1. Trend in penetration rates of top 5 countries (ranking based on 2008)

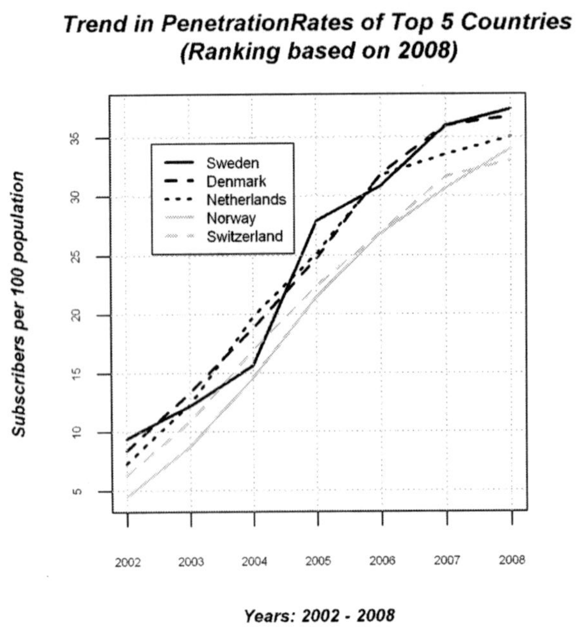

If we compare the ranking of the top 5 countries by penetration rate in 2002 with that in 2008, we find that not a single country is common between the two lists. The trend in penetration rates for these 10 countries over the last 7 years are shown in the Figures 1 and 2.

The two plots in Figures 1 and 2 clearly show a significant difference in the growth for penetration in the last 7 years among the two sets of countries. The first set of countries that were ranked within the top 5 in 2002 was advanced in BB penetration in the early part of this decade. However, our estimates show that their growth rates slowed down relative to the countries that are ranked within the top 5 in 2008. The second set of countries experienced a higher growth in penetration for BB access technology in the recent years. The statistical analysis of the BB data in the next section will explain this phenomenon in detail.

It is important to mention at this point that these estimated BB penetration rates are based on the definition of BB subscriber and the penetration rate used by ITU in building their database[16].

Dynamics of Fixed Broadband in Countries around the World

The underlying dynamics of the observed phenomena in the last section will be clear if we analyze farther the growth pattern of BB subscribers in various countries around the world. A matrix plot of observed BB penetration for 217 different countries between years 1998 and 2008 is shown in Figure 3. To accommodate the variability associated with the penetration levels across different countries, we transform the data using a logarithmic transformation.

The trend displayed by the temporal shape and the level of the curve by country clearly indicates the heterogeneous nature of the BB adoption dynamics across all countries around the world. Also significant differences in diffusion rates are observed among continents at the aggregate level and among various countries within a given continent. There could be several underlying reasons for this variability. In addition to that, the available time series is incomplete for many countries due to missing data. Thus given the complicated nature of the data for BB adoption

Figure 2. Trend in penetration rates of top 5 countries (ranking based on 2002)

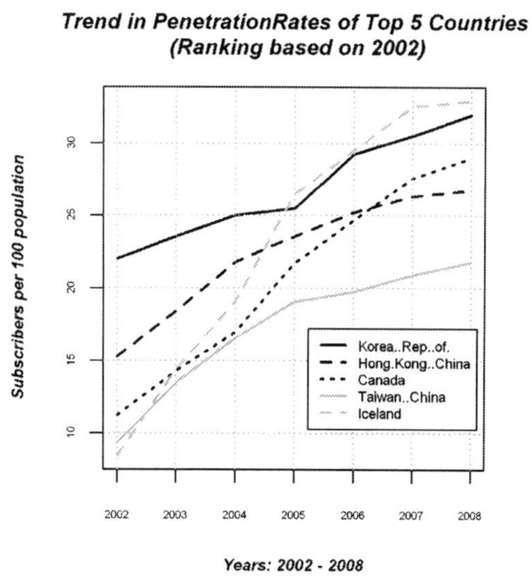

Figure 3. Global dynamics for broadband (fixed) subscribers

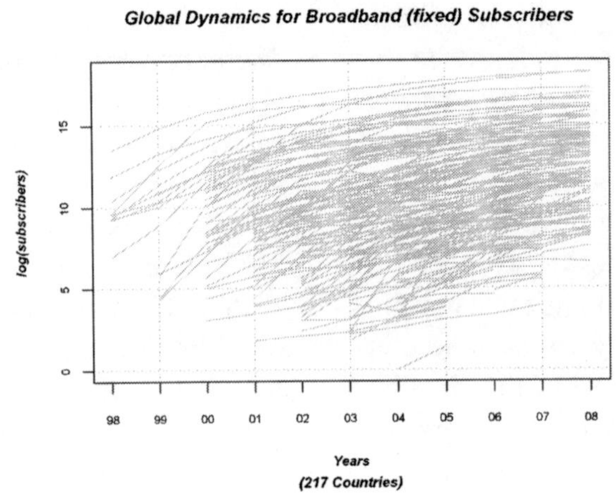

around the world, a disaggregate level study of individual countries appears to be more appropriate to understand the underlying dynamics and the causes behind the variation. However, since such detail study is beyond the scope of this chapter, we focus our analysis of penetration behavior to a small subset of selected countries.

To facilitate our understanding of the phenomena observed in the last section (Figures 1 to 2), this chapter examines 10 countries, 5 of which were ranked in the top 5 for penetration rate in 2002 and the rest ranked in the top 5 in 2008.

Using the time series data from 1999-2007, we use functional data analysis (Ramsay and Silverman, 1997) to estimate the velocity and acceleration rates for BB penetration (per 100 inhabitants). This analysis is restricted to the ten countries discussed above. The Figure 3 display the plots of estimated velocities and accelerations of penetration per 100 inhabitants for selected 10 countries where the 'velocity' implies the "velocity for growth" or rate of change in penetration with respect to time and 'acceleration' measures the rate of change of growth velocity or growth rate of penetration (in short, the change in momentum). These velocity and acceleration rates are estimated by first fitting a smooth function to the observed time series of BB penetration data and then differentiating the function twice (see appendix B for details).. The subtle features (e.g., the bumps and dips) observed in the derivative plots are not visible in the time series plots shown in Figures 1 and 2.

More specifically, Figure 4 and 5 display the estimated velocities and Figure 6 display the estimated acceleration rates of penetration for 10 selected countries. Due to missing data, for countries that are ranked in the top five for penetration rate in 2002(namely Hong Kong, Korea, Taiwan, Iceland and Canada), we used data for the time period 1999-2007. For countries that are ranked in the top five for penetration rate in 2008 (Norway, Denmark, Switzerland, Sweden and Netherland), we used data from 2000-2007. It is clear from the Figure 4 that in spite of the year over year increase in BB penetration for countries Korea, Taiwan, Iceland, Canada and Hong Kong (who were pioneers in BB penetration early in this decade) their penetration velocity is on a declining path. However, the change points for these five countries are different. Korea, Taiwan and Hong Kong, experienced significant declines in their penetration velocity beginning the very first year of the sample period used in our estimation whereas Taiwan and Iceland experienced growth in velocity until 2002 and 2006 respectively. In contrast, for Sweden, Switzerland, Norway, Netherland and Denmark the penetration velocities started to increase from the beginning of this decade exceeding the rates of the other set of 5 countries. For Sweden and Switzerland, we observe an increasing trend in the velocity as of 2007. Same phenomenon is revealed in Figure 6 that displays the acceleration of penetration rates for all 10 countries. Thus the Figures 4 - 6 clearly show that for the time period 2002 – 2007, the rate of growth of the penetration and its acceleration declined for Korea, Taiwan, Iceland, Canada and Hong Kong whereas for Sweden, Denmark, Netherlands, Norway and Switzerland the momentum for penetration increased.

The observed changes in velocities and acceleration rates of BB penetration across various countries could be influenced by political, social or economic factors. Various demand and supply side factors are playing important roles (Nandi & Chakraborty, 2008, Papacharissi and Zaks, 2006) in affecting these rates which can be influenced both by government policies and by the economic environment. The size of the country and the urbanization structure also play important role in achieving higher growth rate of penetration. It is possible that lack of stimulus from demand and supply side factors have slowed the growth rate in countries such as Korea, Taiwan and Hong Kong whereas positive stimulus from demand and supply sides have encouraged further deployment of BB in the countries such as Denmark, Sweden, Switzerland and some others. Other possible reason behind the observed scenario could be that countries like Korea, Taiwan and Hong Kong who were pioneers in the deployment of fixed BB are also pioneers in the deployment of wireless BB.The demand for BB subscription in these countries could be gradually declining due to migration of users from fixed BB to wireless BB. This phenomenon will be discussed in more detail in the next section where we focus primarily on analyzing the deployment of wireless broadband around the world.

Summary Statistics for Mobile Broadband Access

Over the last few years, wireless BB is getting deployed in many countries around the world.

Figure 4. Dynamics of total fixed BB subscription (velocity)

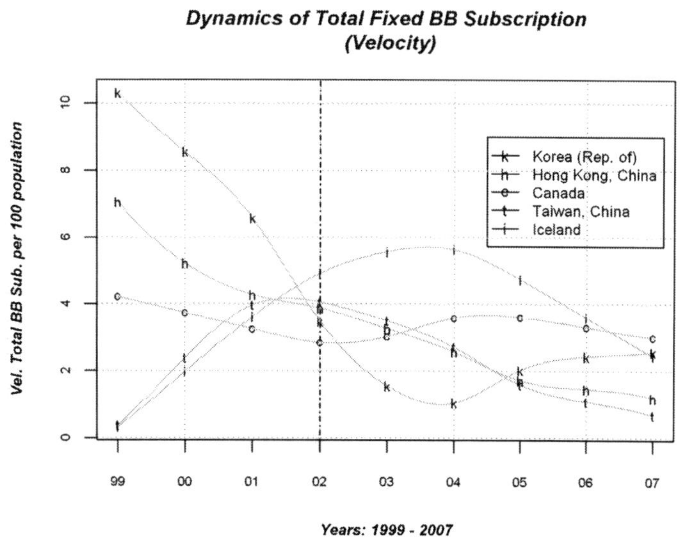

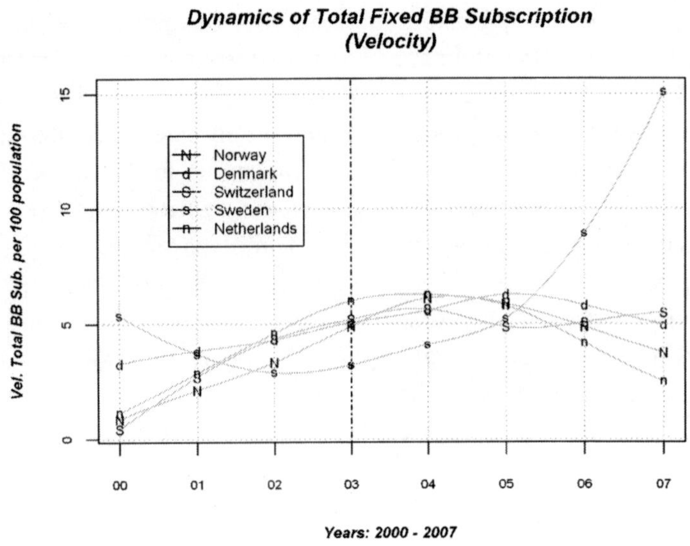

Figure 5. Dynamics of total fixed BB subscription (velocity)

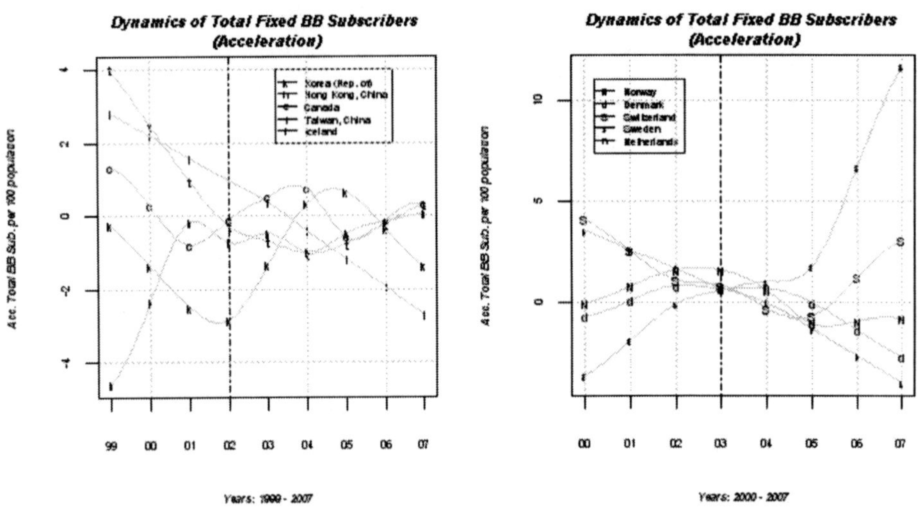

Figure 6. Dynamics of total fixed BB subscribers (acceleration)

In this section we will analyze the growth rate for mobile BB subscribers[17] and its share in the total BB subscriber base by using time series data from various countries around the world. Note that ITU refers to the count for mobile broadband as "subscriptions" and for fixed broadband as "subscribers". In this paper, we will be using the term subscribers for both, although it has to be noted that the term "subscriber" in the case of fixed broadband refers to an individual subscriber whereas the term "subscriber" in the case of mobile broadband means subscriptions. For mobile BB subscribers, the time series data is available since year 2001. Analysis of the mobile BB subscriber data shows that Japan and USA are leading the other countries in subscriber base.

Based on ITU reported data for global mobile BB subscribers, the top 10 countries for 2008 are listed in Table 5 below together with their trend in growth of subscribers over the last 3 years.

The growth rate displayed in the Table 5 indicates that in year 2007, most countries experienced significant growth in their number of mobile subscribers i.e., the year over year changes in growth rates for 2007 over 2006 are in general very high. The same year-over-year growth rates in 2008 are lot less in comparison to previous year. One of the reasons for these declines in the estimated growth rates is due to increased base of mobile broadband subscribers in 2007 for many countries in comparison to that in 2006.

Even though USA has the second largest number of mobile BB subscribers in the world, this ranking changes when we study the penetration rate rather than absolute number of mobile BB subscribers in various countries around the world[18]. The 3-years trend in mobile broadband penetration rate is displayed in Table 6 below for the period 2006-2008.

Numbers in the above table show that Japan is leading the world by penetration rate of mobile subscriptions followed by Korea in second position. It is important to mention that the observed trends of mobile BB subscription base and penetration rates for various countries in Table 5 and 6 are tied to the definition of mobile BB subscription used by ITU in its data collection process. Member countries are aware of difficulties in measuring mobile BB subscriber/subscription and country representatives are providing input to ITU for improving the definition and measurement methodology for obtaining better and unbiased statistics of number of subscribers/subscriptions across countries[19].

The mobile subscription behavior observed in above two tables clearly reflects the recent experience of rapid growth in demand for mobile broadband access in many countries around the world. This is partly a result of the continuous advancement in wireless communication technology, enabling increase in the speed of data transmission together with improved quality and stability of network and thus satisfying the growing demand for mobile BB connections.

Table 5. 3 Years Trend in number of Mobile Broadband Subscribers for Top 10 Countries 2006-2008

Country Name	2006	2007	2008	% Growth 2007/2006	% Growth 2008/2007
Japan	50,594,200	72,690,896	92,613,480	43.61%	27.41%
USA	11,015,968	53,065,192	81,899,720	381.71%	54.34%
Korea	15,512,027	23,426,932	34,061,960	51.02%	45.40%
Italy	17,091,000	24,548,000	30,329,836	43.63%	23.55%
U.K.	7,820,072	12,514,000	20,732,364	60.02%	65.67%
Spain	3,718,116	10,031,050	17,275,184	169.79%	72.22%
France	4,169,474	8,448,387	14,626,233	102.62%	73.12%
Germany	4,500,000	8,700,000	11,500,000	93.33%	32.18%
Australia	2,999,000	6,809,000	11,316,013	127.04%	66.19%
Taiwan, China	3,267,987	6,529,566	8,886,875	99.80%	36.10%

Source: ITU World Telecommunications Indicators Database, June, 2009

*Table 6. 3Year Trend in Mobile Broadband Penetration Rate** for Top 10 Countries 2006-2008*

Country Name	Penetration rate, 2006	Penetration rate, 2007	Penetration Rate, 2008	% Growth 2007/2006	% Growth 2008/2007
Japan	39.46	56.80	72.39	43.94%	27.45%
Korea (Rep.)	32.33	48.58	70.39	50.26%	44.90%
Singapore	24.69	43.31	67.13	75.42%	55.00%
Luxembourg	28.77	43.18	66.72	50.09%	54.52%
Australia	14.72	32.83	54.01	123.03%	64.51%
Italy	29.39	41.69	51.45	41.85%	23.41%
Macao, China	NA	9.55*	47.45	NA	396.86%*
New Zealand	13.12	28.01	45.40	113.49%	62.08%
Austria	20.37?	29.48	41.90	44.72%	42.13%
Portugal	20.51	29.25	40.51	42.61%	38.50%

Source: ITU World Telecommunications Indicators Database, June, 2009

Note: * Could be some outlier in data, ** Penetration rate is defined by number of subscribers per 100 inhabitants

FUTURE ROLE OF FIXED AND MOBILE BROADBAND TECHNOLOGY

Trend in Traditional Landline vs. Mobile Cellular Subscriptions

The scenario described in previous section indicates the possibility that, the current experience of subscription dynamics observed in the voice telephony space between landline and wireless around the world will be replicated in the behavior of fixed versus wireless BB subscriptions. The trend in landline versus mobile cellular subscriptions at the global level for the time period 1980 – 2007 is shown in Figure 7.

The above figure indicates that at the aggregate level, the total number of subscribers for mobile phone in the world has exceeded the number of landline (fixed telephone line) subscriptions in year 2002. However, in reality, the cross over time point is different for different countries depending on their rates of deployment of mobile networks, and other demand and supply side factors (see Appendix-C for observed behavior of landline versus mobile cellular subscriptions for 8 selected countries).

Trend in Fixed vs. Mobile Broadband Diffusion

The trend for fixed and mobile BB subscribers over the last 8 years is summarized in Table 7.

The numbers in the Table 7 clearly illustrate that the growth rates for mobile BB subscribers are relatively higher than that for fixed BB subscribers[20] in the recent years. The above table also shows that at the aggregate level, in 2008, the number of fixed and mobile BB subscribers is very close indicating the possibility of mobile BB subscriptions to exceed that of fixed BB connections in many parts of the world. However the time of crossover could be different for different countries depending on country specific demand and supply side stimulus. Therefore, country level study of the BB adoption behavior will be more interesting. However, due to limitation in time series data, we select 5 countries for analysis. These 5 countries are Japan, USA, Korea, Italy and UK. Behavior of the fixed and mobile broadband subscribers for these countries for the period of 2006-2008 is displayed in Table 8 below.

From Table 8, it is clear that by 2006, the number of mobile BB subscribers had exceeded the number of fixed BB subscribers in Japan,

Figure 7. Trend in Landline versus Mobile Telephone in the World: Aggregate View of 1980-2007

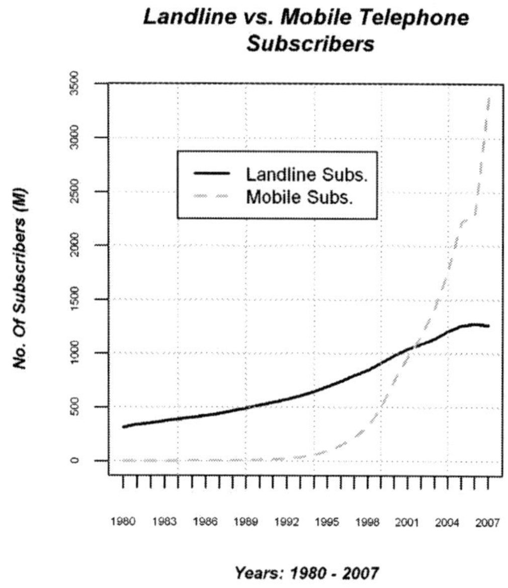

Table 7. Total fixed and mobile broadband subscribers in world

Year	Total Fix Broadband Subscribers	Total Mobile Broadband Subscribers
2001	37,049,893	27,200
2002	66,020,264	326,200
2003	103,772,705	7,171,823
2004	157,924,043	30,201,513
2005	216,327,590	72,972,286
2006	283,195,363	139,406,661
2007	349,228,249	264,753,929
2008*	406,000,000	405,753,929

Source: ITU Telecommunications Indicator Database, June 2009
Note: * End of 2008 data is estimated data

Table 8. Fixed and mobile broadband subscribers for Japan, USA, Korea and Italy, 2006-2008

Country Name	2006 Subscribers	2007 Subscribers	2008 Subscribers	2006 Subscribers	2007 Subscrib-ers	2008 Subscribers
	Fixed BB	Fixed BB	Fixed Bb	Mobile BB	Mobile BB	Mobile BB
Japan	26,438,868	28,287,000	30,107,328	50,594,200	72,690,896	92,613,480
USA	60,522,096	70,187,400	79,014,120	11,015,968*	53,065,192	81,899,720
Korea (Rep. of)	14,042,698	14,709,998	15,474,931	15,512,027	23,426,932	34,061,960
Italy	8,497,422	10,122,126	11,283,000	17,091,000	24,548,000	30,329,836
United Kingdom	13,013,000	15,606,000	17,276,000	7,820,072	12,514,000	20,732,364

Source: ITU World Telecommunications Indicators Database, June, 2009

NOTE:* FCC reported mobile wireless high-speed lines in US in 2006 was 22,287,749 where according to FCC definition, high speed lines are connections to end users locations that deliver services at speed exceeding 200 kbit/s at least in one direction (whereas ITU defined BB speed is 256 kbit/s in at least one direction) (FCC, 2009).

Korea and Italy whereas United States and United Kingdom have just reached that cross over point in 2008.

TREND IN SUBSCRIPTIONS OF VARIOUS BB PLATFORMS AND FUTURE FORECASTS

In this section, we analyze the observed trend in subscriber's behavior for various BB platforms for selected countries and predict the future trend in such behavior. Due to limitation of time series data, we select 3 countries for detail analysis. These countries are USA, Japan and Korea. It is clear from the Table 8 that the observed growth rate of total fixed BB subscribers in USA is relatively higher in comparison to Japan and Korea. If we analyze the composition of subscribers, we find that in all 3 countries, the share of subscribers in "other Technology" category has increased whereas that for DSL and cable technology has declined over the last few years. Here "other technology" includes satellite, Fiber cable, LAN, Ethernet and Power lines. On the other hand, the trend for number of mobile subscribers in the three countries is experiencing a very rapid year over year growth. This phenomenon is displayed in Figure 8 to 10.

Based on such rapid growth in mobile BB adoption, we can expect that mobile BB technology will play an important role as BB access technology and will enhance the scope of innovation and adoption of many new applications. To see this future trend of mobile BB diffusion, we developed a simple time series forecasting model and estimate the model using the data for Japan, USA and Korea to predict the future behavior for fixed versus mobile BB subscribers.. More specifically, based on the available time series (2002 – 2007) for fixed and mobile subscribers, we choose an appropriate forecast trajectory using smoothing splines to predict future trend in subscription for these three countries. The Actual and forecasted results are displayed in Figure 11 to 13 with forecast period ranging from 2009 to 2013.

Figure 8. USA subscription for fixed bb (t) cable modem (c), DSL (d), and wireless BB (w)

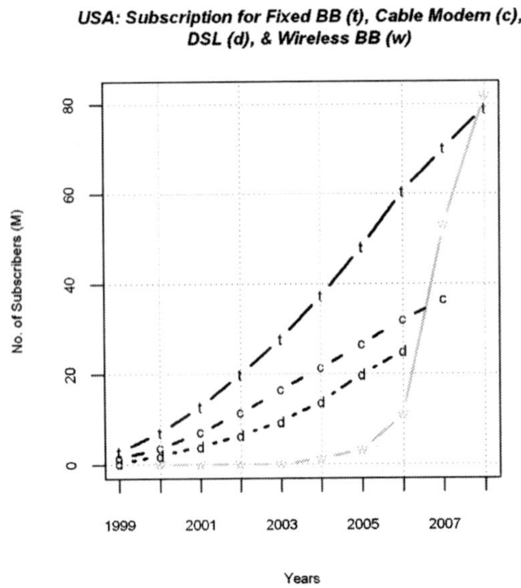

Figure 9. Korea subscription for fixed BB (t), cable modem (c), DSL (d), and wireless BB (w)

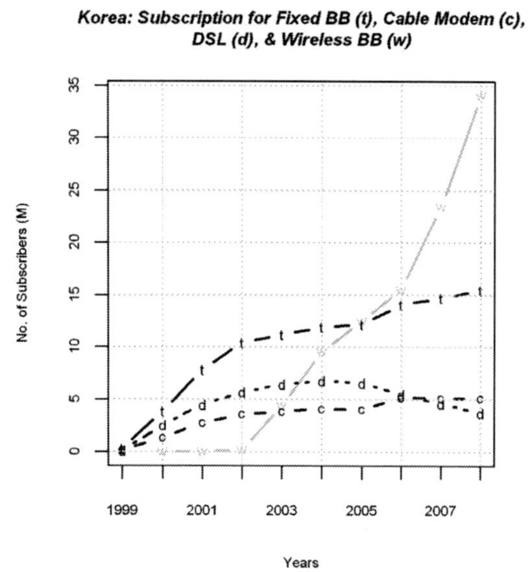

Figure 10. Japan subscription for fixed BB (t), cable modem (c), DSL (d), and wireless BB (w)

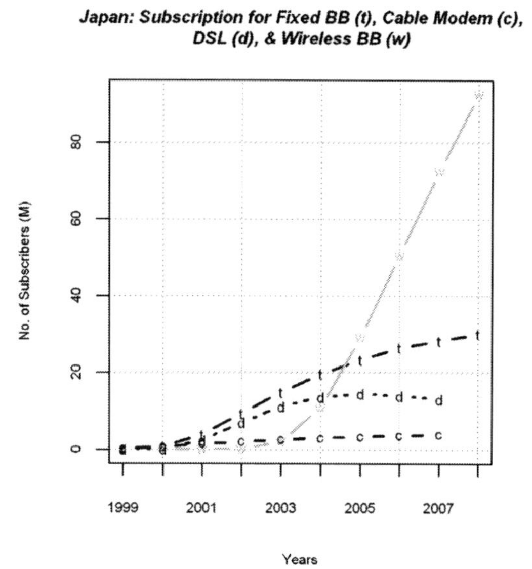

The model predicts that the mobile BB subscriptions will reach to 194M for Japan, 189M for USA and 75M for Korea by 2013. The forecasts for fixed BB subscriber numbers are expected to approach an asymptote around 37 million, 113 million and 19 million in Japan, USA and in Koreas respectively. Detail of the forecasted numbers for three selected countries is shown in Table-D in Appendix-D.

Figure 11. USA fixed (f) vs mobile (w) BB subscribers history and forecast

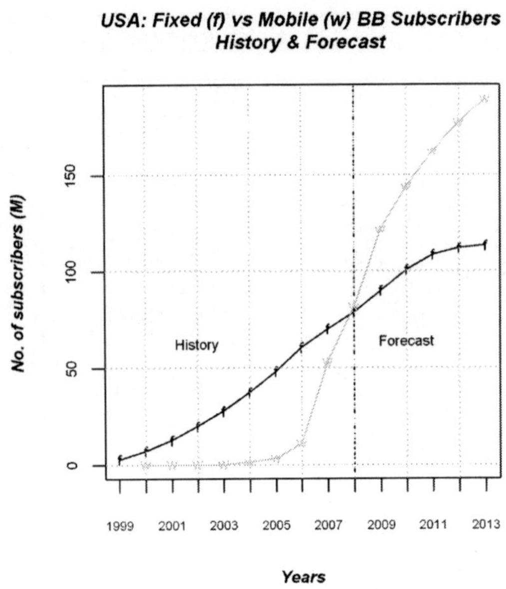

Figure 12. Korea fixed (f) vs mobile (w) BB subscribers history and forecast

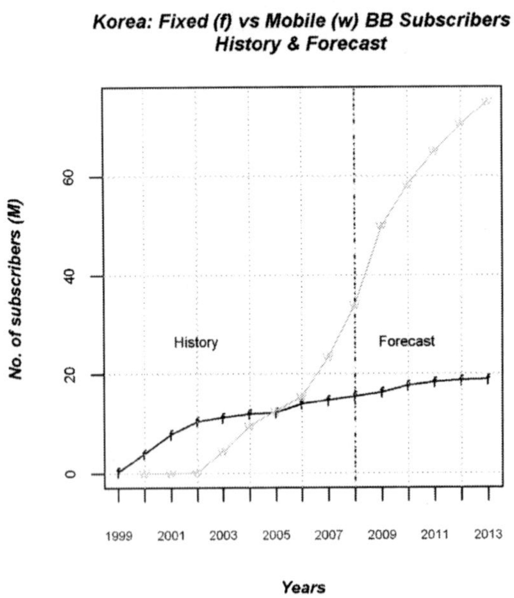

It is important to note at this point that the US smart phone industry is exploding with the introduction of devices like iPhone in 2007, the Netbooks, Droid phone and Palm Pre in 2009 and more recently the iPad in 2010. These events have dramatically transformed the smart phone

Figure 13. Japan fixed (f) vs mobile (w) BB subscribers history and forecast

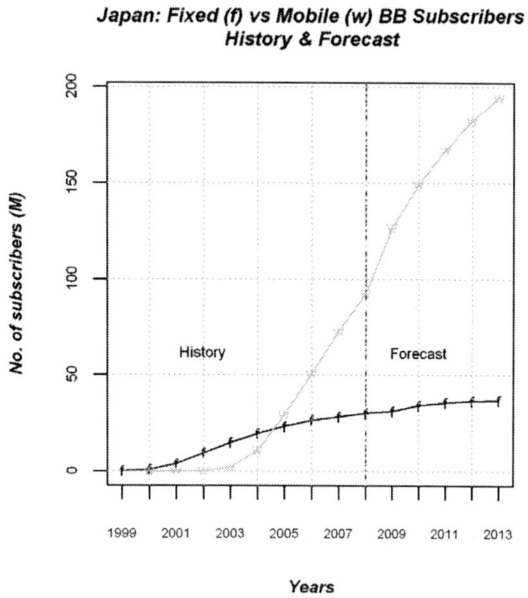

landscape. However, since the last actual in the time series used in the forecasting is year 2008, the impact of these events is not represented in the history. This could be a factor in accelerating the mobile broadband growth in the USA possibly surpassing subscriptions in Japan and Korea. Thus, we expect the US mobile BB subscription forecast of 189M for 2013 to be substantially higher.

It would have been very interesting to study the current and future trend in behavior of fixed versus mobile BB adoption for large emerging developing countries such as India and China. However, due to limited data, we could not include them in the analysis.

CONCLUSION

This chapter begins with the description of primary characteristics of various platforms of BB access technology followed by a brief outline of the benefits of deploying such technology. The focus of the chapter is to analyze and explain the diffusion behavior of BB access technology in different countries around the world. Special emphasis is given to study the evolving role of various technologies in developing high speed BB access to the Internet. Specifically, the focus is to analyze the dynamics of subscription behavior of fixed versus mobile BB connections over the last few years and to predict a directional path of that future trend.

To understand the dynamics behind the observed difference in penetration rates of Fixed BB access connections in various countries, we study selects few countries for which reasonable time series data are available. Functional data analysis methodology is then applied to estimate the velocity and acceleration of BB penetration rates for those selected countries. Estimated results show very differing temporal shapes for the velocity and acceleration curves pointing to distinct growth spurts associated with different countries. This phenomenon is influenced by country specific information policy initiatives and their underlying structure of demand and supply sides. Our observation regarding BB diffusion behavior for various countries clearly reveals that many countries are promoting their supply and demand for BB service for improving their BB adoption but the amount of effort and the timing of efforts differ (Yun, Lee, & Lim, 2002; Tajiri & Okazaki, 2006; Nandi & Chakraborty, 2008). Country specific analysis will reveal the role of various such factors in influencing the deployment and adoption of BB access and provide useful information for policy makers of various countries. However, such study is beyond the scope of this chapter.

Second aspect is discussed in this chapter is the dynamics of trend in subscription behavior of various BB technologies. More broadly, this chapter describes the dynamic behavior of subscription of fixed versus mobile BB access connections. The observed behavior clearly indicates the increasing role of wireless technology in provisioning BB access to users.

This observation is also substantiated with the prediction of future path of fixed versus mobile BB subscribers for selected countries based on simple model.

At this point it is important to highlight that the above described behavior of fixed versus mobile BB subscriptions is very much linked to the innovation and deployment of various applications and devices that determine the need for various bandwidth regarding access connections. In fact, the innovation in information technology and various applications of such technology are influencing each other to shape the evolving nature of BB access and its adoption.

To conclude, the analysis and discussion on broadband diffusion and the observed evolving trend in adoption of fixed versus mobile BB technologies have broader implications on the current state, future trend of technological applications and market condition in the world. Similar to voice telephone service, role of mobile BB access platforms are increasing over the last few years. This mobile technology is becoming very popular in developing countries and in remote areas due its cost advantage. In addition to that, convergence in computer, communications and broadcasting technologies are contributing towards shaping the new integrated communication and information platform for future generations. Consequently the successful adoption and use of such a changing technology in any country depends on the ability of its government and people to see the benefits of the evolving technology and the readiness of the government to design appropriate flexible policy choices to accommodate such changes to optimizing the outcome.

REFERENCES

Barro, R. J. (1990). Government Spending in a Simple Model of Endogenous Growth. *The Journal of Political Economy*, *98*(5), 103–125. doi:10.1086/261726

Biagioni, A., Fantacci, R., Marabissi, D., & Tarchi, D. (2009). Adaptive Subscriber allocation schemes for wireless OFDMA systems in WiMAX networks. *IEEE Journal on Selected Areas in Communications*, *27*(2), 217–225. doi:10.1109/JSAC.2009.090212

Crandall, R.W., & Jackson, C.L. (2001, July). The $500 Billion Opportunity: The Political Economic Benefits of Widespread Diffusion of Broadband Internet Access. *Criterion Economics.*

Crandall, R. W., Lehr, R., & Litan, R. (2007). The effects of Broadband Deployment on Output and employment: A Cross-Sectional Analysis of U.S. Data. *Issues in Economic Policy, 6*. The Brookings Institute. Retrieved from http://www.brookings.edu/ papers/2007/06labor_crandall.aspx

David, P. A. (2001, June). *Knowledge, capabilities and Human Capital Formation in Economic Growth.* New Zealand Treasury Working Paper, 01/13.

Federal Communication Commission (FCC). (2009). *High Speed Service for Internet Access: Statistics as of June 30, 2008*. Industry Analysis and Technology Division, Wire line Competition Bureau. Retrieved from http://www.fcc.gov/wcb/stats.

Firth, L., & Mellor, D. (2005, March-April). Broadband: Benefits and Problems. *Telecommunications Policy, 29*(2-3), 223–236. doi:10.1016/j.telpol.2004.11.004

Gasser, T., Kneip, A., & Koehler, W. (1991). A flexible and fast method for automatic smoothing. *Journal of the American Statistical Association, 86*, 643–652. doi:10.2307/2290393

Gasser, Th., & Muller, H. G. (1984). Estimating Regression Functions and Their Derivatives by the Kernal Method. *Scandinavian Journal of Statistics, 11*, 171–185.

Gera, S. Wulong, Su & Lee F.C. (1998). *Information Technology and Productivity Growth: An empirical analysis for Canada and the United States.* Microeconomic Analysis Working Paper, No 20, Canada.

Greenstien, S. M., & Spiller, P. T. (1996). *Estimating the Welfare Effects of Digital Infrastructure*. NBER Working Paper, No. 5770, National Bureau of Economic Research, Inc., Cambridge, MA. Chlamtac, I., Gumaste, A., & Szabo, C. (Eds.) (2005). *Broadband Services: Business Models and Technologies for Community Networks*. John Wiley & Sons Ltd.

International Telecommunications Union. ITU (2006). *Trends in Telecommunications Reform 2006: Regulation in the Broadband World* (p. 77). Geneva, Switzerland: ITU.

International Telecommunications Union. ITU (June, 2009). World Telecommunications Indicator Database. Retrieved from http://www.itu.int

Lehr, W., & McKnight, L. W. (2003). Wireless Internet Access: 3G vs. WiFi. *Telecommunications Policy, 27*(5-6), 351–370. doi:10.1016/S0308-5961(03)00004-1

Loveman, G. (1994). An Assessment of the Productivity Impact of Information Technologies. In Allen, T. J., & Scott Morton, M. S. (Eds.), *Information Technology and Corporation of the 1990s: Research Studies* (pp. 84–110). Oxford, UK: Oxford University Press.

Lucas, R. E. (1998). On Mechanics of Economic Development. *Journal of Monetary Economics, 22*, 3–42. doi:10.1016/0304-3932(88)90168-7

Mankiew, N. G., Romer, D., & Weil, D. (1992). A Contribution to the Empirics of Economic growth. *The Quarterly Journal of Economics, 107*(2), 407–437. doi:10.2307/2118477

Nadiri, M. I., & Nandi, B. (2001). Benefits of Communications Infrastructure Capital in U.S. Economy. *Economics of Innovation and New Technology, 10*(2-3), 89–107. doi:10.1080/10438590100000005

Nandi, B., & Chakraborty, C. (2008). Broadband Diffusion and its Driving Forces. In Dwivedi, Y. K., & Choudrie, J. (Eds.), *Handbook of Research on Global Diffusion of Broadband Data Transmission* (*Vol. 2*). Hershey, PA: IGI Global.

National Research Council. (2002). *Broadband: Bringing Home the Bits* (pp. 62–81). Report of the Committee of Broadband Last Mile Technology, Computer Science and Telecommunications Board, National Academy Press.

Nee, R. V., & Prasad, R. (2000). *OFDM for Wireless Multimedia Communications* (1st ed.). Norwood, MA: Artech House, Inc.

Niyato, D., & Hossain, E. (2007). Wireless Broadband Access: WiMAX and Beyond-Integration of WiMAX and WiFi: Optimal Pricing for Bandwidth Sharing. *Communication Magazine*, *45*(5), 140–146. doi:10.1109/MCOM.2007.358861

Odinma, A., Oborkhale, L., Eng, M. & Kah, M. (2007). The Trends in Broadband Wireless Networks Technologies. *The Pacific Journal of Science and Technology, 8*(1).

Papacharissi, Z., & Zaks, A. (2006). Is Broadband the Future? An analysis of Broadband Technology Potential and Diffusion. *Telecommunications Policy, 30*(1), 64–75. doi:10.1016/j.telpol.2005.08.001

R Development Core Team. (2005). *R: A Language and Environment for Statistical, R Foundation for Statistical Computing.* Paper presented in R Development Core Team Meeting, Vienna, Austria. Retrieved from http://www.R-project.org

Ramsay, J. O., & Silverman, B. W. (1997). *Functional Data Analysis*. New York: Springer.

Roller, L. H., & Waverman, L. (1996). The Impact of Telecommunications Infrastructure and Economic Development. In Howitt, P. (Ed.), *The Implication of knowledge based Growth of Micro-Economic Policies*. Calgary: University of Calgary Press.

Romer, P. M. (1989). Capital accumulation in the Theory of long-run Growth. In Barro, R. J. (Ed.), *Modern Business Cycle Theory*. Cambridge, MA: Harvard University Press.

Romer, P. M. (1990). Endogenous Technological Change. *The Journal of Political Economy*, *98*(5), 71–102. doi:10.1086/261725

Sawyer, A., Allen, J. P., & Lee, H. (2003). Broadband and Mobile Opportunities: A socio-Technical Perspective. *Journal of Information Technology*, *18*(2), 121–136. doi:10.1080/0268396032000101171

Sein, M. K., & Harindranath, G. (2004). Toward Understanding the Role of ICT in National Development. *The Information Society*, *20*(1), 15–24. doi:10.1080/01972240490269942

Stiroh, K. J. (2002). Are ICT Spillovers Driving the New Economy? *Review of Income and Wealth*, *I*(48), 33–57. doi:10.1111/1475-4991.00039

Subramaniam, G., & Varadhan, R. (2008). Borrowing Strength in Time Series Data Mining. In *Proceedings of American Statistical Association Conference, JSM, 2008.*

Subramaniam, G., & Varadhan, R. (2009). *UseR!2009*. Automatic Numerical Differentiation of Noisy, Time-Ordered Data. In R.

Tajiri, N., & Okazaki, T. (2006). *Policies for Broadband Diffusion in Developing Countries*. Working Paper, Waseda University, GITS, Japan.

Vaughan-Nichols, S. J. (2004). Achieving Wireless Broadband with WiMAX. *Computer*, *37*(6), 10–13. doi:10.1109/MC.2004.1266286

Wildman, S. (1992). *Information Technology, Productivity and Trade Implications*. Working Paper No 521, Columbia Institute for Tele-Information.

Yun, K., Lee, H., & Lim, So-Hye (2002). *The Growth of Broadband Internet Connections in South Korea: Contributing Factors.* Working paper, Asia-Pacific Research Center, Stanford University. Retrieved from http://APARC.stanford.edu

KEY TERMS AND DEFINITIONS

ICT (Information and Communications Technology): Technology of information retrieval, conversion, storage, protection, processing and transmission.

BB Technology (Broadband Technology): Refers to transmission capacity with sufficient bandwidth to permit combined provisioning of voice, data and video. Broadband does not refer to any specific lower limit of speed of transmission or bit-rate as it is evolving with technological dynamics and consumer demand.

BB Access Networks (Broadband Access Networks): Broadband access network, often shortened to 'Broadband Internet', is a high data transmission rate Internet connection. DSL and Cable modem, which are popular consumer broadband access technologies, are typically capable of transmitting faster than a dial-up modem.

BB Penetration (Broadband Penetration): A measure for extent of access to broadband communications within the population of a particular location where location could refer to a region, a country, or the entire world. Normally, it is computed as a ratio between number of subscribers and total population for a particular location. Often, this ratio is multiplied by 100 to represent broadband penetration per100 inhabitants.

Broadband Diffusion: Refers to the spread of deployment and adoption of broadband technology.

Broadband Deployment: Refers to development of broadband networks or infrastructure through which broadband services can be delivered.

Mobile Wireless Broadband: Mobile broadband wireless access which may be defined as wireless access application(s) in which the location of end user termination is mobile and speed of data rate is 1-4 mbit/s.

Broadband Subscribers (BB subscribers): Fixed broadband Internet subscribers refers to a subscriber who pays for high-speed access to the public Internet (a TCP/IP connection), at speeds equal to, or greater than, 256 kbit/s, in one or both directions. This can include cable modem, DSL, leased lines, fibre-to-the-home/business and other broadband subscribers. This total is measured irrespective of the method of payment.

ENDNOTES

1 The views expressed in this chapter (without any reference) are solely those of authors and do not represent the views of AT&T. We express our sincere thanks to Carroll W. Creswell, Richard N. Clarke and Pradeep Bansal of AT&T for their helpful suggestions and comments

2 See Stuckmann and Zimmermann, 2007

3 See the book Broadband Services:Business Models and Technologies for Community Networks (editors:
I. Chlamtac, A. Gumaste and C. A. Szabo, 2005, John Wiley & Sons Ltd.

4 With Broadband 'dial-up' Internet access, users plug a computer into their phone line via modem and dial some specific telephone number to access Internet connections. With its 'always on' characteristics, broadband internet connection on the other hand, eliminates the need for dial-up service altogether.

5 Due to high fixed cost, BB access providers' choice of technology is dependent on the existing network which they can use for provisioning BB service with minimum cost

6 BPL successfully applied in Iceland and Japan and BB over Railway infrastructure was tested by trial project in India.

7 In USA, by deploying Fiber cable network, AT&T and Verizon became notable competitors for TV market by offering their U-Verse and FIOS service respectively.

8 Few of them are Odinma, Oborkhale, Eng and Kah, 2007, Niyato and Hossain, 2007. Nee and Prasad, 2000, Biagioni, Fantacci, Marabissi and Tarchi, 2009

9 See Lehr and McKnight, 2003

10 As documented in the literature on new growth theory (Romer, 1989, 1990; Barro, 1990, Mankiew, 1992; and Lucas, 1998, David, 2001), human capital and accumulated stock of knowledge accelerate economic growth by inducing technical change in

the growth process. Access to knowledge, however, is dependent on availability of information and communications infrastructure systems. While a well-developed information infrastructure can serve as a catalyst to the growth process by facilitating rapid diffusion of knowledge, inadequacy of such an infrastructure can constrain growth by limiting access to sources of technical change

11 The availability of BB infrastructure in the host countries also attract FDI and encourages job migration from foreign companies to host nations (out Sourcing)

12 Estimate obtained from ITU website, ICT Statistics News-blog

13 ITU database of June 2009 has incomplete data for 2008. Therefore, we used the data for 2007

14 On March, 2008, in ICT Statistics Newlog hosted by ITU (www.itu.in), a report was posted with distribution of world BB subscribers by technology for Q4 of 2007. According that report, 65.19% of world BB subscribers use DSL, 21.95% use Cable Modem and 12.86% use other technology for BB connections.

15 It is important to highlight here is that most of the countries listed in Table 4 are mall countries in compare to Western country like USA or Asian country such as India or China. In large countries, individual states or sections may have broadband penetration that exceeds that of these similarly sized small country leaders. Extensive rural areas in large countries may also make deployment of nationwide infrastructure a bigger job than that in a small country. In addition to that, nation's population and urbanization structure also play important role towards achieving high BB penetration rate.

16 As a fixed BB subscriber (based on ITU definition) refers to a connection between a physical location and a point of presence providing access to the Internet, there could be multiple users at that location (in the case of multi-users HH and multi-users business). Hence one can expect that the number of users will be far higher than the number of subscribers counted in this manner. Further the fixed BB speed includes an open range with minimum speed of 256 Kbit/s in at least one direction.

17 Mobile BB subscribers represent here number of mobile BB subscriptions and ITU' defined mobile BB subscriptions as number of subscriptions to mobile cellular networks with access to data communications (e.g. the Internet) at broadband speeds (here defined as greater than or equal to 256 kbit/s in one or both directions)

18 One important reason for that is the size of USA in comparison of other countries in the list of top 10.

19 One such issue is the subscription to multiple SIM cards by a single subscriber which is much dependant on the country specific roaming price structure. In addition to that, in some countries Pre-paid subscriptions dominate the Post-paid subscriptions. Such differences in subscription behavior can generate bias in measuring count of mobile BB subscriptions across countries. Many discussions took place in 7th World Telecommunications/ICT indicators meeting at Egypt in March, 2009 to highlight such issues and challenges related to measuring mobile broadband access and mobile broadband subscribers/subscriptions.

20 The observed behavior in growth rate is based on the underling definition of fixed and mobile BB subscribers used in ITU database and any modification of either fixed or mobile BB subscribers can change the interpretation of the behavior.

APPENDIX A

Data definitions provided by ITU (2009)

Cable Modem Internet Subscribers: Internet subscribers using modems attached to cable television networks. Speed should be equal to, or greater than, 256 kbit/s, in one or both directions

DSL Internet Subscribers: Internet subscribers using Digital Subscriber Line (DSL) technology. Speed should be equal to, or greater than, 256 kbit/s, in one or both directions

Fixed Broadband Internet Subscribers: Total fixed broadband Internet subscribers refers to a subscriber who pays for high-speed access to the public Internet (a TCP/IP connection), at speeds equal to, or greater than, 256 kbit/s, in one or both directions. This can include cable modem, DSL, leased lines, fibre-to-the-home/business and other broadband subscribers. This total is measured irrespective of the method of payment.

Total Fixed Broadband Subscribers Per 100 Inhabitants Number of total broadband subscribers divided by the population and multiplied by 100.

Mobile Broadband Subscriptions: Number of subscriptions to mobile cellular networks with access to data communications at broadband speeds greater than or equal to 256 kbit/s in one or both directions.

Mobile Broadband Subscriptions Per 100 Inhabitants: Calculated by dividing the number of mobile broadband subscriptions by the population and multiplying by 100.

Main (Fixed) Telephone Lines in Operation: A main line is a (fixed) telephone line connecting the subscriber's terminal equipment to the public switched network and which has a dedicated port in the telephone exchange equipment. It may not be the same as an access line or a subscriber. The number of ISDN channels is included.

Mobile Cellular Telephone Subscriptions - (Post-Paid + Pre-Paid): Refers to the subscriptions to a public mobile telephone service and provides access to Public Switched Telephone Network (PSTN) using cellular technology. This can include analogue and digital cellular systems. This also includes subscriptions to IMT-2000 (Third Generation, 3G).

Note: For farther detail definition, see ITU (2009).

Appendix B

Functional data arise from data consisting of repeated measurements of objects over time. Functional data analysis or FDA basically involves estimating individual curves often from noisy data. Let us consider a large array or a N x p matrix of time series Y (tij) for i = 1,...N where N is the "number" of time series and discrete sampling points j = 1,... p and p < N. We use the functional data analysis (FDA) approach to analyze the broadband subscription data for both fixed and wireless access. FDA is a fairly new technique, due to Ramsay and Silverman (1997). The basic philosophy of FDA is that we should think of observed data as functions as opposed to a vector of observations in discrete time.

In the ITU telecommunications indicator database, we have times series of subscription data for 217 countries by various wireline and wireless technologies like landline, wireless telephony, and internet access using several technologies like dial-up, DSL, Cable, ISDN and wireless broadband. Our focus is in particular, understanding the dynamics of fixed broadband subscription and its impact on growth of wireless broadband. We use functional analysis to model adoption for fixed broadband data for different countries and to understand if the trend in these growth curves varies in longitudinal or spatial aspects across different countries or if products exhibit different magnitudes of dynamics. We refer to these subscription data over time as a set of functional observations and the first step in FDA involves converting raw data into functional form. This is called "recovering, from the observed data, the underlying functional object.

Ramsay and Silverman recommend the use of penalized smoothing splines for smoothing. Subramaniam and Varadhan (2008) and Varadhan and Subramaniam (2009) proposed several alternative approaches for e.g., using semi-parametric models, local polynomial smoothing with plug-in bandwidth or the Gasser-Muller kernel global plug-in bandwidth for smoothing. These methods allow for automatic estimation of the smoothing or bandwidth parameter. This parameter provides the tradeoff between bias and variance for the estimated models. Our analysis utilized the Gasser-Muller kernel global plug-in bandwidth method (T. Gasser et.al., 1991) to smooth the time series of BB penetration. *R* (2005) was used for all the statistical computations, and in particular, the "lokerns" library in *R* was used to fit these models. The advantage of using Gasser-Muller method is that it automatically estimates the bandwidth or the smoothing parameterusing a data driven procedure which estimates asymptotically optimal bandwidth from the data. For this, it uses a mathematical expression that describes the asymptotic value of mean integrated square error and that expression contain unknown functionals. We "plug-in" values for the unknown which is estimated from the sample data. The asymptotically optimal bandwidth b_A: is obtained by minimizing a large sample approximation to mean integrated square error (MISE):

$$b_A = \left(\frac{1}{n} \cdot \frac{c_1}{c_2} \cdot \frac{\sigma^2}{\int_0^1 v(t) r''(t)^2 \, dt} \right)^{\frac{1}{5}}$$

with constants and a known function. The residual variance σ^2 and the functional $\int_0^1 v(t) r''(t)^2 dt$ are unknown and estimators have to be plugged into the expression for to make it an estimator for the optimal bandwidth (plug-in estimator). The details are given in T. Gasser et.al. (1991) and Gasser, Th et. al (1984).

Once the data are represented by functional objects we extend the analysis to extract features from the functional. Once such feature extraction we use here are the estimated first and second derivatives of the functions. The derivatives give further insights on rate of change in adoption and identify the different growth spurts associated with the curves.

Appendix C

These paper select eight countries for describing the experience of cross over points of landline versus mobile telephone: These countries are USA, Canada, Korea, Hong Kong, Denmark, Netherland, China and India. We selected these 8 countries to represent the observed diversion in landline to wireless migration patter in voice telephone across various countries around the world. Using historical data for the time period 1999 – 2007, we plot the subscribers of landline vs. wireless phone for all these 8 countries which are shown in Figure-C below.

For each graph shown above, the vertical axis is displaying number of subscribers in million and the horizontal axis is displaying corresponding time period. The graphs show that in all these countries, number of wireless subscriptions crossed the number of landline subscriptions by 2006. However, different country reached this cross point at different times between the sample periods 1999-2007. Countries such as Korea and Hong Kong, being pioneer in telecommunications advancement, have reached the cross over point in early 2002 whereas other European countries such as Denmark and Netherlands reached that cross over point during 2005-2006. On the other hand, even though most of the technologies were available in USA, still the wireless-landline crosses over point took place in lot later than other many other countries. Same is the scenario observed in Canada. The primary reason behind this could be that landline infrastructures are so highly developed in these two countries and the landline penetration rate is so high in each of these countries that phone subscriber's need of using wireless phone is less in comparison to other countries with poor landline infrastructure. On the other hand, the country like India and China, the landline penetration was not that high at the time when people around the world started to use wireless phones due to lack of sufficient landline infrastructure and affordable price for service. However, in recent past these countries are very rapidly increasing their wireless infrastructure

Figure 14. Number of subscribers for landline and mobile phone and the total (Year = 1999 to 2007)

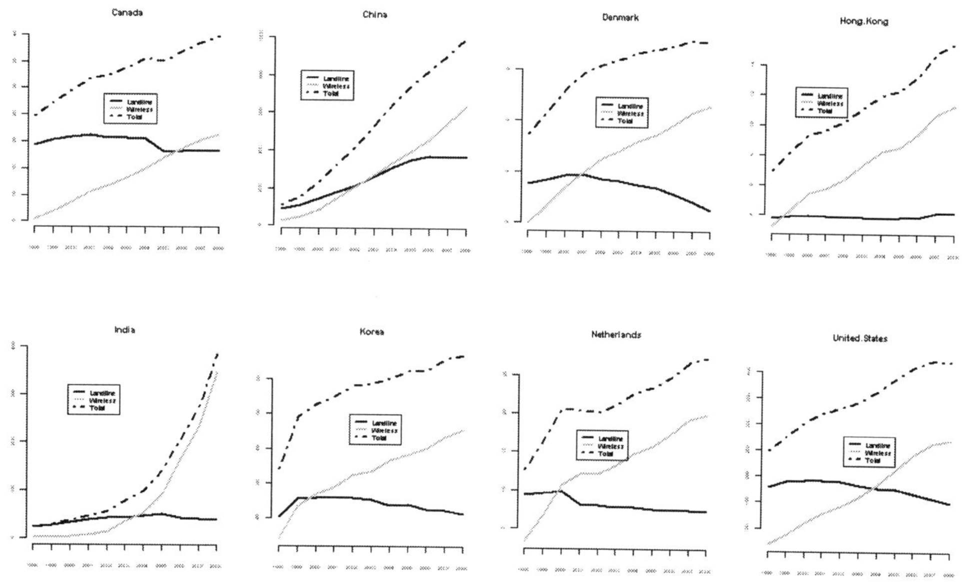

compared to their landline infrastructure in recent years due to cost advantage of deploying wireless infrastructure. Economic conditions also improving for many of these countries. Thus for the case of India and China, we not only observe the early cross over points but also the majority of the growth in telephone subscribers belong to wireless subscribers pool.

Appendix D

Table 9. Forecasts for number of fixed and mobile subscribers (millions)

Forecasts\Country	Korea		Japan		USA	
Subscribers in millions	Fixed BB	Mobile BB	Fixed BB	Mobile BB	Fixed BB	Mobile BB
2009	16.3	50.0	31.1	126.7	90.0	121.7
2010	17.7	58.3	34.1	149.2	100.9	144.2
2011	18.4	65.0	35.7	167.4	108.7	162.4
2012	18.7	70.6	36.5	182.5	112.0	177.5
2013	18.9	75.0	36.8	194.4	113.4	189.3

Chapter 15
Strategic Moves Related to Broadband Diffusion on the French VoD and TV Market

Nabyla Daidj
Telecom Business School, France

Pierre Vialle
Telecom Business School, France

ABSTRACT

The launch of triple-play services, including IPTV can be considered as a "critical event" having a strong structural impact on the market. In this paper, we analyse the evolution of the VoD supply in France, and in particular the effect of the launch of IPTV systems. Firstly, the launch of attractive triple-play offers has induced a strong development of the French broadband Internet access market, which in turn has resulted in the entry of numerous players in the VoD market. Secondly, the adoption of IPTV system has changed the competitive and strategic positioning of main players in the VoD value-chain, thus inducing or amplifying strategic moves along this value-chain. In particular, we show how the diffusion of IPTV systems has induced a switch from complementary convergence to competitive convergence

INTRODUCTION

The emergence of the so-called multimedia industry, based on the expected merging of three vertical industries (telecommunications, computing and the media) into a new horizontally layered one, raised tremendous interest at the end of the 1990s. There was a great deal of conjecture about how this new industry would be structured and what types of players would be successful in positioning themselves in this emerging value chain (Collis et al., 1997; Vialle, 1998). In fact, it is only recently that the actual patterns of convergence can be observed and analysed (i.e. Vialle, 2007). According to Greenstein & Khanna (1997), convergence may include competitive as well as complementary dimensions. Different views on this issue have been expressed. For example, while Collis et al. (1997) have evoked a competition scenario based on substitution between actors and services, Chan-Olmsted & Khanna (1997) have focused on a complementary convergence scenario based on differentiated services and infrastructures.

DOI: 10.4018/978-1-60960-011-2.ch015

Video on Demand (VoD) probably one of the best publicised examples of a multimedia service, as VoD can be considered as exhibiting a symbolic significance for convergence. The supply of VoD services requires the combination of resources initially possessed by players from telecommunications, computing and audiovisual industry. It also exemplifies the disrupting character of convergence, as this is a substitute for more traditional distribution channels, such as movie theatres, broadcasting TV, or DVDs. In this paper, we identify a "critical event" which strongly impacts VoD business as well as other services: the launch of "triple-play" services in France and in particular of IPTV.

In order to analyse the effect of the launch of IPTV systems, we draw from the competences and resources approach (i.e. Barney, 1991; Prahalad & Hamel, 1990; Teece et al., 1997), combined with an overall evolutionist perspective (i.e. Nelson & Winter, 1982). Therefore, we put the emphasis on the critical resources and competences which are at the heart of actors' competitive position. We also consider the learning processes and trajectories, as well as the changing amount and value of resources, which allow them to operate strategic moves.

This critical event also allows us to broadly distinguish between two phases of convergence. During the first phase, major players are strongly constrained by their past trajectories and their current stock of resources and competence. They are subject to strong information asymmetries as they have a low level of understanding of other industries' business and tend to remain on their core resources and capabilities. Moreover, the business prospects of services such as VoD remain uncertain, so that they adopt a rather cautious approach in their strategic moves. This is a phase of rather complementary convergence. After the launch of triple-play services, VoD also exemplifies a second stage of convergence in which players do not remain on their initial resources and competences bundle, move along the value chain, and eventually enter each other's business. It is a more competitive convergence. Players from the various industries have become more familiar with other companies' business by interacting with them for the production and marketing of new services. The more sizeable business prospects also contribute to reduce the perceived risk of more adventurous strategies. This is particularly true of telecom operators which benefit from an increased market power. In the case of Orange, the learning accumulated with VoD business, the extension of a critical resource (the installed base of ADSL subscribers), as well as the blurring of service borders have induced a strategic move into the TV business.

In this paper, after presenting the VoD business and context, we first analyse the initial situation of the French broadband Internet and VoD market in the pre-IPTV complementary convergence period, and introduce the "critical event" represented by the launch of triple-play services. Then we analyse the entries of different companies and the inter-organisational arrangements used. Finally, we show how the change in the balance of power induced by IPTV has amplified the strategic moves of major market players, inducing the emergence of a competitive convergence dimension. We conclude by outlining competitive and regulatory implications.

BACKGROUND: VIDEO ON DEMAND BUSINESS AND CONTEXT

VoD is a service providing video programs (typically movies, but other programs as well) as requested by the customer in a catalogue and delivered at a time chosen by the customer.

Movies are distributed through multiple channels in sequential release windows in a specific order[1]. The majority of the revenues generated by a film are received within 18 months of a film's distribution cycle (Table 1). This cycle is based on potential revenue along the different market

Table 1. Release windows in France for movie distribution channel (Source Adapted from Kuhr (2008), Olivennes (2007))

Release authorisation	Exception	Distribution mode
From release approval		Movie theatre
6 months		Video cassette or DVD, authorised VoD platforms
9 months		Pay-per-view
12 months		Encrypted Pay TV
24 months	18 months	Broadcast TV (if movie co-producer)
36 months		Broadcast TV

segments, according to territoriality (by country and by linguistic zone) and time (duration and distribution rights) agreements (Debande & Chetrit, 2001). The release window timeline is designed to maximise revenues across the value chain. It may be expected that release windows will be shorter in the future, with significant effects on the VoD business. Rights holders look to digital distribution as the next source of revenue growth.

The VoD Value Chain and Critical Resources

The VoD value chain will depend on the nature of technical architecture used and on the complexity of the network. While this framework can vary according the nature of the technology used, the general structure is quite similar. Table 2 presents the VoD value-chain, structured around three main activities involving several players: content provision, service provision and distribution, and commercial distribution. The business model related to VoD services (Berne, 2003; Berne et al., 2003) relying on the sharing of revenue across the value chain is still a key issue and differs from country to country[2].

From a conceptual perspective, the Actors-Resources-Activities model (Axelsson & Easton, 1995) can be useful in order to analyse a value-chain, which are also a network. In the case of VOD content delivery, the detailed functions for each activity show the critical role of service provision in the value chain, comprising technical, marketing, and IPR management functions. Some players can be positioned on more than one activity. Each layer requires critical resources and competences.

THE TWO PERIODS OF VOD: PRE-IPTV AND POST-IPTV

The Pre-IPTV Situation and the Launch of Triple-Play Services

Moviesystem's Quasi-Monopoly

During the pre-IPTV period, VoD services were marketed via websites or Portals, and were usually viewed on a PC. Customers had to engage into specific transactions with the VoD service providers in order to view movies, i.e. consult a movie catalogue online, select a movie, pay for it, and download the movie (or view a streaming movie). This context conforms to the complementary convergence scenario presented by Chan-Olmsted & Khanna (2003), in which PCs and TVs distribute differentiated services.

The French VoD market started with the quasi-monopoly of a firm called Moviesystem, later bought by Canal+ and renamed Canalplay. The case of the first mover on the French market

Table 2. The VoD value chain (Source Adapted from Guillou (2004))

Activity	Function	Critical resources and competences	Actors
Content provision	produce movies, manage rights, market programs	Access to creative inputs Exclusive rights catalogue	Content providers, movie studios, film companies, broadcasters
Aggregation and services provision*	Stock films on servers (based on technologies for encoding such as digitalisation and compression) and make them available for distribution. Multiple functions, i.e.: negotiate rights with content providers, ensure the protection of these programs by appropriate encryption, produce program guides and graphic interfaces, set up and implement a commercial policy, manage invoicing and payment, negotiate agreements with physical distributors (network operators) for integration in the platform offer, either under the service provider's own name, end distributors name, or using "co-branding."	Selective access to content Content market knowledge Technical system integration and management expertise Access to distributors	Various players (aggregators, video service providers)
"Physical" transport and/or commercial distribution	Construct a server-client architecture dedicated to VoD (including a billing system) provide the infrastructure and access networks to carry VoD	Access to customers Efficient infrastructure Innovation capability	Access network providers, network operators, ISPs

is particularly relevant for understanding the specificities of the VoD business. Created in March 2000, Moviesystem has been the pioneer of VoD in France. Starting from a traditional activity of purchasing the exclusive rights for movies and wholesale distribution of VHS cassettes to professionals, Moviesystem entered the VoD market with the acquisition of a VoD Internet service, NetCiné.com. Moviesystem was initially supported by Pathé and Europa Films, which provided funds and the VoD rights for their catalogues. Consequently supported by innovation funds, Moviesystem signed numerous agreements with content providers and claimed to manage at the end 2004 a catalogue of 1,500 films from 70 production companies. Moviesystem's three main activities in VoD business were the purchase and management of exclusive rights, the development and operation of a specialised middleware platform and the commercial operation of VoD services.

Moviesystem addressed the VoD market through both direct and indirect distribution. With regard to direct distribution, Moviesystem operated its own service directed at Internet users, called NetCine. NetCine's web site offered films for streaming that are grouped into various categories (such as news, comedy, drama, children, adult, etc).

With regard to indirect distribution, Moviesystem mainly developed partnerships with ISPs. Under each ISP's brand name, NetCiné's services were distributed on the main French ISPs portals, whether focused on ADSL, such as France Telecom's Wanadoo, Club Internet, 9 Télécom and AOL, or on cable TV (Noos). The indirect distribution channel allowed Moviesystem to address a potential market of roughly two-thirds of broadband access subscribers. Moviesystem was also able to provide its distributors with a complete and customised solution. Moviesystem was in charge of the updated provision of movies, the analysis of customers' data, the customisation of the user interface, and the management of digital rights whereas distributors were responsible for providing the adequate infrastructure, including servers located near end-users and for billing users. The revenues were split as follows: rights owners got 50% of revenues, an average of 5% was spent for the billing platform used and the remaining 45% was more or less equally shared between Moviesystem and ISPs.

The initial success of Moviesystem is representative of the first phase of convergence, during which the capabilities of players were significantly constrained by their past experience. Moviesystem has been able to bridge two different industries by significantly reducing transaction costs resulting from strong uncertainties and from information asymmetries (Daidj et al., 2004). On the one hand, firms in the media industry traditionally distributed their films via movie theatres, TV channels or video rental outlets. They were therefore not familiar with dealing interactively with end customers, such as analysing their behaviour and preferences and managing direct or indirect transactions with them. They were also not accustomed to the technical and management aspects of computing and communications systems. ISPs and network operators, on the other hand, lacked experience in managing a catalogue of films. It requires artistic and marketing capabilities, such as the ability to assess the commercial value of films on this complementary market, to package them, and to put them together to provide an attractive offer. However, the launch of triple-play services has triggered a second phase of convergence, marked by the emergence of competitive convergence.

The Critical Event: The Launch of Triple-Play Offers on the French Market

Triple play services have been increasingly popular in France due to the initiative taken by the ISP "Free" (Iliad Group) to launch the first triple-play offer in Winter 2003, including ADSL2 Broadband Internet, several IPTV channels, and free telephone calls to fixed subscribers (now to 70 countries), for a monthly subscription of 29.90 Euro[3]. Free was subsequently imitated by the other ADSL services providers proposing similar offers. The high level of competition and innovation in the French market has resulted in strong growth, leading to a penetration rate exceeding 50% of TV households since 2006, and reaching a total of 15.55 million subscribers at the end of 2007; a market which is largely dominated by ADSL based broadband services (Table 3)[4]. With 4.1 million IPTV subscribers, the French market exhibits the highest number of subscribers in Europe (NPA Conseil, 2006, 2008).

The IPTV Period: Market Change and Strategic Issues

The launch of IPTV systems and the high market growth resulting from the provision of inexpensive triple-play services have induced significant

Table 3. Broadband Internet subscribers in France(Source ARCEP (2008))

Subscribers (million)	2000	2001	2002	2003	2004	2005	2006	2007
Broadband Internet	0.196	0.602	1.655	3.569	6.561	9.471	12.695	15.550
ADSL	0.068	0.408	1.368	3.172	6.103	8.902	12.019	14.800

changes on the French market. Numerous actors have entered the marked and followed a trajectory according to their initial stock of resources and competences. Therefore, they rely, to a variable extent, on partnerships and alliances in order to complement their own resources.

Market Entries on the French VoD Market

The launch of attractive triple-play offers has not only resulted in a strong market growth, but also induced numerous market entries. Since 2005, the French VoD market has been characterised by multiple entries from players. The French VoD market has become very competitive, exhibiting the highest number of competing services in Europe, with 32 different services offered (NPA Conseil, 2008) (Table 4).

Table 4 shows that the high number of VoD services in France is partly due to the high number of VoD services linked to IPTV. IPTV systems allow offering a much more user-friendly way of distributing VoD services. They allow consumers to view content on their TV, using the IPTV graphic interface and remote control. Consumers are then billed by their ISP. Therefore, IPTV has become the main distribution channel for VoD: 80% of VoD content is now viewed via an IPTV system (NPA Conseil, 2008). Attracted by the high growth of the French residential broadband market, numerous players have entered the VoD service market from 2005 onwards (Table 5). However, the various players have adopted different entry strategies (Daidj & Vialle, 2007).

Telecommunications companies and ADSL service providers are naturally leading the market due to their hold on their subscribers with their IPTV box which allows them to control the most appropriate viewing terminal for VoD, the television. The competitive pricing of ADSL services leads them to push for the adoption of other services such as VoD in order to improve their profitability. According to their resources, they can either provide their own services by negotiating with rights owners (such as France Telecom/Orange with Pathé), or distribute the services of another service provider (such as Alice with VirginMega).

Content aggregators can be divided into two categories. TV companies try to take advantage of their TV program catalogue (reports, documentaries, games, and series) that they complement with movie catalogues. They benefit from their traditional relations with producers, from their market knowledge and their brand image. The other category is composed of niche players that are specialized in a specific topic such as Vodeo (reports, documentaries), Imineo (education), and Cinezine (independent producers). They distribute their VoD services on their portals, but may also distribute their services via other service providers, such as TF1Vision with Neuf Cegetel, or Vodeo with Noos.

Retailers have expanded their operations to the electronic distribution of movies, taking advantage of their brand image and on their market

Table 4. Commercial VoD services in selected European countries (Source NPA Conseil, 2008)

Country	Number of Suppliers	Internet Services	IPTV Services	Cable Services	Satellite Services	DTTB Services
France	32	26	13	1		
Netherlands	30	27	3	1		
UK	16	11	4	4	1	1
Germany	26	20	4	2	2	

Table 5. Entries on the French VoD market (Source DAIDJ & VIALLE (2007))

2002	2006
Moviesystem (Bought by Canal+ in 2004)	Arte VOD
2003-2004	VirginMega (Retailer)
France Telecom: Ma Ligne TV (ADSL operator)	FNAC
2005	Neuf Cegetel (ADSL operator)
Canal+: Canalplay (Ex Moviesystem) (Pay-TV)	Alice (ADSL operator)
Vodéo (Content Aggregator)	MSN (Search engine/Portal)
TF1: TF1vision (Broadcast TV)	Google (Search engine/Portal)
TPS: TPSVOD (Pay-TV)	Yahoo (Search engine/Portal)
France Télévision (Broadcast TV)	Cinezime (Content aggregator)
M6 Video (Broadcast TV)	Universcine (Content aggregator)
Editions Montparnasse (Producer)	Toondra (Content aggregator)
Free (ADSL operator)	Glowria (Retailer)
AOL (ADSL operator)	Imineo (Content aggregator)
Club Internet (ADSL operator)	Paris Première video (Cable TV channel)
	Orange (24/24vidéo) (ADSL operator)

knowledge. FNAC has chosen to subcontract its VoD service to Glowria, while its main competitor VirginMega has decided to develop its own platform. Glowria is another type of retailer initially specialized in renting DVDs by Post, following the model of Netflix in the USA. This company has invested in a platform and a catalogue of movies, and distributes its services directly through its website, through agreements with distributors reselling its services under their own brand, or providing its technical platform without providing movies.

Global players such as Apple or Amazon also threaten to become competitors. Amazon has launched its VoD services in the USA, Amazon_Unbox, and may expand its target market to other countries, such as France. Apple has launched its iTune Video Store in the USA and plans its extension to Europe.

Foreign search engine and portal companies have also entered this market with a different strategy that we can qualify as "disruptive" by analogy with the concept of disruptive innovation by Christensen (1997). Google (with YouTube), MSN and Yahoo provide free of charge VoD services which cannot compete in either terms of quality or choice with paid services. However, they outperform traditional services in two dimensions, namely price and the participation of users. By attracting masses of users viewing and producing videos, they may pave the way for more commercial services.

None of these actors possess all the necessary resources that are necessary for the VoD value chain. Therefore, they rely, to a variable extent, on partnerships and alliances (Le Goff & Mouline, 2003). As already mentioned, services providers, for example, have to partner with media companies and content providers (Table 6). These alliances will enable the service providers to deliver not only attractive digital content to the mass market, but also customized content to specific customer segments.

Table 6. Partnerships in the French VoD industry (Source Daidj & Vialle (2007))

Telecom operators/ADSL operators	
France Telecom (Orange)	Partnerships with Broadcast TV (Arte, TF1, France Télévisions), aggregators (Zooloo Kids) and content providers (Warner Bros)
Neuf Cegetel	Partnerships with Glowria (platform) and TF1 Vision (content)
Alice	Partnership with VirginMega (platform)
Content aggregators	
Vodeo	Partnerships with other service providers (e.g. Noos)
Retailers	
FNAC	Partnership with Glowria (replacing Moviesystem)
Glowria	Partnerships with several service providers
Search engines and portals	
MSN	Partnerships with producers (Endemol), TV challens (i>TELE, Eurosports, Canal J, Filles TV) and Music majors (Sony and EMI)
Google	Partnerships with INA, CanalPlay, Vodeo, Arte
Yahoo	Partnerships with Dailymotion, TV5, INA, Vodeo

The Emerging Competitive Convergence Dimension

One significant impact of the diffusion of IPTV systems is the progressive emergence of a competitive convergence situation. This phenomenon is accelerated by the blurring of borders between services, because of innovation such as Subscription VoD (SVoD), and Catch-up TV. Major players on the French market have made significant strategic moves, in particular by incorporating or developing new activities in stages of the value-chain different than the ones they were initially positioned on.

VoD therefore exemplifies a second stage of convergence in which players do not remain on their initial resources and competences bundle due to strong information asymmetries. Players from the various industries have become more familiar with other companies' business by interacting with them for the production and marketing of new services. We present here some significant strategic moves.

From Packaging (TV) to VoD Service Provision

A first type of strategic move observed on the French market is TV companies, located on the "TV packaging" or "TV edition" stage of the value chain, moving to another packaging position as a VoD service provider, by acquisition or development (Figure 1). The case of Canal+ is particularly representative of this type of strategic move.

Canal+

The first historical strategic move has been the acquisition of Moviesystem by Canal+, which marks the start of a strategic waltz between major players. In May 2004, Moviesystem was bought by Canal+ group, the main French Pay-TV company and changed its brand names (Moviesystem and NetCiné) to CanalPlay, which proved to be a mixed blessing. On the one hand, its business could benefit from the resources of a large company and from various synergy opportunities, however, on the other hand, it was losing the advantage of being an independent player on the market.

Figure 1. From TV channel to VoD services move

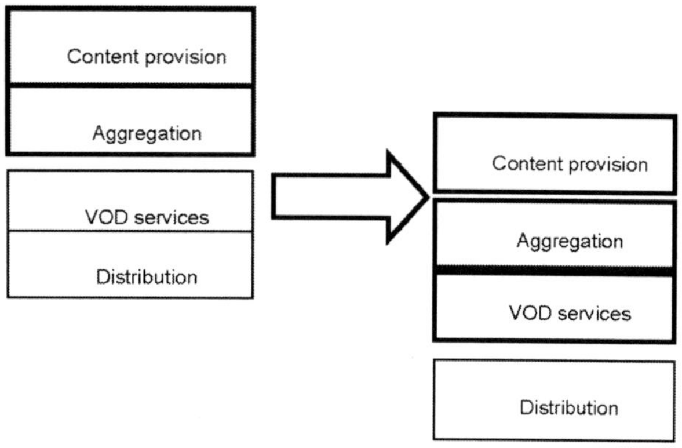

In particular, the fact that Canal+'s main shareholder, Vivendi Universal, was also a major shareholder of France Telecom's major competitors on the fixed and mobile markets, influenced the decision by France Telecom to lower its dependence on CanalPlay for VoD services. When France Telecom launched its IPTV service, "Ma ligne TV" in 2003/4, it was mainly relying on Moviesystem's services. Progressively, France Telecom has implemented a policy of negotiating directly with rights owners, first for its IPTV VoD services, then for its portal-based VoD service. CanalPlay is still a significant provider to France Telecom/Orange among others but it is difficult to assess its position. Currently, CanalPlay offers more than 2,000 videos on demand, 1,000 of which are films made for cinema. It claims that 2.4 million videos were purchased via CanalPlay in 2006 and, since January 2007, customers can download permanently and burn films on DVDs. In 2005, Canal Play became a provider of VoD services for Free IPTV services.

Since September 2009, the three privale and competitors channels Canal+, TF1 and M6 have started negociations to reach an agreement to propose a common catch-up TV platform. This "coopetitive strategy" (Bengtsson & Kock, 2003) adopted by these TV networks, has two main goals. The first one is to create entry barriers because of the threat of new entrants such as Hulu (the American platform managed by NBC Universal, News Corp. and The Walt Disney Company who own rights on many famous series among them: The Office, Desperate Housewifes, Dr House). The second objective is to create a sustainable business model based on online advertisements at higher prices.

TF1

Another move was made by TF1, the French leading broadcaster, which followed a different strategy. In 2005, TF1 launched a streaming VOD service (TF1 Vision). TF1 strategy relies on purchasing popular contents' rights and on partnerships with the major American and European studios (among them Warner, Universal, Sony Columbia, Europa Corp, Pathé). The site TF1 Vision proposes also catch-up TV with the series from TF1 Broadcasting (*Lost, Koh-Lanta...*) and Premium VOD with the season 2 of Heroes practically at the same time as its American distribution. TF1 is referenced on IPTV through the main operators: Alice, Free, and Neuf. Since July 2009, after alliances with all ISPs and several Telco's (among them: SFR, Free, Numericâble), TF1 has signed an agreement with Orange to make TF1 Vision content available on Orange's 24/24 video VOD platform in France. Until the end of 2009, 24/24 video will be offering Orange TV subscribers as well as Internet users a selection of premium TF1 Vision content including American series with the latest seasons of Fringe, Lost, Grey's Anatomy, Ugly Betty and TF1 dramas and series. The videos will be available to rent and customers will be able to watch them during a 24-hour period.

From Distribution to VoD Services and Eventually TV Packaging and Content

Another major strategic move is broadband access providers, such as telecom operators and ISPs, initially only supplying a distribution channel to VoD service providers, moving to upward stages of the value-chain (Figure 2). As for TV companies, there are some differences in their strategic approach.

IPTV systems significantly increase the market power of broadband access suppliers, which also supply the IPTV box giving exclusively access to the services they have selected (Figure 3). Via the IPTV box, they *de facto* control what is offered to subscribers. Of course, subscribers still have the possibility to go to their PC, and access any website providing VoD services, but they will have to stand higher transaction costs than if they remain in front of their TV, use their remote control and get billed by their broadband access provider. Even if subscribers have technically access to all the services offered on the Internet, they are subject to what we call "potato couch switching cost". The strength of this switching cost is already reflected in the high market share of IPTV related VoD services (80% of VoD content according to (NPA Conseil, 2008).

Orange (France Telecom)

Orange has a very offensive strategy concerning VoD. This strategy has also extended to audiovisual services, such as TV channel and content production. Orange VoD strategy must also be understood in the context of its overall strategy of "convergence operator", aiming at providing seamless services across IPTV, Internet and mobile channels.

Initially relying on Moviesystem for its portal-based VoD service, France Telecom has progressively developed its own service, "24/24 video", by negotiating directly with content providers. At the end of 2007, Orange had 7.3 billion subscribers, of which around 1.2 million have access to IPTV programs. Its "24/24 Video" service uses its catalogue of 3,000 films and 6,000 TV programmes to

Figure 2. Strategic move from distribution to VoD services

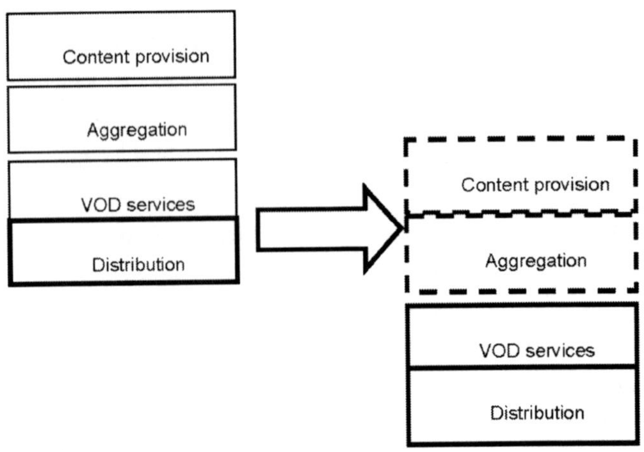

Figure 3. The competitive role of IPTV systems

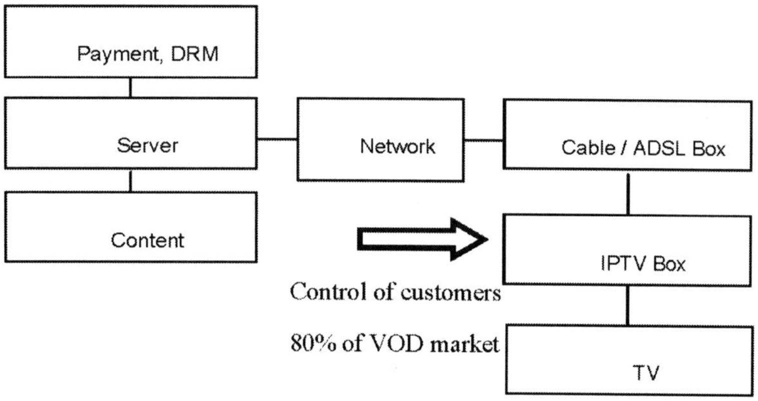

update its offer every week. In November 2006, Orange has launched a new SVoD service "24/24 Jeunesse" allowing access to series for children for a monthly fee of Euro 4.90, followed later by the launch of "24/24 Musique" and "24/24 Séries".

However the most disruptive strategic move of Orange has been the entry into TV business, resulting in a blurring of borders between VoD and TV, as well as with fixed and mobile channels. Four main events can be outlined. The first was the launch in 2007 of "Orange sport TV" for its IPTV subscribers. The second was the acquisition of three shares (out of twelve) of the soccer league championship diffusion rights, for both live and VoD diffusion, on Internet, TV and mobile. The third one is the announcement of the launch of an exclusive catch-up TV service[6] of the French public TV channels, "Rewind TV". The last (and not the least), was the announcement of an exclusive deal with Warner Bros, HBO, International Television, Gaumont, and Fidélité Films. Orange has negotiated exclusive rights for their catalogue for duration of at least 5 years, for direct diffusion and VoD, for IPTV, Internet and mobile channels. Orange has also announced the launch of six exclusive Orange new pay-TV channels diffusing this catalogue. Orange is also marginally involved in movie production, via a subsidiary called "studio 37", and created in 2007.

Free

Free (Iliad Group) is the third ADSL operator in France with 2.9 million subscribers, but the first for IPTV with around 2.4 million customers with IPTV access[7]. This is due to the fact that the triple play package has been the standard offer by Free, while other competitors offered the choice between single, double and triple play. Free initially had a strategy of giving an access to the main VoD service providers on the market: Canalplay (Canal+), TF1 Vision (TF1), M6 Video (M6), and Vodeo.tv. However, Free has also made a move towards providing its own services. After negotiating directly with Warner and Disney-ABC International Television, Free has launched a SVoD service, "Free Home Video", giving access to 50 films and 100 series every week for a monthly fee of Euro 5.90. Free has also announced the launch of a High Definition premium service for a monthly fee of Euro 9.99.

Neuf

Neuf (Neuf Cegetel) is the second ADSL operator with 3.2 million customers at the end of 2007. Resulting from the merging of several ISPs, Neuf Cegetel only launched its VoD service in December 2006. Its main offer, "Neuf VoD" mainly distributes the services from its partner Glowria. Neuf also gives access to TF1 Vision and M6 Video. Neuf represent an unknown factor on the VoD and related services market. First, one of its major indirect shareholders is also the major shareholder of Canal+, Vivendi Universal. Neuf Cegetel has been recently integrated in Vivendi Universal's mobile telecommunications subsidiary, SFR. Until now, no integrated strategy has emerged from the different components of Vivendi Universal. However, Orange's initiative in TV business is considered as a frontal attack for Canal+, and may induce significant strategic change.

CONCLUSION: COMPETITIVE AND REGULATORY IMPLICATION

In Europe, until 2004, the VOD business model was not viable for the pioneers for several reasons: high initial capital cost per end-user, high operating costs, difficulties to propose an attractive catalogue of video content, low revenue per end-user. Now with the emergence of the IPTV system, the situation is changing and one of the more significant cases is the French market.

The IPTV system implemented by broadband Internet access providers is challenging both the supply of traditional services, such as broadcast TV, but also the supply of on-line services such as VoD. The diffusion of IPTV systems and the resulting "potato couch" switching cost, combined with the blurring of borders between services and business models may challenge established regulatory approaches.

The "Potato Couch" Switching Cost

Regulatory concerns have been traditionally focused on issues of access (i.e. unbundling) and interconnection for current and new generation networks. Under the "old" Internet access way using a PC, content issues were not a concern, as subscribers can access any content supplied on the Internet. However, the "potato couch switching cost" that we have outlined shed new light on content provision issues. To some extent, the problem becomes similar to the "closed wall garden" effect identified in the case of mobile Internet.

Service and business model evolution may increase operator's hold on the VoD business. Until recently, with the exception of Orange, broadband Internet access providers were giving access to a variety of VoD service providers. However, changes in business models for VoD may well strongly affect the current situation. In line with similar initiative for music provision, operators are now launching SVoD services based on subscription. For a monthly fee, subscribers have unlimited access to a catalogue of movies and TV series. It may be expected that this service will be attractive for consumers. From a competitive point of view, customers switch from a position of occasional customer billed on a per use basis, to fully owned subscribers who are more captive. Until now, Orange only provides TV series on SVoD, but Free already provides movies.

In the new Audiovisual Media Services Directive, which modernises the European framework for content regulation, distinction is made between linear services (i.e. TV service), and non-linear services (i.e. VoD)[8]. Service and business model evolution also tends to challenge these distinctions, by inducing overlaps of service categories. Firstly, SVoD services have adopted the pay-TV subscription model, offering a choice of movies and other programs for a monthly fee. Some VoD initiative may also adopt a financing mode relying on advertising. Secondly, TV channels adopt a "on demand" feature with catch-up TV services.

Current services don't include movies, but it may change in the future. From the customer point of view, the difference between a VoD service on subscription and a pay-TV channel, which program can also be viewed on demand may not be so important.

Finally, because of technological innovation (i.e. 3G) and of increased system integration capabilities by telecommunications operators, the same content can be viewed on IPTV, Internet and mobile channels, apparently providing an advantage to convergence-based operators. This challenges approaches based on the analysis of single channels.

REFERENCES

Amit, R., & Zott, C. (2001). Value creation in e-business. *Strategic Management Journal*, *22*(6/7), 493–520. doi:10.1002/smj.187

ARCEP (2008). *L'Observatoire de l'Internet haut débit*. Retrieved October 13, 2008, from http://www.arcep.fr/i ndex.php?id=8764

Arino, M. (2007). Content Regulation and New Media: A Case Study of Online Video Portals. *Communications & Strategies*, *66*(2), 115–135.

Axelsson, B., & Easton, G. (1992). *Industrial Networks - A new View of Reality*. London: Routledge, Inc.

Barney, J. B. (1991). Firm Resources and Sustained Competitive Advantage. *Journal of Management*, *17*, 99–120. doi:10.1177/014920639101700108

Bengtsson, M., & Kock, S. (2003, May). *Tension in Co-opetition*. Paper presented at the Academy of Marketing Science Annual Conference, Washington D.C.

Berne, M. (2003). *Business models for Services: the case of Video of Demand*. Working paper, TELECOM & Management SudParis.

Berne, M., Daidj, N., & Epinette, O. (2003). IP Services Billing. *Working paper*, TELECOM & Management SudParis.

Blazquez Cabrera, F. J. (2007). Systèmes de gestion des droits numériques: dernières évolutions en Europe. *IRIS Plus*, *1*, 1–8.

Christensen, C. M. (1997). *The Innovator's Dilemma. When New Technologies Cause Great Firms to Fail*. Boston, MA: Harvard Business School Press.

Collis, D. J., Bane, P. W., & Bradley, S. P. (1997). Winners and Losers. Industry Structure in the Converging World of Telecommunications, Computing, and Entertainment. In Yoffie, D. B. (Ed.), *Competing at the age of digital convergence*. Boston, MA: Harvard Business School Press.

Conseil, N. (2006). *Le développement de la vidéo à la demande en France et en Europe,* Paris.

Conseil, N. (2008). *Vidéo à la demande en Europe: second recensement des services de VOD.* Paris.

Daidj, N., Epinette, O., & Vialle, P. (2004). Positioning of Firms on an E-business Value Chain: the Case of Video on Demand. In J. Chen (Ed.), *The Fourth International Conference on Electronic Business.* Beijing: *Academic Publishers/World Publishing Corporation.*

Daidj, N., & Vialle, P. (2007). Le développement d'une concurrence multiforme pour la VOD en France. In P. Vialle (Ed.), *Mutation des STICS. Acteurs, Ressources, Activités*. Paris: Hermes Science Publishing.

Debande, O., & Chetrit, G. (2001). The European Audiovisual Industry: An overview. *European Investment Bank (EIB) Sector Papers*, September.

Diericks, I., & Cool, K. (1989). Asset stock accumulation and sustainability of competitive advantage. *Management Science*, *35*, 1504–1511. doi:10.1287/mnsc.35.12.1504

Dosi, G., Teece, D., & Winter, S. (1990). Understanding Corporate Coherence: Theory and evidence. *Journal of Economic Behavior & Organization*, *23*, 1–30.

Guillou, B. (2004). *Online Distribution of Films*. European Audiovisual Observatory, May.

Hoarau, Ch. (2003). *Le repositionnement de l'industrie électronique et du multimédia comme support de services, nouvelles sources de création de valeur*. Observatoire des Stratégies Industrielles, Ministère de l'Economie, des Finances et de l'Industrie, septembre.

Isacowitz, R. (2003). VOD and the case for Content Protection. *Converge Network Digest*, November 17.

Kuhr, M.(2008). Media Windows in Flux. Challenges for Audiovisual Media Chronology. *IRIS plus*, 4.

Le Goff, J., & Mouline, A. (2003).*Les stratégies des firmes dans les industries du multimédia, 10 ans d'opérations de rapprochement (1993-2002)*. Paris: Ministère de l'Economie, des Finances et de l'Industrie.

Lieberman, M., & Montgomery, D. (1998). First mover (Dis)Advantage: retrospective and link with the Resource-Based View. *Strategic Management Journal*, *19*(12), 1111–1125. doi:10.1002/(SICI)1097-0266(1998120)19:12<1111::AID-SMJ21>3.0.CO;2-W

Nelson, R. (1991). Why Do Firms Differ and How Does it Matter? *Strategic Management Journal*, *12*, 61–74. doi:10.1002/smj.4250121006

Nelson, R., & Winter, S. (1982). *An Evolutionary Theory of Economic Change*. Boston, MA: Harvard University Press.

Olivennes, D. (2007).*Le développement de la protection des oeuvres culturelles sur les nouveaux réseaux*. Retrieved November, 2008 from http://www.culture.gouv.fr /culture /actualites/index-olivennes231107.htm

Prahalad, C. K., & Hamel, G. (1990). The core competencies of the corporation. *Harvard Business Review*, (May-June): 29–44.

Rallet, A. (1996). Convergence technologique et organisation industrielle de l'audiovisuel, de l'information et des télécommunications. In Brousseau, E., Petit, P., & Phan, D. (Eds.), *Mutations des Télécommunications, des Industries et des Marchés*. Paris: ENSPTT-Economica.

Scheuer, A. (2006). Traditional paradigms for new services? The Commission Proposal for a 'Audiovisual Media Services Directive'. *Communications & Strategies*, *62*(2), 71–91.

Srivastava, R. K., Fahey, L., & Christensen, K. (2001). The Resource-Based View and Marketing: the Role of Market-Based Assets in Gaining Competitive Advantage. *Journal of Management*, *27*, 777–802. doi:10.1177/014920630102700610

Teece, D. J., Pisano, G., & Shuen, A. (1997). Dynamic Capabilities and Strategic Management. *Strategic Management Journal*, *18*, 509–533. doi:10.1002/(SICI)1097-0266(199708)18:7<509::AID-SMJ882>3.0.CO;2-Z

Vialle, P. (1998). *Stratégies des opérateurs de télécoms*. Paris: Editions Hermès.

Vialle, P. (Ed.). (2007). *Mutation des STICS. Acteurs, Ressources, Activités*. Paris: Hermes Science Publishing.

Zhu, K. (2001). Internet-based Distribution of Digital Videos: the Economic Impacts of Digitization of the Motion Picture Industry. *Electronic Markets*, *11*(4), 273–280. doi:10.1080/101967801753405562

ENDNOTES

1 On 20 December 2005, for VOD, representatives of the industries involved agreed a 33-week blocking period for the first time in France. On 23 November 2007, an agreement was reached between producers, ISPs and the Government, under which films may be made available via VoD six months after they are first shown in movies theatres. In October 2009, the Hadopi 2 Law has been adopted, reducing the delay between the main phases of the media chronology. Consequently, movies will be released via DVD six months and via VOD four months after cinema release. Kuhr (2008) describes how rules on media chronology have developed in Europe and analyses the reasons why there are different models and their implications. (http://www.obs.coe.int/oea_publ/iris/iris_plus/iplus4_2008.pdf.en)

2 The share-out often quoted is 50/25/25 for content providers, aggregators and service providers respectively.

3 Convergence has many facets among them are: convergence of industries, networks, services & applications, media & content, devices but also pricing models. Consequently, triple play services are often linked with a bundled pricing approach of a discounted monthly subscription.

4 The number of TV households in France has been estimated at 24.219 million households in 2005 by the European Audiovisual Observatory.

5 Excluding free of charge services, video clip services and adult services. The same service provider can distribute services through several channels. Therefore the number of service providers may be inferior to the sum of the services offered through the various electronic distribution channels.

6 Catch-up TV services allow consumers to view selected past TV programs during a given period after their diffusion. Most catch-up TV services don't include movies.

7 However, the exact number of subscribers effectively using IPTV service is unknown.

8 See ARINO (2007) and SCHEUER (2006) for a discussion of the new Directive.

Chapter 16
Media Platform Competition:
The Displacement Effect of IPTV on Cable TV in Taiwan

Yu-li Liu
National Chengchi University, Taiwan

Wen-yi Hsu
National Chengchi University, Taiwan

ABSTRACT

In Taiwan, there are three major TV platforms: terrestrial TV, cable TV, and IPTV. When faced with various TV platforms, the consumers in Taiwan will be able to choose from many broadcasts. When a new medium emerges, there are always concerns of its displacement effects on existing media. Many studies have explored the displacement effects of newly-emerging media, and have come up with a variety of sometimes conflicting findings (Lee & Leung, 2006). This chapter aims to analyze the media platform competition, namely, the displacement effect of IPTV on cable TV. The research methods used include a literature review and a telephone survey. With regard to the telephone survey, 612 of CHT's IPTV users were selected systematically based on the name lists provided by the service provider. In this study, while the platform displacement effect of IPTV on cable TV is clear, the statistics reveal a different result for the time displacement effect.

INTRODUCTION

In recent years, Internet Protocol TV (IPTV) has become a new service in many developed countries. According to the Multimedia Research Group (MRG)'s forecast released in November 2009, the number of global IPTV subscribers will grow from 28 million in 2009 to 83 million in 2013, reflecting a compounded annual growth rate of 31% (http://www.mrgco.com/iptv/gf1109.html). IPTV is a way of getting TV transmitted over high-speed Internet connections. It can be stored on a server and sent to computers or straight to special set-top boxes, either over asymmetric digital subscriber lines (ADSL) or cable lines. IPTV can be in the form of Video on Demand (VOD) or live TV. To watch IPTV, a person sitting at home needs to have access to broadband and a set-top box that is plugged into the TV. (http://news.bbc.co.uk/2/hi/technology/4318331.stm).

DOI: 10.4018/978-1-60960-011-2.ch016

According to the Point-Topic report of February 2005, delivering IPTV is now a part of many operators' plans. However, providing this service is a high-cost and low-margin business. In Taiwan, Chunghwa Telecom (CHT) is the major IPTV service provider. It was issued a license in February 2004 by the former broadcast regulator the Government Information Office (GIO) to offer IPTV service to its ADSL subscribers. CHT dominated the telecommunications services market until 2000 when the government opened up three new fixed networks. It still dominates the broadband market, because it owns most of the local loops in Taiwan. CHT, however, is not the only telecommunications company that provides video service in the world. Fastweb TV in Italy, Verizon in the United States, and PCCW in Hong Kong have all entered the video service market. Foley found four main reasons why telecommunications carriers wish to enter video markets: (1) the local phone revenue does not increase; (2) they wish to fully utilize the extra broadband; (3) the home video market may have the potential for more revenue; and (4) they can more fully exploit the economies of scale (Foley, 1992). The above four reasons can all explain why CHT is eager to provide IPTV service.

In Taiwan, the cable TV penetration rate is over 80%. Most viewers watch terrestrial TV through cable TV. The rate of DBS penetration is only 2%. Since CHT started to provide IPTV service in 2004, the Taiwanese people have had another alternative to watching TV. The government and the public welcome the competition, but the cable operators are greatly concerned about the new competitor. They have complained to the regulator that CHT is allegedly a dominant telecommunications operator and that its IPTV platform is a national service, while the cable systems are local services. They are afraid that they might lose many subscribers because of this giant TV platform. Therefore, they have been trying every means to deter the development of CHT's IPTV service. They have even threatened the channel operators not to provide content to CHT's IPTV platform.

CHT's IPTV service has also been confronted with another obstacle. In December 2003, the Legislative Yuan revised three broadcasting-related laws requiring all the government, military, and party shares to be withdrawn from the broadcast, cable TV and satellite TV by December 2005. Since more than one third of CHT's shares are still owned by the government, the newly-established regulator, the National Communications Commission (NCC), ruled in January 2007 that CHT could only act as an open platform and that it could not produce programs and aggregate channels. It could, however, provide VOD and interactive services, because the subscribers had full autonomy to choose their interested content. CHT's IPTV platform could also provide content through third parties, namely, other content providers and aggregators (www.ncc.gov.tw). Even though CHT's IPTV is not similar to IPTV services worldwide, it is still very interesting and important to know whether there will be any displacement effect of IPTV on cable TV.

In fact, most studies that compare the relationship between new and existing media focus on the competitive displacement effect (Dutta-Bergman, 2004), by which is meant that different kinds of media will compete for the same resources such as consumers' time, place, and attention. However, most of the displacement effect studies compare different media such as the Internet's impact on traditional media (newspaper, radio and TV). Very few studies compare the competition among different TV platforms, not to mention the IPTV's displacement effect on cable TV.

The purpose of this chapter is to study the IPTV's impact on cable TV and find out whether IPTV exerts a displacement effect on cable TV. If there is time displacement, it then needs to be asked what is the relationship between the displacement effect and other variables such as functional consideration for subscription/adoption, frequently used services, the viewing hours, degrees of satisfaction, media use and demographics. In addition, this chapter will analyze whether there is platform displacement of IPTV on cable TV. If there is, then what are the factors involved?

THEORIES OF DISPLACEMENT EFFECTS

People usually have a limited amount of time to spend on the media. The more time they spend on one medium, the less time they will spend on others (McComb, 1972; Lee & Leung, 2006). In the past, whenever a new medium arose, there were concerns about whether displacement effects would occur and whether they would impact the existing media, such as radio (Lazarsfeld, 1940), television (Himmelweit, Oppenheim & Vince, 1958; Belson, 1961; Mendelsohn, 1964; Weiss, 1968; Rubenstein et al., 1973; Williams, 1986), cable television (Kaplan, 1978; Grotta & Newsom, 1982; Sparkes, 1983), VCR (Harvey & Rothe, 1985; Henke & Donohue, 1989) and the Internet (Finhoult & Sproull, 1990; Kayany & Yelsma, 2000; Dimmick, Kline & Stafford, 2000; Ferguson & Perse, 2000; Nie, N, & Ebring, L., 2000; Chen, Boase, & Wellman, 2002; Kaye & Johnson, 2002, 2003; Dutta-Bergman, 2004; Tsao & Sibley, 2004; Lee & Leung, 2006; Nguyen & Western, 2006).

Displacement theory is based on the argument that different media will compete with each other for resources such as time and money. Numerous studies have explored the displacement effects in relation to existing media, but they have given rise to varied, and sometimes contradictory, findings (Lee & Leung, 2006). The conflicting findings have led to three different schools of thought (Kayany and Yelsma, 2000; Tsao & Sibley, 2004). One school depicts an asymmetrical relationship (increase-decrease) in regard to viewer use between the new and existing media, while the second school supports a complementary relationship (increase-increase). The third perspective in regard to the effects of media displacement focuses on functional displacement (Tsao & Sibley, 2004). Although Dutta-Bergman's (2004) paper proposes a shift in the research focus by investigating the content complementarity issue, it does not disapprove of competition-based theories.

These schools of thought originated from the bioecological theory of the niche. Dimmick & Rothenbuhler (1984) used niche theory to study competition between the coexistence of media industries. The expectation was that there would be an increase-decrease relationship between the new and existing media when the displacement effects occurred (Kayany and Yelsma, 2000; Tsao & Sibley, 2004). However, the process may require increase-increase for both parties to succeed (Dimmick & Rothenbuhler, 1984).

Lee & Leung (2006) explored two approaches to study the displacement effects of new media. Under the medium-centric approach, the more time that users spend on a new medium as a result of new features, the less time they will spend on existing media. Thus, a "more-less" hypothesis was proposed. A user-centric approach focuses on the needs and gratifications of media users. No new medium can substitute all the uses and gratification of existing media. If a new medium serves similar needs of the existing media, no displacement is to be expected. Therefore, a "more-more" hypothesis was proposed.

Although the research findings are not consistent, the major focuses of displacement effects of new media studies have usually involved time and functional displacement (Lee & Leung, 2006).

In Taiwan, only a few studies have applied the theories of displacement effects. The authors of those studies were interested in finding out whether the Internet had displacement effects on traditional media (Jiang, 2005; Hu, 2008). Jiang (2005) found that only heavy Internet users had a displacement effect on the usage of television and magazine (Jiang, 2005). Hu's study suggested that Internet usage is increasing. Although people spend less time watching TV, reading newspapers and magazines, and listening to the radio, traditional media will not be replaced by new media because different media have their own characteristics and they also have their own supporters (Hu, 2008).

Time Displacement

The emphasis of displacement theory is the amount of time people spend on different media types (Dutta-Bergman, 2004), and so the concept of time displacement means that the amount of time available for the consumption of various types of media is limited. The basic concept of time displacement assumes "a zero-sum relationship for the amount of time invested in various activities" (Kayany & Yelsma, 2000). In other words, people only have certain amounts of time and money to spend. If they spend more on one medium, then they will spend less on others (Lee & Leung, 2006).

The zero-sum hypothesis was rejected by many scholars but Nguyen & Western (2006) suggested that these findings do not totally discredit the displacement hypothesis. However, until now most of the displacement effect studies have focused on comparisons among different types of media, and very few studies have compared the competition among different TV platforms. There are also studies that focus on cable TV's impact on local television channels or local television stations. Since these advantages of cable TV had been widely known during the 1970's, quite a few studies demonstrated that cable TV greatly reduced the number of viewers of local television stations. Kaplan (1978) found that cable television decreased local network television viewing: 36% respondents reported decreased viewing of local network stations. The effect of cable service on local television stations is quite considerable. Grotta & Newsom's findings (1982) are generally consistent with those in Kaplan's study, and they concluded that the greatest impact of cable TV appears to be on local television stations, with newspapers and magazines having the least impact. On this basis, the following hypothesis is proposed:

H1: IPTV has a time displacement effect on the users who are also cable TV subscribers.

In the last two years, there have been few studies on IPTV users' characteristics. However, there seems to be more room for discussion. For instance, with regard to the correlation between IPTV subscriptions and demographics, some studies have pointed out that several personal characteristics are unable to provide a more credible explanation as to the reasons why the consumers subscribe to IPTV service. Furthermore, it is quite certain in the analysis that once there are more members in the family, the possibility of subscribing to IPTV will be higher. When the consumers are older, the possibility of subscribing to IPTV is also higher. Because the respondents may not be those who decide to subscribe to the IPTV services, the influence of the age factor should be further studied (Lin, 2006). Furthermore, it could be an interesting subject to discuss in the future.

Meanwhile, Lin (2006) suggested that the "life style" has a considerably positive effect in terms of "subscribing to IPTV services", and its effect is much more important than "education" or "income" based on the statistical results. Furthermore, those people who frequently use the Internet will prefer to subscribe to IPTV as well.

Liu (2006) found in her study that the main reason why people subscribe to IPTV is that it is "easy to subscribe to", while the reason why they do not subscribe is that they "do not understand IPTV". This chapter not only examines the user behavior of the IPTV subscribers, but also looks at the effect of time displacement on cable TV. Hence, the following research questions are proposed:

Q1: What are the main reasons why IPTV users adopt IPTV? What are the functions and services that the IPTV subscribers frequently avail themselves of?

Q2: Is there a correlation between the IPTV users' time displacement in relation to cable TV and the degree of satisfaction, and demographics?

Functional Displacement

Time displacement suggests that the more time that users spend on a new medium, the less time they will spend on existing media. However, the concept of functional displacement is that a new medium will displace the existing media if they can serve similar function in a better or more effective way (Schramm et al., 1961; Lee & Leung, 2006).

Henke & Donohue (1989) suggested that functional displacement at an individual level recognizes that patterns of media use depend on individual needs and motivations; and which medium one uses to meet the needs can change rapidly as new media or technological innovations emerge. When the new medium serves the same functions as the existing media, such competition can "result in either of them becoming irrelevant or secondary" in satisfying the needs of the people (Kayany & Yelsma, 2000).

In the development of media, some studies have pointed out that when the users have different choices for similar function, they might choose alternative media for similar functions. Therefore, functional displacement will occur. Himmelweit, Oppenheim & Vince (1958) found that television displaced radio listening, movie attendance and comic reading because they all had a similar "escape" function. Perse & Courtright (1993) argued that cable TV and the VCR were alternatives to broadcast television because they fulfilled the same functions of relaxation and entertainment. Kayany & Yelsma (2000) concluded that the functional equivalence perspective could explain Kaplan's findings, namely, that when users consider the TV function of cable TV, terrestrial TV and IPTV, functional displacement will occur. Therefore, theoretically it is also possible to have time displacement. According to the previous studies, the functions that TV provide include the information function and entertainment function (Kayany & Yelsma, 2000). Since CHT's IPTV also provides VOD and interactive services, this chapter also tries to find out whether there is a correlation between time displacement and the IPTV subscribers' use of the functions. In addition, in terms of the time spent on media, it is interesting to know whether there is a significant difference between the different functions and the time spent on IPTV. The following research questions are thus proposed:

Q3: Is there a correlation between the IPTV users' time displacement on cable TV and the functions IPTV provides?

Q4: Is there a significant difference between IPTV's functional approach and IPTV viewing?

Platform Displacement

No studies in the literature have mentioned the term "platform displacement." The authors are thus very interested in finding out whether IPTV will have a "platform displacement" effect on cable TV, namely, whether IPTV users who are also cable subscribers will stop subscribing to cable TV or digital cable TV. Terrestrial TV, cable TV and IPTV all provide TV programs in the form of TV channels, and they are all TV services. However, when they go digital, they need to have a set-top-box to receive the programs (Only DTTV can also be received through a digital tuner built into the TV). The three above-mentioned TV services usually promote channels via the TV platform. Digital cable TV and IPTV not only provide TV programs, but also interactive services. CHT's IPTV even provides VOD service. Although digital cable TV can also provide VOD service, Taiwan's cable operators still hesitate to provide VOD services. Therefore, platform displacement can be defined by stating that subscribers will stop subscribing to the existing platform (such as cable TV) because they feel they can be served by the new platform (such as IPTV). Since DTTV users do not have to pay subscription fee, this chapter only focuses on whether there is "platform displacement" of IPTV on cable TV.

In recent years, there have been studies that have analyzed the competition between cable TV and IPTV. Liu (2006) analyzed the advantages and disadvantages of CHT's IPTV and cable TV. In studying the impact of IPTV on cable TV, she concluded that IPTV lacked channels and its programs were not timely. However, because CHT had a lot of resources and its IPTV provided VOD and interactive services, it was believed that it would serve as a threat to the development of cable TV, if it provided more channels. Lee (2007) also demonstrated that IPTV service would compete directly with cable companies in Taiwan. Therefore, in addition to time displacement and functional displacement, this chapter also analyzes whether IPTV subscribers who also subscribe to cable TV will stop subscribing to cable TV. What, then, are the main reasons why IPTV subscribers stop subscribing to cable TV? This chapter also analyzes whether there is a platform displacement effect of IPTV on cable TV. If there is, then what are the factors that are related to IPTV subscribers? Hence, the following research question is proposed:

Q5: Since competition exists among different TV platforms, will there be a platform displacement effect of IPTV on cable TV subscribers? If there is, on what will their considerations be based?

Methods

Data

The research methods adopted include a literature review and telephone survey. With regard to the telephone survey, the research team outsourced the interviews to a survey company in May 2007. It dialed 2,814 numbers selected systematically based on the name lists provided by the service provider. After subtracting the non-human factors, 1,301 households were asked to answer the questions. Finally 612 of CHT's IPTV users were successfully interviewed. The acceptance rate was 47%.

Measurement

Use of Different TV Platforms

This study focused on the use on different TV platforms (such as cable TV and IPTV). Kaplan (1978) and Grotta & Newsom (1982) suggested that future studies should measure the use both before and after the users adopt the new media (such as cable TV). In this study, the respondents were asked to estimate their total viewing time (in minutes) in relation to each TV platform per day. They were asked to state whether the time they spent on TV increased, remained the same, or decreased after they subscribed to IPTV.

Time Displacement

When users spend more time on a new medium and they also tend to spend less time on existing media, then a time displacement effect is established. We can also ask: "For the users who have more than two TV platforms, if they spend more time on a new platform, do they also tend to spend less time on existing platform?" Therefore, if the total viewing time of the IPTV users who are also cable TV subscribers does not change or decrease, the hypothesis of "more-less" is supported, i.e., the time displacement effect is deemed to exist. However, since we want to know whether there is a time displacement of IPTV on cable TV subscribers, if the cable viewing time of the IPTV users who are also cable TV subscribers does not change or decrease, the hypothesis of IPTV's time displacement effect on cable TV is not supported.

Functional Displacement

With regard to the functional displacement effect, the respondents were asked what their major concerns were when they decided to subscribe to the new TV platform. For digital cable TV users, a third item, "interactive function," was added. For IPTV users, a fourth item, "video on demand (VOD)," was added.

Platform Displacement

The IPTV users were asked whether they also subscribed to cable TV, whether they stopped subscribing to cable TV after they subscribed to IPTV, or whether they had ever subscribed to cable TV before. If the respondents terminated their cable TV subscription after subscribing to IPTV, the platform displacement effect of IPTV on cable TV is deemed to exist.

Results

1. IPTV Respondents' Profile

The data consisted of 333 females and 279 males. The educational level of the respondents, of which 44.4% were college graduates and 8.4% had received graduate degrees, was above the national average. As for the age level, 34.9% were 40-49 years old, and 21.5% were 30-39 years old. The household income of 42.57% of the respondents was NT$80,000 (USD 2,352) or higher, 51.2% had more than two TV sets, and 50.9% of the families had more than two children under 20.

2. Time Displacement Effects

On average, IPTV users watched TV for 159.15 minutes every day. They watched programs on the IPTV platform for 105.56 minutes every day. They also spent 11.35 minutes on non-TV purposes (such as interactive services) every day.

i. IPTV does not have a time displacement effect on cable TV subscribers

H1 hypothesizes that IPTV has a time displacement effect on the users who are also cable TV subscribers. After users subscribed to IPTV, 67.7% (n=408) of the IPTV users maintained the same number of viewing hours per day, 23.1% (n=139) of the IPTV users decreased their viewing time, and 9.2% (n=55) of the IPTV users increased their viewing time. In other words, more than 90 percent of the IPTV users do not increase their TV viewing time because of their IPTV subscription.

Among the IPTV subscribers, only 18.8% (n=111) of users subscribed to cable TV. For those who subscribed to both IPTV and cable TV, 82.4% (n=89) maintained the same total viewing time, 11.1% (n=12) decreased their total viewing time, while 6.5% (n=7) increased their total viewing time. When asked to respond to cable viewing time, 87% (n=96) maintained the same viewing time on cable TV, 10.3% (n=11) decreased their cable viewing time, and 2.7% (n=3) increased their cable viewing time after they subscribed to IPTV. Since 87% of the IPTV/cable TV subscribers maintained their cable viewing time and only 10.3% decreased their cable viewing time, the time displacement effect of Chunghwa's IPTV on the cable TV platform was not clear. Therefore, H1 was not supported.

i. Users stopping cable subscription spend more time on IPTV than users who still subscribe to cable TV

According to the survey, 59.1% (n=349) of the IPTV users stopped subscribing to cable TV services after they subscribed to IPTV service. The t-test shows that there was a significant difference between users who stopped their cable TV subscription and users who still subscribed to cable TV in terms of IPTV viewing time per day ($t[445]$ -6.418, $p<.01$). Users who stopped cable TV subscription (M=123.69, n=337) spent more time on IPTV viewing than users who still subscribed to cable TV (M=49.6, n=110)(see Table 1).

ii. Time displacement effect and the users' demographics

The Chi-square analysis showed that there is no correlation between the time displacement effect and the users' demographics.

3. Functional Approach

Kayany & Yelsma (2000) indicated that "the functional niche of any new medium does not occur instantly but over time". It might thus be inappropriate to measure the functional displacement effect on the existing media in the early stage of the new medium's adoption because of the novelty effect of the new medium. Since this study only focuses on TV platforms and IPTV is a new service, it might be too early to analyze the functional displacement effect of IPTV on cable TV. Therefore, this study used the "functional approach" instead of "functional displacement" to perform the following analysis.

i. Functional considerations for subscription

More than forty percent of IPTV users subscribed to IPTV because of the entertainment function (41.7%, n=190) provided by IPTV. 27.1% of users subscribed for the information function and 26.1% for the VOD function. Only 5.1% subscribed for the interactive function. The main reasons why people subscribed to the IPTV subscription included the triple play's (telephone, broadband, and IPTV) special deal (21.1%, n=120), the diversity of programs (14.9%), the reasonable price (11.1%), and other promotions (11.0%). Obviously, forty-three percent subscribed because of economic considerations. With regard to frequently used services, 52.4% (n=216) of IPTV subscribers used the VOD function, 14.3% used the interactive program guide, and 13.7% used the "searching for news and information" service.

IPTV users subscribing to cable TV or digital cable TV all considered the entertainment function to be more important than the information function. As for the IPTV users who also subscribed to digital cable TV, their main concerns included the diversity of programs (27.8%), family interests (16.7%) and more channels (16.7%)(see Table 2).

In analyzing the entertainment and information functions that cable TV and IPTV provide, this study indicates that most of the users considered the entertainment function more than the information function. Cable TV provides more than 100 channels. Digital cable TV provides even more channels and a better quality picture. As for IPTV, even though it only provides thirty more channels, it has VOD services. The users can select the programs at their convenience.

ii. Functional considerations and IPTV viewing hours

After ANOVA was conducted, this study found that there was a significant correlation between functional considerations and IPTV viewing hours ($F[2,243]=3.080$, $p<.05$). The IPTV subscribers who considered more of the information function spent more time on IPTV (see Table 3).

iii. Time displacement effect and the functional approach

This study also tried to examine whether there was any correlation between the time displacement effect on cable TV platform and the functional ap-

Table 1. IPTV viewing for non-subscription and subscription of cable TV

	Still subscribe	Stop cable subscription	t	Sig.
IPTV viewing (min) per day	49.60 (n=110)	123.69 (n=337)	-6.418**	.000

** Significant at the 0.01 level (2-tailed)

Table 2. Reasons for subscription to different TV platforms

Subscription status	Functional consideration for subscription (or adoption)	Main reasons for subscription or adoption
IPTV users subscribing to cable TV	Entertainment function 60.9% (n=61) Information function 39.1% (n=39)	NA
IPTV users subscribing to digital cable TV	Entertainment function 51.8% (n=9) Information function 43.3% (n=8) Interactive function 4.9 (n=1)	Diversity of programs 27.8% (n=5) Family interest 16.7% (n=3) More channels 16.7% (n=3)
IPTV users	Entertainment function 41.7% (n=190) Information function 27.1% (n=123) VOD function 26.1% (n=118) Interactive function 5.1% (n=23)	Special deal for triple play 21.1% (n=120) Diversity of programs 14.9% (n=84) Price is reasonable 11.1% (n=63) Other promotion 11% (n=62)

proach for IPTV subscribers. Through Chi-square analysis, no correlation was found (χ^2=3.453, df=2, p=.178 >.05).

iv. Satisfaction with the provided functions and IPTV viewing hours

The t-test showed that the IPTV users who were more satisfied with the entertainment function would spend more time on IPTV than the low satisfaction users (p<.05). This was also true for the information function (p<.001). Therefore, the more the IPTV users were satisfied with the entertainment and information functions, the more time they spent on IPTV (see Table 4).

v. Satisfaction with the functions provided and demographics

With regard to the IPTV users' demographics, only education (p<.05) variable was found to be correlated with users' satisfaction in the entertainment function. Users who graduated from junior high school were more satisfied with IPTV's entertainment function than the users who had obtained a higher degree. There was an inverse

Table 3. The IPTV subscribers' functional approach and their time spent on IPTV

Media use (min)	Functional approach			Stat.	
	Entertainment function	Information function	VOD	F	Sig.
Time spent on IPTV (per day)	102.72 (n =187)	133.33 (n =120)	102.33 (n =118)	3.080*	.047

** Significant at the 0.01 level (2-tailed). * Significant at the 0.05 level (2-tailed).

Table 4. IPTV users' satisfaction and viewing time

Satisfaction and viewing time		IPTV viewing time (mins per day)	t	Sig.
Entertainment satisfaction	high	112.31 (n=274)	2.003*	.046
	low	94.63 (n=272)		
Information satisfaction	high	121.21 (n=311)	4.033**	.000
	low	86.02 (n=234)		

** Significant at the 0.01 level (2-tailed). * Significant at the 0.05 level (2-tailed).

significant correlation between users' satisfaction with the information function and the users' age ($p<.01$). Younger users were more satisfied with IPTV's information function.

4. PLATFORM DISPLACEMENT EFFECT

This study sought to find out whether there was a platform displacement effect, namely, whether the IPTV users who were also cable TV subscribers stopped subscribing to cable TV after they subscribed to IPTV. The results show that 59.1% (n=349) of the users stopped subscribing to their cable TV service after they had subscribed to the IPTV service. 18.8% (n=111) of the IPTV users still subscribed to cable TV. Therefore, a platform displacement effect did occur.

Table 5 shows that the main reasons for stopping subscribing to cable TV included "too expensive" (35%, n=116), "children have to concentrate on study" (23%), and "no time to watch" (17.8%). Only 7.9% answered that they stopped their cable TV service because they "can watch programs on another digital TV platform." In Taiwan, cable TV subscribers can receive more than 100 channels by paying NT$550 (USD 16) monthly. Before October 2007, CHT's IPTV subscribers could receive thirty five channels by paying NT$150 (USD 4.4). The result proves that the IPTV users are very sensitive to price.

The t-test shows that there was no significant difference for IPTV users who stopped cable TV service and those who still kept cable TV service in terms of the "age" and "income" variables, while "the number of TV sets" variable ($t[455]=5.657$, $p<.01$) was the exception.

As for the 111 IPTV users who still subscribed to cable TV, only 7 users (6.3%) considered dropping the cable service within six months because of price. A total of 78 IPTV users had at least two digital platforms, 22 IPTV users also subscribed to digital cable TV, of which 5 users considered dropping cable service within 6 months, and 63 IPTV users also had DTTV set-top boxes or built-in digital TV sets. Since there was no subscription fee for the DTTV service, if the users said that they would consider dropping it, it meant that they would consider not using DTTV facilities within six months. For the 78 IPTV users who had at least two digital TV platforms, 23 users considered dropping the IPTV service within six months (see Table 6).

The Chi-square analysis shows that the "most frequently used digital platform" is not correlated with "considering stopping the subscription within 6 months" ($\chi^{2}=6.556$, $p=.161>.05$).

i. Satisfaction with the Provided Functions and Cable TV Subscription

IPTV users subscribing to cable TV were found to be more concerned with the medium's entertainment function (60.9%, n=61) than with its information function (39.1%, n=39) (see Table 2). The t-test showed that the IPTV users who stopped cable TV subscription tended to be more

Table 5. Reasons why IPTV users stopped subscribing to cable TV

	Stopped subscribing to cable TV (n=349)
No time to watch	17.8% (n=59)
Children have to study	23.0% (n=76)
Too expensive	35.0% (n=116)
Programs are not interesting	13.3% (n=44)
Can watch other digital TV	7.9% (n=26)

Table 6. Considering Dropping the Digital Service within 6 Months

Subscription status	No. of subscribers	No. considering dropping
IPTV users subscribing to digital cable	22 (6 have DTTV)	5 (22.7%)
IPTV users having DTTV	63 (6 have digital cable)	6 (9.5%)
IPTV users having at least 2 digital TV platforms	78	23 (29.5%) considered dropping IPTV

satisfied with the information function (M=3.16) than those users who still subscribed to cable TV (M=2.83) ($p<.01$). In some ways, the information function provided by IPTV might have had a functional displacement effect on some cable TV subscribers. In reality, CHT does provide more interactive functions than the cable TV services.

CONCLUSION AND DISCUSSION

The findings of this study show that IPTV does not have a time displacement effect on IPTV users subscribing to cable TV. More than 90 percent of the IPTV users in the sample did not increase their TV viewing time because of their IPTV subscription. This proves that people have a fixed amount of time to spend on TV. For those who subscribed to both IPTV and cable TV, 87% of the IPTV/cable TV subscribers maintained their cable viewing time and only 10.3% decreased their cable viewing time. The time displacement effect of CHT's IPTV on the cable TV platform was therefore not clear.

This study also defines the "platform displacement effect" and demonstrates that since almost sixty percent of the IPTV users who were also cable TV subscribers stopped subscribing to cable TV, the platform displacement effect does occur. The main reasons for cable TV's non-subscription included "the subscription is too expensive", "children have to concentrate on study," and "no time to watch TV." This shows that the subscribers were very sensitive to price. The cable TV monthly fee is almost four times that of IPTV in Taiwan. However, 19% of IPTV users still kept their cable TV service, perhaps because they could not feel satisfied with the limited programs that IPTV provides. For the IPTV users who never subscribed to cable TV, their main reason was "no time to watch TV" which also echoes the reference in the literature that people only have a fixed amount of time to spend on TV.

Those IPTV users who terminated their cable TV subscription were found to spend more time

Table 7. Correlation between IPTV users' entertainment and information satisfaction with cable TV subscription

	Stopping cable TV subscription		stat.	
	Still subscribe(n=111)	Stop cable subscription(n=349)	t	Sig.
IPTV users' entertainment satisfaction	2.96(n=110)	2.93(n=339)	.265	.791
IPTV users' information satisfaction	2.83(n=108)	3.16(n=340)	-2.758**	.006

** Significant at the 0.01 level (2-tailed).

on IPTV than those who still subscribed to cable TV. In addition, they became more satisfied with the information function that IPTV provides. There was no significant difference between the IPTV users who terminated their cable TV subscription and those who still subscribed to cable TV in terms of the demographic variables.

About 13% of IPTV users had at least two digital TV platforms. They had more TV sets than the single IPTV platform users. However, there was no significant difference between the two kinds of users in terms of demographics and average viewing time for IPTV and TV as a whole.

This study also tried to explore the factors contributing to IPTV's time displacement effect. The findings show that there is no correlation between the functional consideration and the time displacement effect, although the users' satisfaction factor was an exception. IPTV users who were more satisfied with the entertainment and information functions that IPTV provides tended to spend more time on IPTV.

More than 80% of IPTV users were CHT's broadband customers before they subscribed to CHT's IPTV. They also replied that CHT's special deal for triple play (phone, DSL, and IPTV service) was one of the main reasons why they subscribed to IPTV. More than half of the IPTV subscribers used VOD service frequently, since they liked to watch TV at their convenience. Although technically speaking cable TV can also provide VOD service, the cable operators in Taiwan have not provided VOD service to their subscribers, which makes CHT's IPTV more advantageous than cable TV.

Currently, CHT's IPTV cannot manage channels or produce programs because of Taiwanese laws and regulations that do not allow the government to own any media. However, since the government still owns 35% of CHT, CHT can only play the role of an "open platform," which means that all the content providers, content aggregators, ISPs, and other fixed networks can use CHT's IPTV platform by paying reasonable prices. The regulator only allows CHT's IPTV to play an active role in VOD and interactive services (www.ncc.gov.tw). The survey showed that when IPTV users subscribed to the service, they considered more of its entertainment function. This finding also indicated that the characteristics of CHT's IPTV such as VOD and interactive services were not considered to be very important at the time of subscribing. This shows that CHT has more room to improve in its VOD and interactive services.

According to related studies, if a new medium has a displacement effect on existing media, it is because the new medium can provide a similar function to the existing media that satisfies more of the users' needs. It can also explain why Kaplan's study (1978) found that, after cable TV appeared, the audience would decrease the time spent on terrestrial TV. However, this study found that there was no time displacement effect of Chunghwa's IPTV on the cable TV platform. In other words, the functional niche of the new medium does not appear immediately, and takes time to develop (Kayany & Yelsma, 2000). It is worthwhile continuing to study the time displacement effect and functional displacement when the penetration rate of IPTV is increasing.

This study also echoes that of Liu & Hsu (2007) in that time displacement among the TV platforms is not an absolute displacement. A different TV platform has its advantages and disadvantages, and viewers also wish to have more choices with regard to TV platforms. The conclusion drawn from this study can contribute to the literature on the displacement effect. However, its limitation is that when this study was conducted, the penetration of IPTV was very low (about 17%). Some questions could not generate enough replies from respondents. Therefore, the authors were very cautious in interpreting the findings and drawing conclusions.

Cable TV has a penetration of over 80% in Taiwan, higher than that of broadband network. It will thus be hard for telephone companies to capture the video market, and hence the risk for telephone companies launching IPTV service in Taiwan is high. Since the cable MSOs in Taiwan have warned the cable channels not to come to CHT's IPTV platform, it is hard for CHT to gain video market share quickly if it does not have interesting content to attract subscribers. Therefore, some studies suggest that CHT has to adopt a differentiated strategy and exploit its unlimited channels and VOD service to offer a greater variety of content and price flexibility, as well as develop innovative interactive services in order to gain a competitive advantage (Lee, 2007). Since CHT has a more than 80% market share of the DSL broadband market and our study shows that the IPTV platform displacement effect does occur, it still has more potential to compete with cable TV in Taiwan.

REFERENCES

Belson, W. (1961). Effects of television on the reading and buying of newspapers and magazines. *Public Opinion Quarterly*, *25*, 366–381. doi:10.1086/267033

Chen, W., Boase, J., & Wellman, B. (2002). The global villagers: comparing Internet users and uses around the world. In Wellman, B., & Haythorntwaite, C. (Eds.), *The Internet in Everyday Life* (pp. 74–113). Oxford: Blackwell. doi:10.1002/9780470774298.ch2

Dimmick, J., Kline, S., & Stafford, L. (2000). The gratification niches of personal e-mail and the telephone: Competition, displacement, and complementarity. *Communication Research*, *27*, 227–248. doi:10.1177/009365000027002005

Dimmick, J., & Rothenbuhler, E. W. (1984). The theory of the niche: quantifying competition among media industries. *The Journal of Communication*, *34*(1), 103–119. doi:10.1111/j.1460-2466.1984.tb02988.x

Dutta-Bergman, M. (2004). Complementarity in consumption of news types across traditional and new media. *Journal of Broadcasting & Electronic Media*, *48*(1), 41–61. doi:10.1207/s15506878jobem4801_3

Ferguson, D. A., & Perse, E. M. (2000). The World Wide Web as a functional alternative to television. *Journal of Broadcasting & Electronic Media*, *44*(2), 155–174. doi:10.1207/s15506878jobem4402_1

Finholt, T., & Sproull, L. (1990). Electronic groups at work. *Organization Science*, *1*(1), 41–64. doi:10.1287/orsc.1.1.41

Foley, J. M. (1992). Economic factors underlying telephone company efforts to enter home video distribution. *Journal of Media Economics*, (Fall): 57–68.

Grotta, G. L., & Newsom, D. (1982). How does cable television in the home relate to other media use patterns? *The Journalism Quarterly*, *59*, 588–591, 609.

Harvey, M. G., & Rothe, J. T. (1985). Video casette recorders: Their impact on viewers and advertisers. *Journal of Advertising Research*, *25*(6), 19–27.

Henke, L., & Donohue, T. R. (1989). Functional displacement of tradition TV viewing by VCR owners. *Journal of Advertising Research*, (April/May): 18–23.

Himmelweit, H., Oppenheim, A. N., & Vince, P. (1958). *Television and the child*. London: Oxford University Press.

Hu, H. H. (2008). The tendency of media usage behavior in Taiwan--A case study of 2004-2008 Shih Hsin Communication Database. Shih Hsin University. MA thesis.

Jiang, Y. S. (2005). Displacement and complementary effect of Internet usage and traditional media usage--A case study of 2005 Shih Hsin Communication Database. Shih Hsin University. MA thesis.

Kaplan, S. J. (1978). The impact of cable television services on the use of competing media. *Journal of Broadcasting*, *22*(Spring), 155–165.

Kayany, J. M., & Yelsma, P. (2000). Displacement effects of online media in the socio-technical contexts of households. *Journal of Broadcasting & Electronic Media*, *44*(2), 215–229. doi:10.1207/s15506878jobem4402_4

Kaye, B., & Johnson, T. (2002). Online and in the know: Uses and gratifications of the Web for political information. *Journal of Broadcasting & Electronic Media*, *46*(1), 54–71. doi:10.1207/s15506878jobem4601_4

Kaye, B., & Johnson, T. (2003). From here to obscurity?: Media substitution theory and traditional media in an on-line world. *Journal of the American Society for Information Science and Technology*, *54*(3), 260–273. doi:10.1002/asi.10212

Lazarfeld, P. F. (1940). *Radio and the printed page*. New York: Dell, Sloan and Pearce.

Lee, C. I. (2007). IPTV's competitive strategies and business models for Telcos-- Chunghwa Telecom's IPTV as Case Study. Fu Jen Catholic University. MA thesis.

Lee, P. S., & Leung, L. (2006). Accessing the displacement effects of the Internet. *Telematics and Informatics*. doi:.doi:10.1016/j.tele.2006.08.002

Lin, S. C. (2006). A Study on Factors of Purchasing the Multimedia-on-Demand (IPTV) Service. Department of Applied Economics, National Chung Hsing University. MA thesis.

Liu, L. T. (2006). The Study of the Market for CHT's IPTV. Department and Graduate School of Business Administration, Chung-Hua University. MA thesis.

Liu, Y. L., & Hsu, W. Y. (2007). The *Displacement effect of the digital television platforms*. Paper presented at The Fifth Conference of Media & Communication in Chinese Civilization. National Chengchi University, Taipei.

McComb, M. (1972). Mass media in the marketplace. *Journalism Monographs (Austin, Tex.)*, *24*, 1–102.

Mendelsohn, H. (Ed.). (1964). *Listening to Radio*. New York: Free Press.

Nguyen, A. & Western, M. (2006). The complementary relationship between the internet and graditional mass media: the case of online information. Retrieved from http//informationr.net/ir/11 -3/paper259.html

Nie, N., & Ebring, L. (2000). Study Offers Early Look at How Internet is Changing daily Life. Retrieved from www.stanford.edu/dept/news/pr

Perse, E. M., & Courtright, J. A. (1993). Normative images of communication media: Mass and interpersonal channels in the new media environment. *Human Communication Research*, *19*, 485–503. doi:10.1111/j.1468-2958.1993.tb00310.x

Rubenstein., et al. (1973). *Television and social behavior* (Technical report of the Surgeon General's Advisory Committee, Vol. IV). US Department of Health, Education and Welfare, Washington, DC.

Schramm, W., Lyle, J., & Parker, E. (1961). *Television in the Lives of our Children*. Stanford, CA: Stanford University Press.

Sparkes, V. M. (1983). Public perception of and reaction to multi-channel cable television service. *Journal of Broadcasting*, *27*(Spring), 163–175.

Tsao, J. C., & Sibley, S. D. (2004). Displacement and reinforcement effects of the Internet and other media as sources of advertising information. *Journal of Advertising Research*, (March): 126–142. doi:10.1017/S0021849904040073

Weiss, W. (1968). The impact of mass media. In Lindzey, G., & Aronson, E. (Eds.), *Handbook of Social Psychology* (*Vol. 5*, pp. 77–195). Reading, MA: Addison-Wesley.

Williams, T. (1986). *The Impact of Television*. New York: Academic Press.

KEY TERMS AND DEFINITIONS

Displacement Theory: It is based on the argument that different media will compete with each other for resources such as time and money.

Time Displacement: It suggests that the more time that users spend on a new medium, the less time they will spend on existing media.

Functional Displacement: A new medium will displace the existing media if they can serve similar function in a better or more effective way.

Platform Displacement: Subscribers will stop subscribing to the existing platform because they feel they can be served by the new platform.

Internet Protocol TV (IPTV): IPTV is a way of getting TV transmitted over high-speed Internet connections. It can be stored on a server and sent to computers or straight to special set-top boxes, either over asymmetric digital subscriber lines (ADSL) or cable lines.

Chunghwa Telecom (CHT): the dominant fixed network in Taiwan.

National Communications Commission (NCC): the integrated government agency regulating both telecommunications operators and broadcasting media in Taiwan. It was established in February 2006.

Government Information Office (GIO): the former broadcasting regulator in Taiwan.

ENDNOTE

This chapter is part of a large project supported by the National Science Council in Taiwan. This is the first study to focus on IPTV users' viewing behavior in Taiwan, and is also one of only a few to look into this issue, regardless of the countries involved. The results of the study contribute not only to the scholarly literature but also to the IPTV and cable TV industry.

Compilation of References

AAR. (2009). The American Recovery and Reinvestment Act of 2009. Retrieved January 14, 2010, from http://www.broadband.gov/

Adeye, C. (2002). *ICT and Poverty: a literature review.* Retrieved from http://web.idrc.ca/uploads/user-S/10541291550ICTPovertyBiblio.doc

Aminuzzaman, S., Baldersheim, H., & Jamil, I. (2003). Talking back: Empowerment and mobile phones in rural Bangladesh: A study of the village pay phone of Grameen Bank. *Contemporary South Asia, 12*(3), 327–348. doi:10.1080/0958493032000175879

Amit, R., & Zott, C. (2001). Value creation in e-business. *Strategic Management Journal, 22*(6/7), 493–520. doi:10.1002/smj.187

ANACOM (2009). Consumo de Banda Larga survey (BCS, 2006).

ANACOM (2009). Índice Europeu da Satisfação de Consumidor (ECSI, 2008).

ANACOM (2009). Inquérito ao Consumo das Comunicações Electrónicas (ICSCE, 2008).

Anwer, M. (2006) *ICT for Better Society*. Paper presented at 9th International Conference on Computer and Information Technology. Dhaka, Bangladesh.

ARCEP (2008). *L'Observatoire de l'Internet haut débit*. Retrieved October 13, 2008, from http://www.arcep.fr/index.php?id=8764

Arino, M. (2007). Content Regulation and New Media: A Case Study of Online Video Portals. *Communications & Strategies, 66*(2), 115–135.

Armstrong, M. (2001). Access pricing, bypass and universal service. *The American Economic Review, 91*(2), 297–301. doi:10.1257/aer.91.2.297

Armstrong, M. (1998). Network Interconnection in Telecommunications. *The Economic Journal, 108*, 545–564. doi:10.1111/1468-0297.00304

Arunachalam, S. (2002). *Reaching the unreached: How can we use ICTs to empower the rural poor in the developing world through enhanced access to relevant information?* (p. 11). India: M.S.S.R. Foundation. M S Swaminathan Research Foundation.

Ashraf, M. M., Hanisch, J., & Swatman, P. (2007). *ICT Intervention Towards Development In The Third World Countries: some perspectives of IS research.* Paper presented at 9th International Conference on Computer and Information Technology (ICCIT). Dhaka, Bangladesh.

Atkinson, R. D., Correa, D. K., & Hedlund, J. A. (2008). Explaining international broadband leadership. Retrieved October 13, 2009, from http://www.itif.org/files/ExplainingBBLeadership.pdf

Axelsson, B., & Easton, G. (1992). *Industrial Networks - A new View of Reality*. London: Routledge, Inc.

Bain, J. S. (1959). *Industrial Organization*. Hoboken, NJ: John Wiley & Sons.

Bakalis, S., Abeln, M., & Mante-Meijer, E. (1997). The Adoption and Use of Mobile Telephony in Europe. In L. Haddon (Ed.), *Communications on the Move: The Experience of Mobile Telephony in the 1990s* (pp. 19-28). Farsta: COST248 Report.

Bangsburg, P.T. (2000). Singapore to meld Internet logistics. *Journal of Commerce*, (March),10.

Barney, J. B. (1991). Firm Resources and Sustained Competitive Advantage. *Journal of Management*, *17*, 99–120. doi:10.1177/014920639101700108

Barro, R. J. (1997). Economic growth in a cross-section of countries. *The Quarterly Journal of Economics*, (106): 407–443.

Barro, R. J. (1997). *Determinants of Economic Growth*. Cambridge: The MIT Press.

Barro, R. J. (1990). Government Spending in a Simple Model of Endogenous Growth. *The Journal of Political Economy*, *98*(5), 103–125. doi:10.1086/261726

Batchelor, S., & Norrish, P. (2004). *Framework for the assessment of ICT pilot projects: beyond monitoring and evaluation to applied research. InfoDev*. Washington, DC: World Bank.

Bauer, M. (Ed.). (1995). *Resistance to New Technology – Nuclear Power, Information Technology, Biotechnology*. Cambridge: Cambridge University Press. doi:10.1017/CBO9780511563706

Bayes, A., Vn Braun, J., & Akhteret, R. (1999). *Village pay phones and poverty reduction: insights from a Grameen Bank initiative in Bangladesh. ZEF – Discussion Papers on Development Policy No. 8, Center for Development Research*. Bonn: ZEF.

BBCNews. (2009). UK broadband 'not fit' for future. Retrieved October 1, 2009 from http://news.bbc.co.uk/2/hi/tech nology/8282839.stm

BC. (2009) Broadband Canada: Connecting Rural Canadians, Retrieved January 14, 2010, from http://www.ic.gc.ca/eic/site /719.nsf/eng/home

Beeching, P. (2006). How to win friends and influence people. *Market Leader* [online], 34.

Belson, W. (1961). Effects of television on the reading and buying of newspapers and magazines. *Public Opinion Quarterly*, *25*, 366–381. doi:10.1086/267033

Beneke, J. (2008). Long walk to broadband freedom. In Y. Dwivedi, A. Papazafeiropoulou & J. Choudrie (Eds.), *Handbook of Research on Global Diffusion of Broadband Data Transmission* (pp. 13-28). IGI Global: Information Science Reference.

Bengtsson, M., & Kock, S. (2003, May). *Tension in Co-opetition*. Paper presented at the Academy of Marketing Science Annual Conference, Washington D.C.

Benkler, Y. (2009). Next Generation Connectivity: A review of broadband Internet transitions and policy from around the world. Retrieved November 2, 2009, from http://www.fcc.gov/stage/pdf/Berkman_ Center_Broadband_Study_13Oct09.pdf

Berkman Center for Internet and Society. (2009). *Next Generation Connectivity: A review of broadband Internet transitions and policy from around the world*. Harvard University.

Berne, M. (2003). *Business models for Services: the case of Video of Demand*. Working paper, TELECOM & Management SudParis.

Berne, M., Daidj, N., & Epinette, O. (2003). IP Services Billing. *Working paper*, TELECOM & Management SudParis.

Bhuiyan, A. J. M. S. A. (2004). Universal access in developing countries: a particular focus on Bangladesh. *The Information Society*, *20*(4), 269–278. doi:10.1080/01972240490480983

Biagioni, A., Fantacci, R., Marabissi, D., & Tarchi, D. (2009). Adaptive Subscriber allocation schemes for wireless OFDMA systems in WiMAX networks. *IEEE Journal on Selected Areas in Communications*, *27*(2), 217–225. doi:10.1109/JSAC.2009.090212

Blanchard, C. (1995). Telecommunications regulation in New Zealand: light-handed regulation and the Privy Council's judgement. *Telecommunications Policy*, *19*(6), 456–475. doi:10.1016/0308-5961(95)00024-Z

Blazquez Cabrera, F. J. (2007). Systèmes de gestion des droits numériques: dernières évolutions en Europe. *IRIS Plus*, *1*, 1–8.

Bode, K. (2009, June 12). Small ISPs Want FCC To Ban ESPN 360 Model. *Broadband DSL Reports.* Retrieved from www.dslreprots.com.

Boles de Boer, D., Enright, C., & Evans, L. (2000). The performance of internet service provider (ISP) markets of Australia and New Zealand. *Info*, *2*(5), 487–495. doi:10.1108/14636690010801672

Boles de Boer, D., & Evans, L. (1996). The economic efficiency of telecommunications in a deregulated market: The case ofNew Zealand. *The Economic Record*, *72*(216), 24–39. doi:10.1111/j.1475-4932.1996.tb02606.x

Bridges.org. (2001). South Africa telecommunications overview, commentary, and statistics (policy brief). Retrieved May 2, 2001 from http://www.bridges.org/publications/123 Christian, C. (2005). Telkom's ADSL now officially over 1000% more expensive than other countries. Retrieved November 11, 2005 from http://www.mybroadband.co.za /nephp/?m=show&id=1167

Bridges.Org. (2003). *Case Study Series on ICT-Enabled Development: The Women's Information Resource Electronic Service.* Retrieved from *http://*www.bridges.org/case_studies/135

Brier, N. (2005). Viral Marketing: Making the Right Connection. *Admap Magazine*, 465, October.

Brown, J. & Reingen, P. (1987). Social ties and word of mouth referral behaviour. *Journal of Consumer Research* [online]. [Accessed 12 March 2009].

BSI. (2009). Black Sea Interconnection (BSI) project homepage. Retrieved December 2009 from http://www.blacksea-net.eu/

BT.(2006). *Openreach-Open for business*. News release (DC06-010, January 10, 2006).

BTRC. (2009). Retrieved February 12, 2010 from http://www.btrc.gov.bd

Budde, P. (2002). Australia- Internet wholesale market. Retrieved November 10, 2009, from http://budde.com.au

Burton, J. (2005, March 21). High speed links to the Internet encounter consumer slowdown. *Financial Times London.*

Business Times (2006, March 25). New GEBiz features lets firms post electronic catalogue.

Business Wire (2006, June 26). Singapore iN2015 masterplan offers digital future for everyone

Cable and Satellite Asia (1997). No couch potatoes? *Cable and Satellite Asia,* July, 18.

Campanella, M. (2008). The FEDERICA project - A federated infrastructure for Future Internet research. EURESCOM mess@ge, issue 2/2008. Retrieved December 2009 from http://www.eurescom.eu/ message/messageJul2008/The_FEDERICA _project_A_federated_infrastructure-%20for_Future_Internet_research.asp

Canon, R. (2003). The Legacy of the Federal Communications Commission's Computer Inquiries. *Federal Communications Law Journal*, *55*(2), 167–206.

Cant, M., Brink, A., & Brijball, S. (2006). *Consumer Behaviour* (1st ed.). Cape Town: Juta & Co.

Carroll, A., Barnes, S., Scornavacca, E. & Fletcher, K. (2005). Consumer perceptions and attitudes towards SMS advertising: recent evidence from New Zealand. *Journal of Consumer Psychology* [online] [Accessed 13 March 2009].

Carson, J. (2005) Word of mouth marketing: A New Mandate. *Admap Magazine*, 465, October

Castells, M. (2007). *Mobile Communication and Society*. Cambridge, MA: MIT Press.

Castells, M., & Himanen, P. (2002). *The Information Society and the Welfare State: The Finnish Model*. Oxford: Oxford University Press.

Cathelat, B. (1993). *Socio-Lifestyles Marketing*. Chicago: Probus.

Cave, M. (2006). Six Degrees of Separation: Operational Separation as a Remedy in European Telecommunications Regulation. *Communications & Strategies*, *64*, 89–103.

Cave, M. (2008). *Vertical Integration and the Construction of NGA Networks*. Paper commissioned by Telstra

Cecchini, S.S., C. (2003). Can information and communications technology applications contribute to poverty reduction? Lessons from rural India. *Information Technology for Development, 10*(2), 73–84. doi:10.1002/itdj.1590100203

CEPAL (2008a, April). *The global information society: A statistical view*. Chile: United Nations.

CEPAL (2008b, June). *Digital Review 2007 of Latin America and the Caribbean*. Chile: United Nations.

CERN LHC. (2009). The Large Hadron Collider: Our understanding of the Universe is about to change. Retrieved December 2009 from http://public.web.cern.ch/public/en/LHC/LHC-en.html

CERN WLCG. (2009). Worldwide LHC Computing Grid (WLCG) project homepage. Retrieved December 2009 from http://lcg.web.cern.ch/LCG/public/

Chan, J. W. K., Yung, K. L., & Burns, N. D. (2000). Environment-strategy fit: A study of Hong Kong manufacturing logistics. *Logistics Information Management, 13*(5), 286–300. doi:10.1108/09576050010378513

Chapman, R., Slaymaker, T., & Young, J. (2003). *Livelihoods approaches to information and communication in support of rural poverty elimination and food security*. London: Overseas Development Institute.

Chen, W., Boase, J., & Wellman, B. (2002). The global villagers: comparing Internet users and uses around the world. In Wellman, B., & Haythorntwaite, C. (Eds.), *The Internet in Everyday Life* (pp. 74–113). Oxford: Blackwell. doi:10.1002/9780470774298.ch2

Choudrie, J., & Dwivedi, Y. K. (2004). Towards a Conceptual Model of Broadband Diffusion. *Journal of Computing and Information Technology, 12*(4), 323–338. doi:10.2498/cit.2004.04.06

Choudrie, J., & Dwivedi, Y. K. (2005). The Demographics of Broadband Residential Consumers of a British Local Community: The London Borough of Hillingdon. *Journal of Computer Information Systems, 45*(4), 93–101.

Choudrie, J., & Dwivedi, Y. K. (2006a). A Comparative Study to Examine the Socio-Economic Characteristics of Broadband Adopters and Non Adopters. *Electronic Government -. International Journal (Toronto, Ont.), 3*(3), 272–288.

Choudrie, J., & Dwivedi, Y. K. (2006b). Broadband Diffusion: Usage and Impact on the Consumer-User Behaviour. *Journal of Intelligent Systems, 15*(1-4), 389–408.

Choudrie, J., & Dwivedi, Y. K. (2006c). Investigating Factors Influencing Adoption of Broadband in the Household. *Journal of Computer Information Systems, 46*(4), 25–34.

Choudrie, J., & Dwivedi, Y. K. (2007). Broadband Impact on Household Consumers: Online Habits and Time Allocation Patterns on Daily Life Activities. *International Journal of Mobile Communications, 5*(2), 225–241. doi:10.1504/IJMC.2007.011817

Choudrie, J., & Dwivedi, Y. (2005). The demographics of broadband residential consumers in a British local community: The London Borough of Hillingdon. *Journal of Computer Information Systems, 45*(4), 93–101.

Choudrie, J., & Dwivedi, Y. (2006). Investigating factors influencing adoption of broadband in the household. *Journal of Computer Information Systems, 46*(4), 25–34.

Christensen, C. M. (1997). *The Innovator's Dilemma. When New Technologies Cause Great Firms to Fail*. Boston, MA: Harvard Business School Press.

CIA. (2008). Retrieved from http://www.isilo.com/download /CIAWFB08.htm

Cirami, R., Coretti, I., Marcantonio, P. D., Pucillo, M., & Santin, P. (2008). EVALSO: Enabling Virtual Access to Latin-America Southern Observatories (Poster presentation). SPIE conference on Astronomical Instrumentation, Marseilles, 23-28 June 2008. Retrieved December 2009 from http://www.evalso.eu/evalso/wp- content/uploads/2008/08/jra2_spie08_poster.pdf

CNNIC (The China Internet Network Information Centre). Semiannual survey report on Internet development in China, 1997-2009. Retrieved November 5, 2009, from http://www.cnni.net.cn

Colle, R.D.R., R. (2001). *Modules for training telecentre staffs: An Interim Report with sample modules.* Cornell University.

Collis, D. J., Bane, P. W., & Bradley, S. P. (1997). Winners and Losers. Industry Structure in the Converging World of Telecommunications, Computing, and Entertainment. In Yoffie, D. B. (Ed.), *Competing at the age of digital convergence*. Boston, MA: Harvard Business School Press.

Commerce Commission. (2003). Telecommunications Act 2001 Section 64 review and schedule 3 investigation into unbundling the local loop network and the fixed public data network Final Report. Wellington, New Zealand. Retrieved from http://www.comcom.govt.nz/Industry Regulation/Telecommunications/I nvestigations/LocalLoopUnbundling/ContentFiles/Documents/finalreport.PDF

Commerce, Industry and Technology Bureau (2004). *Digital 21 strategy: Sustainability and opportunities.* Hong Kong: Commerce, Industry and Technology Bureau, The Government of the Hong Kong Special Administrative Region.

Conseil, N. (2006).*Le développement de la vidéo à la demande en France et en Europe,* Paris.

Conseil, N. (2008). *Vidéo à la demande en Europe: second recensement des services de VOD.* Paris.

Copps, M. (2001). Remarks of Michael J. Copps, Commissioner Federal Communications Commission to National Association of State Utility Consumer Advocates 2001 Annual Meeting.

Correa, D. K. (2007). Assessing broadband in America: OECD and ITIF Broadband Rankings. Retrieved November 18, 2009, from http://www.itif.org/files/BroadbandRankings.pdf

Corry, W., & Mundell, J. (2006) Demystifying blogs – embracing objective communication. Market Research Society Annual Conference [accessed on April 27, 2009]

Covec (2004). Benchmarking Telecom's UPC service. Retrieved from http://www.comcom.govt.nz//IndustryRegulation/Telecommunications/Investigations/Unbundling-PartialCircuits/ContentFiles/Documents/Final%20Draft%20Report%20(August%204%202004)%20-%20UPC%20Benchmarking.pdf).PDF

Crandall, R., Lehr, W., & Litan, R. (2007). The Effects of Broadband Deployment on Output and Employment: A Cross-Sectional Analysis of U.S. Data. *Issues in Economic Policy*, *6*, 1–34.

Crandall, R., Sidak, G., & Singer, H. (2002). The empirical case against asymmetric regulation of broadband Internet access. *Berkeley Law and Technology Journal*, *17*(1), 953–987.

Crandall, R. W., Lehr, R., & Litan, R. (2007). The effects of Broadband Deployment on Output and employment: A Cross-Sectional Analysis of U.S. Data. *Issues in Economic Policy, 6*. The Brookings Institute. Retrieved from http://www.brookings.edu/ papers/2007/06labor_crandall.aspx

Crandall, R.W., & Jackson, C.L. (2001, July). The $500 Billion Opportunity: The Political Economic Benefits of Widespread Diffusion of Broadband Internet Access. *Criterion Economics.*

Cummings, R., Harrison, G., & Rutström, E. (1995). Homegrown Values and Hypothetical Surveys: Is the Dichotomous Choice Approach Incentive Compatible? *The American Economic Review*, *85*(1), 260–266.

Daidj, N., & Vialle, P. (2007). Le développement d'une concurrence multiforme pour la VOD en France. In P. Vialle (Ed.), *Mutation des STICS. Acteurs, Ressources, Activités*. Paris: Hermes Science Publishing.

Daidj, N., Epinette, O., & Vialle, P. (2004). Positioning of Firms on an E-business Value Chain: the Case of Video on Demand. In J. Chen (Ed.), *The Fourth International Conference on Electronic Business.* Beijing: *Academic Publishers/World Publishing Corporation.*

Dart, J., & Davies, R. (2003). A Dialogical. Story-Based Evaluation Tool: the most significant change technique. *The American Journal of Evaluation*, *24*(3).

David, P. A. (2001, June). *Knowledge, capabilities and Human Capital Formation in Economic Growth*. New Zealand Treasury Working Paper, 01/13.

Davies, J.R. (2006, July). Korea broadband market December 2005. *eCommerce Innovation Centre*. Cardiff University.

DB. (2009). Digital Britain: The Final Report, Retrieved January 14, 2010, from http://www.culture.gov.uk/what_we_do/broadcasting/6216.aspx

Debande, O., & Chetrit, G. (2001). The European Audiovisual Industry: An overview. *European Investment Bank (EIB) Sector Papers*, September.

DEISA. (2009). Distributed European Infrastructure for Supercomputing Applications (DEISA) project homepage. Retrieved December 2009 from http://www.deisa.eu/

Dierickx, I., & Cool, K. (1989). Asset stock accumulation and sustainability of competitive advantage. *Management Science*, *35*, 1504–1511. doi:10.1287/mnsc.35.12.1504

Dimmick, J., Kline, S., & Stafford, L. (2000). The gratification niches of personal e-mail and the telephone: Competition, displacement, and complementarity. *Communication Research*, *27*, 227–248. doi:10.1177/009365000027002005

Dimmick, J., & Rothenbuhler, E. W. (1984). The theory of the niche: quantifying competition among media industries. *The Journal of Communication*, *34*(1), 103–119. doi:10.1111/j.1460-2466.1984.tb02988.x

Dixit, A., & Nalebuff, B. (1991). *Thinking strategically: the competitive edge in business, politics and everyday life*. New York, London: W. W. Norton & Company.

DOGC. (2009). Digital opportunities for global community. Retrieved November 11, 2009, from http://www.dosite.go.jp

Dosi, G., Teece, D., & Winter, S. (1990). Understanding Corporate Coherence: Theory and evidence. *Journal of Economic Behavior & Organization*, *23*, 1–30.

du Plessis, M. (2005). Tit for tat. Retrieved September 15, 2005 from http://www.iweek.co.za/View Story.asp?StoryID=155165

Dutta-Bergman, M. (2004). Complementarity in consumption of news types across traditional and new media. *Journal of Broadcasting & Electronic Media*, *48*(1), 41–61. doi:10.1207/s15506878jobem4801_3

Dwivedi, Y. K., & Irani, Z. (2009). Understanding the Adopters and Non-adopters of Broadband. *Communications of the ACM*, *52*(1), 122–125. doi:10.1145/1435417.1435445

Dwivedi, Y. K., Choudrie, J., & Brinkman, W. P. (2006a). Consumer Usage of Broadband in British Households. *International Journal of Services and Standards*, *2*(4), 400–416. doi:10.1504/IJSS.2006.010472

Dwivedi, Y. K., Choudrie, J., & Brinkman, W. P. (2006b). Development of a Survey Instrument to Examine Consumer Adoption of Broadband. *Industrial Management & Data Systems*, *106*(5), 700–718. doi:10.1108/02635570610666458

Dwivedi, Y. K., & Irani, Z. (2009). Understanding the Adopters and Non-adopters of Broadband. *Communications of the ACM*, *52*(1), 122–125. doi:10.1145/1435417.1435445

Dwivedi, Y. K., Khan, N., & Papazafeiropoulou, A. (2007a). Consumer adoption and usage of broadband in Bangladesh. *Electronic Government: An International Journal*, *4*(3), 299–313. doi:10.1504/EG.2007.014164

Dwivedi, Y. K., Khoumbati, K., Williams, M. D., & Lal, B. (2007b). Factors affecting consumers' behavioural intention to adopt broadband in Pakistan. *Transforming Government People. Process and Policy*, *1*(3), 285–297.

Dwivedi, Y. K., & Lal, B. (2007). Examining the Socio-economic Determinants of Broadband Adoption in the United Kingdom. *Industrial Management & Data Systems*, *107*(5), 654–671. doi:10.1108/02635570710750417

Dwivedi, Y. K., Lal, B., & Williams, M. D. (2009). Managing Consumer Adoption of Broadband: Examining Drivers and Barriers. *Industrial Management & Data Systems*, *109*(3), 357–369. doi:10.1108/02635570910939380

Dwivedi, Y. K., Papazafeiropoulou, A., Gharavi, H., & Khoumbati, K. (2006c). Examining the Socio-economic Determinants of Adoption of an e-Government Initiative 'Government Gateway'. *Electronic Government -. International Journal (Toronto, Ont.)*, *3*(4), 404–419.

Dwivedi, Y. K., & Weerakkody, V. (2007). Examining the Factors Affecting the Adoption of Broadband in the Kingdom of Saudi Arabia. *Electronic Government: An International Journal*, *4*(1), 43–58. doi:10.1504/EG.2007.012178

Dwivedi, Y. K., & Williams, M. D. (2008). The Influence of Demographic Variables on Citizens' Adoption of e-Government. *Electronic Government: An International Journal*, *5*(3), 261–274. doi:10.1504/EG.2008.018874

Dwivedi, Y. K., Williams, M. D., Weerakkody, V., Lal, B., & Bhatt, S. (2008b). Consumer Adoption of Broadband in India: A Pilot Study. *Journal of Cases on Information Technology*, *10*(3), 39–52.

Dwivedi, Y. K., Papazafeiropoulou, A., Willem-Paul, B., & Lal, B. (2008a). Examining the Influence of Service Quality and Secondary Influence on the Behavioral Intention to Change Internet Service Provider. *Information Systems Frontiers*. Retrieved January 7, 2010 from http://www.springerlink.com /content/aw80021039441235/

Easterly, W., King, R., Levine, R., & Rebelo, S. (1994). Policy, Technology Adoption and Growth. *NBER Working Paper No. 4681*. Cambridge: National Bureau of Economic Research.

EC. (2000). Regulation (EC) No 2887/2000 of the European Parliament And Of The Council of 18 December 2000 on unbundled access to the local loop.

EC. (2006). Communication from the commission to the council, the European parliament, the European economic and social committee and the committee of the regions on the Review of the EU Regulatory Framework for electronic communications networks and services, *COM(2006) 334 final, 29.6.2006*

EC. (2007a). Proposal for a directive of the European parliament and of the council of … 2007 amending Directives 2002/21/EC on a common regulatory framework for electronic communications networks and services, 2002/19/EC on access to, and interconnection of, electronic communications networks and services, and 2002/20/EC on the authorisation of electronic communications networks and services.

EC. (1996). *Directive 96/19amending Directive 90/388 with regard to the implementation of□full competition in telecommunications markets*. Full Competition Directive.

EC Press Release. (2008). Europe's GÉANT, the world's highest speed computer network, goes global. Retrieved December 2009 from http://europa.eu/rapid/pressReleases Action.do?reference=IP/08/354&

EC. (2005) i2010 – A European Information Society for growth and employment Retrieved January 14, 2010, from http://europa.eu/rapid/pressReleases Action.do?reference=MEMO/05/184& format=HTML&aged=0&language=EN&guiLanguage=en

EC. (2007b). *Commission staff working document, impact assessment*. Accompanying document to the Commission proposal for a Directive of the European Parliament and the Council, {COM (2007) 697, COM (2007)698, COM (2007)699, SEC (2007)1473}, 13 November 2007.

EC. (2009) Digital economy can lift Europe out of crisis, says Commission report, European Commission, Retrieved January 14, 2010, from http://europa.eu/rapid/pressReleases Action.do?reference=IP/09/1221& format=HTML&aged=0&language=EN&guiLanguage=fr

Economides, N., & White, L. (1995). Access and interconnection pricing? How efficient is the "efficient component pricing rule"? *Antitrust Bulletin*, *40*(3), 557–579.

Economist, (2000, June 24). Government and the Internet: A survey. *The Economist*, 17.

ECTA. (2009) *Broadband Scorecards*. Retrieved October 15, 2009, from http://www.ectaportal.com/en/basic650.html.

EGEE. (2009). Enabling Grids for E-sciencE (EGEE) project homepage. Retrieved December 2009 from http://www.eu-egee.org

Engel, J., Blackwell, R., & Miniard, P. (1993). *Consumer Behavior* (8th ed.). Fort Worth: Dryden Press.

Enright, C. (2000). *Strategic behaviour of internet service providers in New Zealand and the performance of this market.* Unpublished paper based on Masters thesis, New Zealand Institute for the Study of Competition and Regulation, Wellington, New Zealand. Retrieved from http://www.iscr.org.nz/f240,4833 /4833_enright_210600.pdf

ERG. (2007). *ERG Opinion on Functional Separation (ERG (07)44).*

Esbin, B., & Marcus, A. (2009). "The Law is Whatever the Nobles Do": Undue Process at the FCC, 17. *CommLaw Conspectus, 1*, Retrieved from http://commlaw.cua.edu// artic les/v17/17.2/Esbin-Marcus-Revised.pdf.

E-Smith. J. (2009). Internet future in SA. Retrieved November 5, 2009 from http://mybroadband.co.za/news/Internet/10332.html

EU. (2005). i2010 - A European Information Society for growth and employment. European Commission. Retrieved November 11, 2009, from http://ec.europa.eu/information _society/eeurope/i2010/index_en.htm

EU. (2008). Commission casts ICT in green role. Reference IP/08/733, European Commission. Retrieved November 11, 2009, from http://europa.eu/rapid/press ReleasesAction.do?reference=IP/08/733&format= HTML&aged=0 &language=EN&guiLanguage=en

European Broadband Portal. (2009). Check-list of actions for public authorities considering broadband interventions in under-served territories. Draft (1st Version, April 2009). Retrieved November 11, 2009, from http://www.broadband-europe.eu/ Documents/EBP-CHECK-LIST-Broaband%20Actions% 20of%20Public%20Authorities%20v%201-4-2009.pdf

European Commission. (2006). *Safer Internet.* Special Eurobarometer 250/Wave 64.4. Luxemburg: European Commission. Retrieved October 1, 2006, from http://europa.eu.int/information_society/ac tivities/sip/docs/eurobarometer/euro barometer_2005_25_ms.pdf.

European Commission. (2007). *E-Communications Household Survey.* Special Eurobarometer 274/Wave 66.3. Luxembourg: European Commission. Retrieved November 3, 2008, from http://ec.europa.eu/public_opinion/ archives/eb_special_280_260_en.htm

European Commission. (2008). *E-Communications Household Survey.* Special Eurobarometer 293/Wave 68.2. Luxembourg: European Commission. Retrieved November 3, 2008, from http://ec.europa.eu/public_opinion /archives/ebs/ebs_293_full_en.pdf.

Eurostat (2007). *ICT Impact Assessment by Data Linking.* Luxembourg: Working Group on Information Society Statistics (ISS), Doc.F6/ISS-WG/Oct07/08.

EVALSO. (2009). Enabling Virtual Access to Latin-America Southern Observatories (EVALSO) project homepage. Retrieved December 2009 from http://www.evalso.eu/evalso/

Evans, L., Grimes, A., Wilkinson, B., & Teece, D. (1996, December). Economic reforms in New Zealand 1984-1995: the pursuit of efficiency. *Journal of Economic Literature, 34*, 1856–1902.

Evans, L., & Quigley, N. (2000). Contracting, incentives for breach and the impact of competition. *Journal of World Competition, 23*(2).

Evert, G. (2003). All research is interpretive! *Journal of Business and Industrial Marketing, 18*(6/7), 482–492. doi:10.1108/08858620310492365

Faiola, A. (2006, May 27). When escape seems just a mouse-click away: Stress-driven addiction to online games spikes in S. Korea. *Washington Post.*

Fairlie, R. W. (2004). Race and the digital divide. *Contributions to Economic Analysis & Policy, 3*(1).

Fan, X., & Dickie, P. (2000). The contribution of foreign direct investment to growth and stability: A post crisis ASEAN-5 review. *ASEAN Economic Bulletin, 17*(3), 312–323. doi:10.1355/AE17-3F

Fan, Q. (2008). Analysis of the Influence of Australia's Government Policy on Broadband Internet Access. In Dwivedi, Y. K., Papazafeiropoulou, A., & Choudrie, J. (Eds.), *Handbook of Research on Global Diffusion of Broadband Data Transmission* (pp. 109–125). Hershey, PA: IGI Global.

Farrell, J. (1996). Creating local competition. Retrieved from http://law.indiana.edu/fclj/pubs/v 49/no1/farrell.html

FCC. (2003). Review of the Section 251 Unbundling Obligations of the Telecommunications Act of 1996; Deployment of Wireline Services Offering Advanced Telecommunications Capability; Appropriate Framework for Broadband Access to the Internet over Wireline Facilities. *(FCC03-36), 20 February 2003.*

FCC. (1996). *The First Report & Order on the local Competition Provisions of the Telecommunications Act* (FCC 96-325).

FCC. (2004). *FCC Releases New Rule for Network Unbundling Obligations of Incumbent Local Phone Carriers*, Order on remand (FCC 04-290), 15 December 2004.

FCC. (2005). *FCC Eliminates Mandated Sharing Requirement on Incumbents' Wireline Broadband Internet Access Services, Report and order and Notice of Proposed Rulemaking*, (FCC 05-150), 23 September 2005.

FCC. (2010). *Internet Access Services: Status as of June 30, 2009.* Retrieved October 15, 2010, from http://www.fcc.gov/Daily_Releases/Daily_ Business/2010/db0902/DOC-301294A1.pdf

Federal Communication Commission (2002). *Inquiry Concerning High-Speed Access to the Internet Over Cable and Other Facilities; Internet Over Cable Declaratory Ruling; Appropriate Regulatory Treatment for Broadband Access to the Internet Over Cable Facilities.* Declaratory Ruling, 17 FCC Rcd 4798.

Federal Communication Commission (2003). Review of the Section 251 Unbundling Obligations of Incumbent Local Exchange Carriers, Implementation of the Location Competition Provisions of the Telecommunications Act of 1996, Deployment of Wireline Services Offering Advanced Telecommunications Capability, CC Docket Nos. 01-338, 96-98, 98-147, Report and Order and Order on Remand and Further Notice of Proposed Rulemaking, 18 FCC Record 16978.

Federal Communications Commission. (2005a). *Appropriate Framework for Broadband Access to the Internet over Wireline Facilities.* Report and Order and Notice of Proposed Rulemaking.

Federal Communications Commission (2005b, September 23). *In the Matter of Appropriate Framework for Broadband Access to the Internet Over Wireline Facilities.* Policy Statement, 20 FCC Rcd 14986.

Federal Communications Commission (2008, August 20). *Formal Complaint of Free Press and Public Knowledge Against Comcast Corp. for Secretly Degrading Peer-to-Peer Applications; Broadband Industry Practices; Petition of Free Press et al. for Declaratory Ruling that Degrading an Internet Application Violates the FCC's Internet Policy Statement and Does Not Meet an Exception for "Reasonable Network Management."* Memorandum Opinion and Order, 23 FCC Rcd 13028.

Federal Communications Commission (2009, April 8). *In the Matter of A National Broadband Plan for Our Future.* GN Docket No. 09-51, Notice of Inquiry.

Federal Communication Commission (FCC). (2009). *High Speed Service for Internet Access: Statistics as of June 30, 2008.* Industry Analysis and Technology Division, Wire line Competition Bureau. Retrieved from http://www.fcc.gov/wcb/stats.

FEDERICA. (2009). Federated E-infrastructure Dedicated to European Researchers Innovating in Computing network Architectures (FEDERICA) project homepage. Retrieved December 2009 from http://www.fp7-federica.eu/

Ferguson, D. A., & Perse, E. M. (2000). The World Wide Web as a functional alternative to television. *Journal of Broadcasting & Electronic Media*, *44*(2), 155–174. doi:10.1207/s15506878jobem4402_1

Finholt, T., & Sproull, L. (1990). Electronic groups at work. *Organization Science*, *1*(1), 41–64. doi:10.1287/orsc.1.1.41

Firth, L., & Mellor, D. (2005, March-April). Broadband: Benefits and Problems. *Telecommunications Policy*, *29*(2-3), 223–236. doi:10.1016/j.telpol.2004.11.004

Flamm, K., & Chaudhuri, A. (2007). An analysis of the determinants of broadband access. *Telecommunications Policy*, *31*, 312–326. doi:10.1016/j.telpol.2007.05.006

Flamm, K., & Chaudhuri, A. (2007). An analysis of the determinants of broadband access. *Telecommunications Policy*, *31*, 312–326. doi:10.1016/j.telpol.2007.05.006

Foley, J. M. (1992). Economic factors underlying telephone company efforts to enter home video distribution. *Journal of Media Economics*, (Fall): 57–68.

Foster, I., Kesselman, C., & Tuecke, S. (2001). The anatomy of the grid: enabling scalable virtual organizations. *International Journal of High Performance Computing Applications*, *15*(3), 200–222. doi:10.1177/109434200101500302

Foster, I., & Kesselman, C. (2004). Concepts and architecture. In Foster, I., & Kesselman, C. (Eds.), *The grid: blueprint for a new computing infrastructure* (2nd ed.). San Francisco, CA: Morgan Kaufmann.

Franz, L. (2005). Pushing the city limits. iWeek. Retrieved December 1, 2005 from http://www.iweek.co.za/ViewStory.asp?StoryID=157791

Free Press(2009, September4). In the Matter of Inquiry Concerning the Deployment of Advanced Telecommunications Capability to All Americans in a Reasonable and Timely Fashion, and Possible Steps to Accelerate Such Deployment Pursuant to Section 706 of the Telecommunications Act of 1996, as Amended by the Broadband Data Improvement Act, A National Broadband Plan for Our Future. *Comments of Free Press (p. 48).*

Frieden, R. (2005). Lessons from broadband development in Canada, Japan, Korea and the United States. *Telecommunications Policy*, *29*, 595–613. doi:10.1016/j.telpol.2005.06.002

Fuke, H. (2000). *Structural changes and deregulation in the telecommunications industry*. Tokyo: NTT Publishing. (in Japanese)

Fuke, H. (2003). The Spectacular Growth of ADSL in Japan and its Implications. *Communications & Strategies*, *52*, 175–191.

Fuke, H. (2005). Structural Changes in Telecommunications and Regulatory Policy in Japan. *Communications & Strategies*, *59*, 113–139.

Fuke, H. (2007). *Info-communications policy in the broadband era*. Tokyo: NTT Publishing. (in Japanese)

Garbacz, C., & Thompson, H. G. Jr. (2005). Universal telecommunications service: a world perspective. *Information Economics and Policy*, *17*, 495–512. doi:10.1016/j.infoecopol.2005.03.001

Garcia-Murillo, M. (2005). International Broadband Deployment: The Impact of Unbundling. *Communications & Strategies*, *57*, 83–108.

Gasser, T., Kneip, A., & Koehler, W. (1991). A flexible and fast method for automatic smoothing. *Journal of the American Statistical Association*, *86*, 643–652. doi:10.2307/2290393

Gasser, Th., & Muller, H. G. (1984). Estimating Regression Functions and Their Derivatives by the Kernal Method. *Scandinavian Journal of Statistics*, *11*, 171–185.

GÉANT Press Release. (2009). Third Generation GÉANT project launched to extend Pan-European research and collaboration capabilities. Retrieved December 2009 from http://www.geant.net/Media_Centre/PressReleases/Pages/3rd Generationofprojectlaunched.aspx

GÉANT. (2009a). GÉANT project homepage. Retrieved December 2009 from http://www.geant.net

GÉANT. (2009b). Global Connectivity - Connecting research communities across the world. Retrieved December 2009 from http://www.geant.net/ Network/ GlobalConnectivity

GÉANT. (2009c). GÉANT Partners. Retrieved December 2009 from http://www.geant.net/About_GEANT/Partners

Genachowski, J. (2009, September 21). Preserving a Free and Open Internet: A Platform for Innovation, Opportunity, and Prosperity. The Brookings Institution, Washington, D.C.

General Accountability Office (2006). *Broadband Deployment Is Extensive throughout the United States, but It Is Difficult to Assess the Extent of Deployment Gaps in Rural Areas*, GAO-06-426, May.

Gera, S. Wulong, Su & Lee F.C. (1998). *Information Technology and Productivity Growth: An empirical analysis for Canada and the United States.* Microeconomic Analysis Working Paper, No 20, Canada.

Gholami, R., Sang-Young, T. L., & Heshmati, A. (2006). The Causal Relationship Between Information and Communication Technology and Foreign Direct Investment. *World Economy*, *29*(1), 43–62. doi:10.1111/j.1467-9701.2006.00757.x

Gilroy, A. (2009, June 1). Access to Broadband Networks: The Net Neutrality Debate. Congressional Research Service.

Goggin, G. (2006). *Cell Phone Culture: Mobile technology in everyday life*. New York: Routledge.

Goldfarb, A., & Prince, J. (2008). Internet adoption and usage patterns are different: Implications for the digital divide. *Information Economics and Policy*, *20*, 2–15. doi:10.1016/j.infoecopol.2007.05.001

Goldstuck, A. (2009). *Internet Access in South Africa 2008*. Johannesburg: World Wide Worx.

Goldstuck, A. (2008). Internet Access in South Africa 2007. World Wide Worx: Johannesburg. Retrieved June 2, 2006 from http://free.financialmail.co.za/06/ 0602/ technology/atech.htm

Goolsbee, A., & Klenow, P. J. (2002). Evidence on learning and network externalities in the diffusion of home computers. *The Journal of Law & Economics*, *45*, 317–343. doi:10.1086/344399

Government Technology. (2009). The Future Is Fibre. Retrieved November 11, 2009, from http://www.govtech.com/gt/692922

Greek Information Society Observatory. (2008). *Project: "Study for measuring the indicators of the i2010 initiatives for the year 2008". Research Findings amongst Individuals-Households*. Retrieved October 20, 2009, from http://www.observe.gr/page/defaul t.asp?la=2&id=2101&pk=161&return=183

Greek Information Society Observatory. (2009). *Measurement of eEurope/i2010 indicators*. July 2009. Retrieved October 25, 2009, from http://www.observatory.gr/page/default.asp?id=4

Greenstein, S., & Prince, J. (2006). *The diffusion of the Internet and the geography of the digital divide in the United States*. Working Paper 123182, National Bureau of Economic Research, USA.

Greenstien, S. M., & Spiller, P. T. (1996). *Estimating the Welfare Effects of Digital Infrastructure*. NBER Working Paper, No. 5770, National Bureau of Economic Research, Inc., Cambridge, MA. Chlamtac, I., Gumaste, A., & Szabo, C. (Eds.) (2005). *Broadband Services: Business Models and Technologies for Community Networks*. John Wiley & Sons Ltd.

Gremler, D., Thorsten, H., Gwinner, K., & Walsh, G. (2004). Electronic word of mouth via consumer-opinion platforms: What motivates consumers to articulate themselves on the Internet? *Journal of Interactive Marketing*, *18*(1).

Grotta, G. L., & Newsom, D. (1982). How does cable television in the home relate to other media use patterns? *The Journalism Quarterly*, *59*, 588–591, 609.

Group, I. T. (Information Technology Group Center for Information Development at Harvard University) (1999). Readiness for the networked world – A guide for developing countries. Retrieved November 12, 2009 from http://www.harvard.edu

Guillou, B. (2004). *Online Distribution of Films*. European Audiovisual Observatory, May.

Gurumurthy, A. (2006). Promoting gender equality? Some development-related uses of ICTs by women. *Development in Practice*, *16*(6), 611–616. doi:10.1080/09614520600958298

Gurumurthy, A. (2003) *Bridging the digital gender divide: Issues and insights on ICT for women's economic empowerment*. U.N.D.F.f. Women.: Impression communications India.

Haddon, L. (2003). Domestication and Mobile Technology. In Katz, J. E. (Ed.), *Machines that Become Us: The Social Context of Personal Communication Technology* (pp. 43–56). New Brunswick, New Jersey: Transaction Publisher.

Haddon, L. (2004, October). *Cultural Differences in Communication: Examining Patterns of Daily Life*. Paper presented at the Conference Mobile Communication and Social Change, Seoul, South Korea.

Haddon, L., de Gournay, C., Lohan, M., Östlund, B., Palombini, I., Sapio, B., & Kilegran, M. (2002). *From Mobile to Mobility: The Consumption of ICTs and Mobility in Everyday Life.* The COST269 Mobility Workgroup. Retrieved October 5, 2008, from http://www.cost269.org

Hardy, A. P. (1980). The role of the telephone in economic development. *Telecommunications Policy*, *4*(4), 278–286. doi:10.1016/0308-5961(80)90044-0

Harper, R., Palen, L., & Taylor, A. (Eds.). (2005). *The Inside Text: Social, Cultural and Design Perspectives on SMS*. Dordrecht: Springer.

Harris, R. R. R. (2006). *Empowering the poor. Information and communications technology for governance and poverty reduction: a study of rural development projects in India*. Bangkok: UNDP-APDIP.

Harris, R., & Bhatarai, M. (2003). *Fact-Finding Review of ICT in Development in a Rural-Urban Setting*, in *Assessing the feasibility of ICT as a development instrument for rural-urban linkages in Nepal*. UN-HABITAT and Roger Harris Associates. Rural-Urban Partnerships Programme (RUPP) under Support Services Policy and Programme Development (SPPD).

Harvey, M. G., & Rothe, J. T. (1985). Video casette recorders: Their impact on viewers and advertisers. *Journal of Advertising Research*, *25*(6), 19–27.

Hausman, J. (2002). Asymmetric regulation of related services. In Crandall, R., & Alleman, J. (Eds.), *Broadband: should we regulate high-speed internet access?* (pp. 129–156). Washington, DC: AEI-Brookings Joint Center for Regulatory Studies.

Heeks, R., Arun, S., & Morgan, S. (2005). *Researching Women's ICT-Based Enterprise for Development: Methods, Tools and Lessons from Fieldwork*. UK: IDPM, University of Manchester.

Heinz, H., Gross, R., & Acquisti, A. (2005). *Proceedings of the 2005 ACM workshop on social network privacy* [online]. Retrieved March 11, 2009 from http://www.heinz.cmu.edu/~ acquisti/papers/privacy-facebook-acquisti-slides.ppt#584,61, Conclusions.htm

Henke, L., & Donohue, T. R. (1989). Functional displacement of tradition TV viewing by VCR owners. *Journal of Advertising Research*, (April/May): 18–23.

Himmelweit, H., Oppenheim, A. N., & Vince, P. (1958). *Television and the child*. London: Oxford University Press.

Hoarau, Ch. (2003). *Le repositionnement de l'industrie électronique et du multimédia comme support de services, nouvelles sources de création de valeur*. Observatoire des Stratégies Industrielles, Ministère de l'Economie, des Finances et de l'Industrie, septembre.

Hofstede, G. (1980). *Culture's Consequences: International Differences in Work-Related Values*. London: Sage.

Holt, L., & Jamison, M. A. (2008). Broadband and contributions to economic growth: Lessons from the U.S. Experience. PURC Working Paper, Department of Economics, University of Florida.

Horrigan, J. (2008). *Broadband adoption, rural America, and economic decisions*. Paper presented at the Economic Research Service's Broadband in the Rural Economy workshop, Washington, DC.

Horrigan, J. (2009, August 25). Broadband Adoption Barriers. Blogband, The Official Blog of the National Broadband Plan.

Howell, B., & Sangekar, M. (2009). Beyond Surface Similarities: Telecommunications Industry Structure Evolution in Finland and New Zealand. *Prometheus*, *27*(2), 99–115. doi:10.1080/08109020902895243

Howell, B. (2008). Competition, Regulation and Broadband Diffusion: the Case of New Zealand. In Dwivedi, Y. K., Papazafeiropoulou, A., & Choudrie, J. (Eds.), *Handbook of Research on Global Diffusion of Broadband Data Transmission* (pp. 139–154). Hershey, PA: IGI Global.

Howell, B. (2003). *Building best practice broadband: bringing infrastructure supply and demand together.* Wellington, New Zealand: New Zealand Institute for the Study of Competition and Regulation. Retrieved from http://www.iscr.org.nz/f213,4379/4379_ building_best_practice_060603.pdf

Howell, B. (2007). *A Pendulous Progression: New Zealand's Telecommunications Regulation 1987-2007*. Wellington, New Zealand: New Zealand Institute for the Study of Competition and Regulation. Retrieved from http://www.iscr.org.nz/research

Howell, B. (2008). *From competition to regulation: New Zealand's telecommunications sector performance 1987-2007.* Paper presented at the International Telecommunications Society European Regional Conference, Rome, Italy. Retrieved from http://www.iscr.org.nz/f439,12724/12724_From _Competition_to_Regulationbh230708.pdf

Howell, B. (2008a). *The role of price structure in telecommunications technology diffusion.* Paper presented at the International Telecommunications Society European Regional Conference, Rome, Italy. Retrieved from http://www.iscr.org.nz/research

Howell, B., & Obren, M. (2003). *Telecommunications usage in New Zealand 1993-2003. Broadband diffusion*. Wellington, New Zealand: New Zealand Institute for the Study of Competition and Regulation. Retrieved from http://www.iscr.org.nz/f212,436 2/4362_nz_telecommunications_usag_090703.pdf

Hu, W., & Prieger, J. (2008). The Timing of Broadband Provision: The Role of Competition and Demographics. In Dwivedi, Y. (Eds.), *Handbook of Research on Global Diffusion of Broadband Data Transmission* (pp. 241–259). Hershey, PA: IGI Global.

Hu, H. H. (2008). The tendency of media usage behavior in Taiwan--A case study of 2004-2008 Shih Hsin Communication Database. Shih Hsin University. MA thesis.

Huyer, S. Hafkin, N., Ertl, H, & Dryburgh, H. (2005). From the Digital Divide to Digital Opportunities: Measuring Infostates for Development. In Sciadas (Ed.), *Women in the Information Society.* Montreal: Orbicom.

IDA. (2009). Infocomm Development Authority of Singapore. Retrieved November 11, 2009, from http://www.ida.gov.sg/

incrEAST. (2009). Project of the month (January): Black Sea Interconnection. Retrieved December 2009 from http://www.increast.eu/en/441.php

Indjikian, R., & Siegel, D. (2005). The Impact of Investment in IT on Economic Performance: Implications for Developing Countries. *World Development*, *33*(5), 681–700. doi:10.1016/j.worlddev.2005.01.004

InfoComm Development Authority. Singapore (2007). *Annual e-Government Perception Survey, 2007.* Retrieved from http://www.ida.gov.sg/Publications /20071001170637.aspx

Intel (2009). The Economic Impact of Broadband in Developing Nations, Best practices enable developing nations to reap economic benefits of broadband. White paper, Intel Corporation, 2009. Retrieved November 11, 2009, from http://download.intel.com/intel/ worldahead/pdf/economic_impact.pdf

International Monetary Fund. (2002). People's Republic of China- Hong Kong Special Administrative Region: Selected Issues. *IMFCountry Report*, 02/99.

International Telecommunications Union. (2003). Broadband Korea: Internet Case Study. Retrieved February 1, 2007 from http://www.itu.int/ITU-/ict/cs/korea/ material/CS_KOR.pdf

International Telecommunications Union. (2006). *World Telecommunication/ICT Development Report 2006: Measuring ICT for social and economic development,* http://www.itu.int/ITU-D/ict/ publications/wtdr_06/index. html.

International Telecommunications Union. ITU (2006). *Trends in Telecommunications Reform 2006: Regulation in the Broadband World* (p. 77). Geneva, Switzerland: ITU.

International Telecommunications Union. ITU (June, 2009). World Telecommunications Indicator Database. Retrieved from http://www.itu.int

Internet.org.za. (2003). Cost of Internet access in South Africa (1993-2003). Retrieved October 5, 2006 from http://www.internet.org.za/costs.html

Irani, Z., Dwivedi, Y. K., & Williams, M. D. (2009). Understanding Consumer Adoption of Broadband: An Extension ofTechnology Acceptance Model. *The Journal of the Operational Research Society*, *60*(10), 1322–1334. doi:10.1057/jors.2008.100

Isacowitz, R. (2003). VOD and the case for Content Protection. *Converge Network Digest*, November 17.

ITU. (1998). *Resolution 7. Gender and Telecommunication Policy in Developing Countries.* World Telecommunication Development Conference, Valletta.

ITU. (2002). *Gender Mainstreaming.* Retrieved September 28, 2009 from http://www.itu.int/ITU-D/pdf/5196-011-en.doc

ITU. *(2006).* World telecommunication/ ICT development report: measuring ICT for social and economic development. *Geneva.*

ITU. (2008). Report on the World Summit on the Information Society Stocktaking. World Summit on the Information Society, Geneva 2003 - Tunis 2005. Retrieved November 11, 2009, from http://www.itu.int/wsis/stocktaking/ docs/2008/WSIS-Stocktaking2008-e.pdf

ITU-T. (2004). Very high speed digital subscriber line transceivers. ITU-T Recommendation G.993.1. Retrieved January 15, 2010, from http://www.itu.int/rec/ T-REC-G.993.1/en

Izugbara, C. O. (2004). Gendered micro-lending schemes and sustainable women's empowerment in Nigeria. *Community Development Journal*, *39*(1), 72–84. doi:10.1093/cdj/39.1.72

JANET. (2009). JANET Services. Retrieved December 2009 from http://www.ja.net/services/index.html

Jenkins, H., Jr. (2009, September 23). Neutering the 'Net. *Wall Street Journal*.

Jerinabi, U., & Arthi, J. (2005). *Impact of Information Technology and Globalisation on Women's career*. International Conference on Information and Communication Technology of Management. Melaka, Malaysia.

Jew, B., & Nicholls, R. (2000). Competition and regulatory issues. In N. Roehrich (Ed.), *Internet carriage, peering and chaos.* Sydney, NSW: Network Insight Group, RMIT.

Jiang, Y. S. (2005). Displacement and complementary effect of Internet usage and traditional media usage--A case study of 2005 Shih Hsin Communication Database. Shih Hsin University. MA thesis.

Joskow, P. L. (2006). *Vertical Integration*. Prepared for the American Bar Association Antitrust Section's "Issues in Competition Law and Policy Project." Retrieved on October15, 2009 from http://econ-www.mit.edu/files/1191

Kamino, A. (2009). Transition of the US telecommunications mega merger reviews and problems-Balancing of efficiency and public interest [in Japanese]. *Journal of Information and Communication Research, 26*(4), 51–66.

Kamino, A., & Fuke, H. (2010). Diffusion of the Broadband Internet and the Structural Separation. In Gentzoglanis, A., & Henten, A. (Eds.), *Regulation and the Evolution of the Global Telecommunications Industry*. Northampton, MA: Edward Elgar Pub.

Kaplan, S. J. (1978). The impact of cable television services on the use of competing media. *Journal of Broadcasting, 22*(Spring), 155–165.

Karel, A. (2003). *The development and implications of free ISPs in New Zealand.* Wellington, New Zealand: New Zealand Institute for the Study of Competition and Regulation. Retrieved from http://www.iscr.org.nz/f209,4313 /4313_Free_ISPs_220903.pdf

Katz, J. E., & Aakhus, R. (Eds.). (2008). *Perpetual Contact: Mobile Communication, Private Talk, Public Performance*. Cambridge: Cambridge University Press.

Kayany, J. M., & Yelsma, P. (2000). Displacement effects of online media in the socio-technical contexts of households. *Journal of Broadcasting & Electronic Media, 44*(2), 215–229. doi:10.1207/s15506878jobem4402_4

Kaye, B., & Johnson, T. (2002). Online and in the know: Uses and gratifications of the Web for political information. *Journal of Broadcasting & Electronic Media, 46*(1), 54–71. doi:10.1207/s15506878jobem4601_4

Kaye, B., & Johnson, T. (2003). From here to obscurity?: Media substitution theory and traditional media in an on-line world. *Journal of the American Society for Information Science and Technology, 54*(3), 260–273. doi:10.1002/asi.10212

KCC. (2009). Korea Communications Commission. Retrieved November 11, 2009, from http://eng.kcc.go.kr

Kennard, W. (1999). Connecting The Globe: A Regulator's Guide to Building A Global Information Community (pp. IX-3 &4). Washington, DC: Federal Communications Commission.

Keogh, D., & Wood, T. (2005). *Village Phone Replication Manual: Creating Sustainable Access to Affordable Telecommunications for the Rural Poor*. New York: The United Nations Information and Communication Technologies Task Force.

KEYMILE. (2008). Ethernet Point-to-Point vs. PON – A comparison of two optical access network technologies and the different impact on operations. Retrieved November 11, 2009, from http://www.eurocomms.com/asset/ 922/White%20Paper%20EP2P%20vs%2 0PON%20(E%20 2009%2008%2017).pdf

Khoumbati, K., Dwivedi, Y. K., Lal, B., & Chen, H. (2007). Broadband Adoption in Pakistan. *Electronic Government: An International Journal, 4*(4), 451–465. doi:10.1504/EG.2007.015038

Kim, A. (2006). Community Building on the Web. We Media [online]. Retrieved May 3, 2009 from http://www.hypergene.net/wemed ia/weblog.php?id=P40

Kim, F. F. (1997). Government IT Infrastructure. International Market Insights. *National Trade Data Base*. U.S. Department of Commerce, December 16 1997.

Kirby, J. (2004). Using "Word of Mouse" in Brand Marketing. *Admap Magazine*, 454.

Klamer, L., Haddon, L., & Ling, R. (2000). *The Qualitative Analysis of ICTs and Mobility, Time Stress and Social Networking.* Heidelberg: A Report to EURESCOM P-903.

Koh, J. (2000, July). E-biz yet to touch SMEs. *Business Times* (p. 14).

Korea Internet and Security Agency. (2009). *Survey of the Internet Usage 2009*. Seoul: Korea Internet and Security Agency.

Kramer, G. (2005). *Ethernet Passive Optical Networks (professional Engineering)*. McGraw-Hill Professional Publishing.

Kridel, D., Rappoport, P., & Taylor, L. (2001). The demand for high-speed access to the Internet: the case of cable modems. In Loomis, D., & Taylor, L. (Eds.), *Forecasting the Internet: Understanding the Explosive Growth of Data Communications* (pp. 11–22). Boston, MA: Kluwer.

Krueger, A. (2003). The digital divide in educating African-American students and Workers. In Conrad, C. (Ed.), *Building Skills for Black Workers* (pp. 51–76). New York: University Press of America.

Kruger, L.G., & Gilroy, A.A. (2001, May 18). IB10045: Broadband Internet Access: Background and Issues. CRS Issue Brief for Congress.

Kuhr, M.(2008). Media Windows in Flux. Challenges for Audiovisual Media Chronology. *IRIS plus*, 4.

Laffont, J.-J., & Tirole, J. (2002). *Competition in telecommunications*. Cambridge, Massachusetts: MIT Press.

Lamanna, M. (2004). The LHC computing grid project at CERN. *Nuclear Instruments & Methods in Physics Research. Section A, Accelerators, Spectrometers, Detectors and Associated Equipment*, *534*(1-2), 1–6. doi:10.1016/j.nima.2004.07.049

Langdale, J. (1997). International competitiveness in East Asia: Broadband telecommunications and interactive multimedia. *Telecommunications Policy*, *21*(3), 235–249. doi:10.1016/S0308-5961(97)00004-9

Laschinger, K., & Goldstuck, A. (2006). *PC users in South Africa 2006*. Johannesburg: World Wide Worx.

Lazarfeld, P. F. (1940). *Radio and the printed page*. New York: Dell, Sloan and Pearce.

Le Goff, J., & Mouline, A. (2003). *Les stratégies des firmes dans les industries du multimédia, 10 ans d'opérations de rapprochement (1993-2002)*. Paris: Ministère de l'Economie, des Finances et de l'Industrie.

Leamer, E. E., & Storper, M. (2001). *The economic geography of the Internet age*. Working Paper 8450, National Bureau of Economic Research, USA.

Lee, C., & Chan-Olmstead, S. M. (2004). Competitive advantage of broadband internet: A comparative study between South Korea and the United States. *Telecommunications Policy*, *28*(9-10), 649–677. doi:10.1016/j.telpol.2004.04.002

Lee, P. S., & Leung, L. (2006). Accessing the displacement effects of the Internet. *Telematics and Informatics*. doi:.doi:10.1016/j.tele.2006.08.002

Lee, C. I. (2007). IPTV's competitive strategies and business models for Telcos-- Chunghwa Telecom's IPTV as Case Study. Fu Jen Catholic University. MA thesis.

Lee, N. C. (2002, February). Broadband Internet Service: Korea's Experience. Retrieved January 28, 2007 from eng.mic.go.kr/eng/secured.tdf?seq=5&idx=2&board_id=E_04_03

Lee, S.S. (2000, April 12). EDB 50% grant to aid switch to e-commerce. *The Straits Times* (p.71).

Leff, N. H. (1984). Externalities, information costs, and social benefit-cost analysis for economic development: an example from telecommunications. *Economic Development and Cultural Change*, *32*(2), 255–276. doi:10.1086/451385

Lehmann, A. (2002). Foreign direct investment in emerging markets: Income, repatriations and financial vulnerabilities. *International Monetary Fund Working Paper Series*, WP/02/47, 1-31.

Lehr, W., & McKnight, L. W. (2003). Wireless Internet Access: 3G vs. WiFi. *Telecommunications Policy*, *27*(5-6), 351–370. doi:10.1016/S0308-5961(03)00004-1

Li, M. (2005). Passive Optical Networks (PON) Protocols. Hakko-opto, presentation. Retrieved January 15, 2010, from http://www.hakko-opto.com/pdf/ Comparison_PON_Protocols.pdf

Lieberman, M., & Montgomery, D. (1998). First mover (Dis)Advantage: retrospective and link with the Resource-Based View. *Strategic Management Journal*, *19*(12), 1111–1125. doi:10.1002/(SICI)1097-0266(1998120)19:12<1111::AID-SMJ21>3.0.CO;2-W

Lim, V., & Teo, T. (2000). To work or not work at home: An empirical investigation of factors affecting attitudes toward teleworking. *Journal of Managerial Psychology*, *15*(6), 560–586. doi:10.1108/02683940010373392

Lim, R. (2006, July 3). Masterplan to transform Info-Comm. *The Business Times*, Singapore.

Lin, S. C. (2006). A Study on Factors of Purchasing the Multimedia-on-Demand (IPTV) Service. Department of Applied Economics, National Chung Hsing University. MA thesis.

Ling, R. (2004). *Mobile connection: The Cell Phone's Impact on Society*. San Francisco, CA: Morgan Kaufmann.

Ling, R., & Donner, J. (2009). *Mobile Phones and Mobile Communication*. Cambridge, MA: Polity.

Ling, R., & Pederson, P. (Eds.). (2005). *Mobile Communication: Re-negotiation of the Social Sphere*. London: Springer.

Liu, L. T. (2006). The Study of the Market for CHT's IPTV. Department and Graduate School of Business Administration, Chung-Hua University. MA thesis.

Liu, Y. L., & Hsu, W. Y. (2007). The *Displacement effect of the digital television platforms*. Paper presented at The Fifth Conference of Media & Communication in Chinese Civilization. National Chengchi University, Taipei.

Lovelock, P. (1999). *The Evolution of China's NII Initiative: A policy making analysis.* University of Hong Kong: HongKong

Loveman, G. (1994). An Assessment of the Productivity Impact of Information Technologies. In Allen, T. J., & Scott Morton, M. S. (Eds.), *Information Technology and Corporation of the 1990s: Research Studies* (pp. 84–110). Oxford, UK: Oxford University Press.

Lucas, R. E. (1998). On Mechanics of Economic Development. *Journal of Monetary Economics*, *22*, 3–42. doi:10.1016/0304-3932(88)90168-7

Lyrintzis, C. (1984). Political Parties in Post-Junta Greece: A Case of Bureaucratic Clientelism. *West European Politics*, *7*(2), 99–118. doi:10.1080/01402388408424473

MAIT Country Intelligence (2005, June). *Hong Kong's ICT Industry, 48*.

Malecki, E. (2008). *The rural digital economy: competition and the geography of telecommunications*. Paper presented at the Economic Research Service's Broadband in the Rural Economy workshop, Washington, DC.

Mankiew, N. G., Romer, D., & Weil, D. (1992). A Contribution to the Empirics of Economic growth. *The Quarterly Journal of Economics*, *107*(2), 407–437. doi:10.2307/2118477

Manning, R. (2008). *Development Co-operation Report (2008): Efforts and policies of the members of the Development Assistance Committee*. OECD Journal of Development.

Mante-Meijer, E. (2002). The Netherlands and the US Compared. In Katz, J., & Aakhus, R. (Eds.), *Perpetual Contact: Mobile Communication, Private Talk, Public Performance* (pp. 110–125). Cambridge: Cambridge University Press. doi:10.1017/CBO9780511489471.010

ManyPossibilities. (2009). African Undersea Cables. Retrieved July 28, 2009 from http://manypossibilities.net/ african-undersea-cables

Marcheck, J. (2006). FTTx: The Many Roads to Last Mile Fiber. Current-Analysis, presentation, Retrieved January 15, 2010, from http://www.currentanalysis.com /t/fiberpremise2006/files/ CurrentAnalysis-fttx.pdf

Marker, P., McNamara, K., & Wallace, L. (2002). *The Significance of Information and Communication Technologies for Reducing Poverty*. London: Department for International Development.

Maslow, A. (1970). *Motivation and Personality* (2nd ed.). New Jersey: Pearson Education.

Mathiason, N. (2008, July 27). Shaping the way we live: A Korean vision of hi-tech future. *The Observer*.

Matthews, J. (1999). Silicon Island of the East: Creating a semiconductor industry in Singapore. *California Management Review*, *41*(2), 55–78.

Mawson, N. (2006). Telkom moves to shorten local loop in high-demand areas. *Engineering News*. Retrieved May 19, 2006 from http://www.engineeringnews.co.za/ eng/ news/today/?show=86554

Mayzlin, D. (2006). Promotional Chat on the Internet. *INFORMS. Marketing Science*, *25*(2). doi:10.1287/mksc.1050.0137

Mbarika, V., Okoli, C., Byrd, T., & Datta, P. (2005). The Neglected Continent of IS Research: A Research Agenda for Sub-Saharan Africa. *Journal of the Association for Information Systems*, *6*(5), 130–170.

McCarthy, C. (2009). Facebook now twice as big as MySpace? Oh boy! [online]. Retrieved January 23, 2009 from http://news.cnet.com/8301-13577_3- 10148855-36.html?part=rss&subj=new s&tag=2547-1023_3-0-5

McComb, M. (1972). Mass media in the marketplace. *Journalism Monographs (Austin, Tex.)*, *24*, 1–102.

McLeod, D. (2006). *Ivy's Extra Mile* [online]. Financial Mail.

Mediagruppe München (2000). *Online-Nutzung in den Sinus-Milieus*. @facts Newsletter.

Melody, W. (2002). *Building the regulatory foundations for growth in network economies*. World dialogue on regulation for network economies discussion paper #0201.

Mendelsohn, H. (Ed.). (1964). *Listening to Radio*. New York: Free Press.

Menou, M.J., & Potvin, J. (2007). *Toward a conceptual framework for learning about ICT's and knowledge in the process of development- A GK LEAP background document*.

MIC. (2006a). *Report of Panel on Frameworks of Communications and Broadcasting* (in Japanese).

MIC. (2006b). *Agreement between the Cabinet and Ruling Parties on Frameworks of Communications and Broadcasting* (in Japanese).

MIIT (Ministry of Industry and Information Technology). (2006). The 2006-2020 National Informationization Development Strategy. Retrieved October 8, 2008, from http://www.miit.gov.cn/n112 93472/index.html

Miles, M. B., & Huberman, A. M. (1994). *Qualitative data analysis: an expanded sourcebook* (2nd ed.). Thousand Oaks, CA: Sage Publications.

Milgrom, P., & Roberts, J. (1992). *Economics, organization and management*. Englewood Cliffs, New Jersey: Prentice-Hall.

Ministry of Commerce & The Treasury. (1995). *Regulation of access to vertically-integrated natural monopolies*. Wellington, New Zealand: Ministry of Commerce; The Treasury. Retrieved from http://www.med.govt.nz/templates /MultipageDocumentPage____4560.aspx

Ministry of Commerce. Industry and Energy, Republic of Korea (2005). *Key figures: FDI.* Ministry of Commerce, Industry and Energy, Republic of Korea. Retrieved January 28, 2007 from http://english.mocie.go.kr/ language/eng/news/news_view2.jsp

Ministry of Economic Development. (2006). *Promoting competition in the market for broadband services*. Wellington, New Zealand: Ministry of Economic Development. Retrieved from http://www.med.govt.nz/uplo ad/36537/promoting-competition.pdf.

Ministry of Information and Communication, Republic of Korea. (2004). *IT 8-3-9 strategy: The road to $20,000 GDP per capita.* Ministry of Information and Communication, Republic of Korea.

Ministry of Information and Communication. (2002, December 16) Republic of Korea. *e-Korea Vision 2006*. Ministry of Information and Communication, Republic of Korea. MSC revitalised for global stage. (2006, April 7) Retrieved January 28, 2007 from http://sso.msc.com.my/index.php?option= com_content&task=view&id=114&Itemid=57

Miravete, E. (2003). Choosing the wrong calling plan? Ignorance and learning. *The American Economic Review*, *93*(1), 287–310. doi:10.1257/000282803321455304

Modzelewski, F. (2000, September 11), Finding a Cure for Viral Marketing. *Direct Marketing News* [online] [accessed on May 1, 2007]

Mokyr, J. (1990). *The lever of riches. Technological creativity and economic progress*. Oxford: Oxford University Press.

Mokyr, J. (1992). Technological inertia in economic history. *The Journal of Economic History*, *52*(2), 325–338. doi:10.1017/S0022050700010767

Monteiro, A. (2006). Education, lack of engineers hamper SA IT [online]. Moneyweb. Retrieved March 29, 2006 from http://www.moneyweb.co.za/s hares/ict_sector/995758.htm - 39k

Mouzelis, N. (1986). *Politics in the Semi-Periphery: Early Parliamentarism and Late Industrialisation in the Balkans and Latin America*. London: Macmillan.

Mouzelis, N. (1995). Modernity, late development and civil society. In Hall, J. (Ed.), *Civil Society: Theory, History, Comparison* (pp. 224–249). Cambridge: Polity.

Mouzelis, N., & Pagoulatos, G. (2002). *Civil Society and Citizenship in Postwar Greece*. Discussion Paper. Athens: Athens University of Economics and Business.

Muller, R. (2006). SA slipping further behind in broadband rankings [online]. Moneyweb. Retrieved April 20, 2006 from http://moneyweb.iac.iafrica.com/ shares/ict_sector/252408.htm

MyBroadband. (2009a). JINX and CINX traffic growth. Retrieved October 7, 2009 from http://mybroadband.co.za/new s/Internet/9909.html

MyBroadband. (2009b). Reducing ADSL and broadband costs. Retrieved August 4, 2009 from http://mybroadband.co.za/n ews/Broadband/9073.html

MyBroadband.co.za. (2006a). Telkom 1000 times more expensive. Retrieved May 8, 2006 from http://www.mybroadband.co.za /nephp/?m=show&id=2680

MyBroadband.co.za. (2006b). iBurst coverage expansion moving forward. Retrieved May 29, 2006 from http://www.mybroadband.co. za/nephp/?m=show&id=2943

Mzolo, S. (2005). Lower telecom costs 'only talk' [online]. Fin24.co.za. Retrieved December 20, 2005 from http://www.fin24.co.za/articles/default/ display_article.asp?ArticleID=1518-1786_1853631. Dated

Nadiri, M. I., & Nandi, B. (2001). Benefits of Communications Infrastructure Capital in U.S. Economy. *Economics of Innovation and New Technology*, *10*(2-3), 89–107. doi:10.1080/10438590100000005

Nandi, B., & Chakraborty, C. (2008). Broadband Diffusion and its Driving Forces. In Dwivedi, Y. K., & Choudrie, J. (Eds.), *Handbook of Research on Global Diffusion of Broadband Data Transmission* (*Vol. 2*). Hershey, PA: IGI Global.

National Cable & Telecommunications Ass'n v. Brand X Internet Services, 545 U.S. 967 (2005).

National Information Agency. Korea (2008). *Informatization White Paper*. National Information Agency, Korea 2008.

National Research Council. (2002). *Broadband: Bringing Home the Bits* (pp. 62–81). Report of the Committee of Broadband Last Mile Technology, Computer Science and Telecommunications Board, National Academy Press.

Nee, R. V., & Prasad, R. (2000). *OFDM for Wireless Multimedia Communications* (1st ed.). Norwood, MA: Artech House, Inc.

Nelson, R. (1991). Why Do Firms Differ and How Does it Matter? *Strategic Management Journal*, *12*, 61–74. doi:10.1002/smj.4250121006

Nelson, R., & Winter, S. (1982). *An Evolutionary Theory of Economic Change*. Boston, MA: Harvard University Press.

Nevile, A. (2007). Amartya K. Sen and social exclusion. *Development in Practice*, *17*(2), 249–255. doi:10.1080/09614520701197200

New Zealand Institute of Economic Research. (2005). *Telecommunications pricing in New Zealand: a comparison with OECD countries*. Wellington, New Zealand: NZIER.

Nguyen, A. & Western, M. (2006). The complementary relationship between the internet and graditional mass media: the case of online information. Retrieved from http//informationr.net/ir/11 -3/paper259.html

Nie, N., & Ebring, L. (2000). Study Offers Early Look at How Internet is Changing daily Life. Retrieved from www.stanford.edu/dept/news/pr

Niyato, D., & Hossain, E. (2007). Wireless Broadband Access: WiMAX and Beyond-Integration of WiMAX and WiFi: Optimal Pricing for Bandwidth Sharing. *Communication Magazine*, *45*(5), 140–146. doi:10.1109/MCOM.2007.358861

NTIA (National Telecommunications and Information Administration & Economics and Statistics Administration). (2004). *A nation online: entering the broadband age*. Washington, DC: U.S. Department of Commerce.

NTIA. (2000). *Falling through the Net: Toward digital inclusion*. Washington, DC: U.S. Department of Commerce.

NTIA. (2002). *A nation online: How Americans are expanding their use of the Internet*. Washington, DC: U.S. Department of Commerce.

Oden, M., & Strover, S. (2002). *Links to the future: The role of information and telecommunications technology in Appalachian economic development*. Washington, DC: Appalachian Regional Commission.

Odinma, A., Oborkhale, L., Eng, M. & Kah, M. (2007). The Trends in Broadband Wireless Networks Technologies. *The Pacific Journal of Science and Technology, 8*(1).

OECD. (1999). *Communications Outlook*. Paris: OECD.

OECD. (2001). *The development of broadband access in OECD countries*. Paris: OECD.

OECD. (2003). *Policies for Broadband Development. Recent OECD work on Broadband Committee for Information*. Computer and Communications Policy.

OECD. (2003). *The benefits and costs of structural separation of the local loop, DSTI/ICCP/TISP (2002)*. Final.

OECD. (2009). *OECD Communications Outlook, 2009*. Geneva: OECD.

OECD. (2007*). Measuring the impacts of ICT using official statistics*, Working Party on Indicators for the Information Society, DSTI/ICCP/IIS(2007)1/FINAL, Paris, http://www.oecd.org/ dataoecd/ 43/25/39869939.pdf

OECD. (2008). *Broadband subscribers per 100 inhabitants, by technology, December 2008*. OECD Broadband Portal. Retrieved October 15, 2009 from http://www.oecd.org/document/54/0,334 3,en_2649_34225_38690102_1_1_1_1,00.html

OECD. (2009). OECD Broadband Statistics: December 2008. Organisation for Economic Co-operation and Development. Retrieved May 20, 2009 http://www.oecd.org/document/54/0,334 3,en_2649_34225_38690102_1_1_1_1,00.html

OECD. (2009). The Role of Communication Infrastructure Investment in Economic Recovery. DSTI/ICCP/CISP(2009)1/FINAL, Organisation for Economic Co-operation and Development. Retrieved November 11, 2009, from http://www.oecd.org/dataoecd /4/43/42799709.pdf

Ofcom. (2004a). *Strategic Review of Telecommunications Phase 1 consultation document,* 28 April 2004.

Ofcom. (2004b). *Strategic Review Telecommunications Phase 2 consultation documents*, 18 November 2004.

Ofcom. (2005d). *Final statements on the Strategic Review of Telecommunications, and undertakings in lieu of a reference Under the Enterprise Act 2002,* Statement, 22 September 2005.

Ofir, Z., & Kriel, L. (2004). *Evaluating policy influence of ICTs for rural areas: the MSSRF information villages research project*. Johannesburg: Evalnet.

Olivennes, D. (2007).*Le développement de la protection des oeuvres culturelles sur les nouveaux réseaux*. Retrieved November, 2008 from http://www.culture.gouv.fr /culture /actualites/index-o livennes231107.htm

OPTA. (2004). *Vertical integration: efficiency & foreclosure*. Economic Policy Note, no.3, May 2004.

Orbicom (2005). *From the digital divide to digital opportunities: measuring infostates for development* (G. Sciadas. Ed.). Montreal, Canada: NRC press-presses du CNRC.

Organization for Economic Co-Operation and Development. (2000). *Local access pricing and e-commerce.* Paris: OECD.

Organization for Economic Co-Operation and Development. (2001). *The development of broadband access in OECD countries.* Paper DSTI/ICCP/TISP(2001)2/FINAL. Paris: OECD. Retrieved from http://www.oecd.org

Organization for Economic Co-Operation and Development. (2007). *Communications Outlook 2007.* Paris: OECD. Retrieved from http://www.oecd.org

Organization for Economic Co-Operation and Development. (2008). *Broadband growth and policies in OECD countries.* Paris: OECD. Retrieved from http://www.oecd.org

Oxman, J. (1999, July). The FCC and the Unregulation of the Internet. OPP Working Paper No. 31.

Ozcan, K. (2004). *Consumer-To-Consumer Interactions In A Networked Society: Word of mouth theory, consumer experiences, and network dynamics* [online]. PhD dissertation. University of Michigan. Retrieved May 4, 2009 from http://www.insna.org/IN SNA/power_inf.html

Papacharissi, Z., & Zaks, A. (2006). Is Broadband the Future? An analysis of Broadband Technology Potential and Diffusion. *Telecommunications Policy, 30*(1), 64–75. doi:10.1016/j.telpol.2005.08.001

Papacharissi, Z., Zaks, A. (2006). Is broadband the future? An analysis of broadband technology potential and diffusion. Telecommunications Policy, *30*(2006), 64–75 70.

Papazafeiropoulou, A., & Dwivedi, Y. K. (2009). A Framework of Broadband Adoption and Diffusion (BA&D): Understanding Factors Relevant to Residential Consumers and Small & Medium Enterprises. In Dwivedi, Y. K. (Eds.), *Handbook of Research in Global Diffusion of Broadband Data Transmission.* Hershey, PA: IGI Global.

Parker, E. (2000). Closing the digital divide in rural America. *Telecommunications Policy, 24*(4), 281–290. doi:10.1016/S0308-5961(00)00018-5

Parker, E., & Hudson, H. (1992). *Electronic byways: state policies for rural development through telecommunications.* Boulder: West View Press.

Pastore, M. (2000). *The value of word of mouth* [online]. [Accessed on April 24, 2009]

Patel, R., & Parmentier, M. J. C. (2006). The persistence of traditional gender roles in the information technology sector: a study of female engineers in India. *Information Technologies and International Development, 2*(30), 29–46.

Pereira, P., & Ribeiro, T. (2006). The Impact on broadband access to the Internet of the dual ownership of telephone and cable networks. NET Institute, working paper #06-10.

Perloff, J. M. (2003). *Microeconomics* (3rd ed.). Pearson-Addison Wesley.

Perse, E. M., & Courtright, J. A. (1993). Normative images of communication media: Mass and interpersonal channels in the new media environment. *Human Communication Research, 19*, 485–503. doi:10.1111/j.1468-2958.1993.tb00310.x

Petmesidou, M. (1996). Social protection in Greece: a brief glimpse of a welfare state. *Social Policy and Administration, 30*(4), 324–347. doi:10.1111/j.1467-9515.1996.tb00564.x

Petmesidou, M. (2000). Social protection in Greece in the nineties: reforming the 'weak' welfare state. In Mitsos, A., & Mossialos, E. (Eds.), *Contemporary Greece and Europe* (pp. 303–330). Aldershot: Ashgate.

PEW Internet & American Life Project. (n.d.).Retrieved from http://www.pewinternet.org/index.asp.

Phelps, J., Lewis, R., Mobilio, L., Perry, D., & Raman, N. (2004). Viral marketing or electronic word of mouth advertising: Examining consumer responses and motivations to pass along e-mail. *Journal of Advertising Research, 44*(4).

Philadelphia, Pennsylvania November 14, 2001 Retrieved January 15, 2010, from http://www.fcc.gov/Speeches / Copps/2001/spmjc108.html

Pillarisetti, J.R.M., M. (1998). Human development and gender empowerment: methodological and measurement issues. *Development Policy Review*, *16*, 197–203. doi:10.1111/1467-7679.00059

Point-Topic. (2009). World Broadband Statistics: Q1 2009. Retrievbed June 1, 2009 from http://point-topic.com/contentDownload/operatorsource/dsl reports/world%20 broadband%20statistics%20q1%202009.pdf

Poley, J. (2008). *Rural distance education*. Paper presented at the Economic Research Service's Broadband in the Rural Economy workshop, Washington, DC.

Porter, L., & Golan, G. (2006). From subservient chickens to brawny men: A comparison of viral advertising to television advertising. *Journal of Interactive Advertising*, *6*(2).

PRACE. (2009). Partnership for Advanced Computing in Europe (PRACE) project homepage. Retrieved December 2009 from http://www.prace-project.eu

Prahalad, C. K., & Hammond, A. (2002). Serving the World's Poor, Profitably. *Harvard Business Review*, *80*(9).

Prahalad, C. K., & Hamel, G. (1990). The core competencies of the corporation. *Harvard Business Review*, (May-June): 29–44.

Prestona, P., & Cawleyb, A. (2008). Broadband development in the European Union to 2012—A virtuous circle scenario. *Futures*..doi:10.1016/j.futures.2008.07.035

Prieger, J. (2003). The supply side of the digital divide: Is there equal availability in the broadband internet access market? *Economic Inquiry*, *41*(2), 346–363. doi:10.1093/ei/cbg013

Prieger, J., & Hu, W. (2008). The Broadband Digital Divide and the Nexus of Race, Competition, and Quality. *Information Economics and Policy*, *20*(2), 150–167. doi:10.1016/j.infoecopol.2008.01.001

Prieger, J. E., & Lee, S. (2008). Regulation and the Deployment of Broadband. In Dwivedi, Y. K., Papazafeiropoulou, A., & Choudrie, J. (Eds.), *Handbook of Research on Global Diffusion of Broadband Data Transmission* (pp. 278–297). Hershey, PA: IGI Global.

R Development Core Team. (2005). *R: A Language and Environment for Statistical, R Foundation for Statistical Computing.* Paper presented in R Development Core Team Meeting, Vienna, Austria. Retrieved from http://www.R-project.org

Raj, C. (2000, April 11). Singapore to pump up SMEs doing e-commerce. *Business Times* (p. 7).

Rallet, A. (1996). Convergence technologique et organisation industrielle de l'audiovisuel, de l'information et des télécommunications. In Brousseau, E., Petit, P., & Phan, D. (Eds.), *Mutations des Télécommunications, des Industries et des Marchés*. Paris: ENSPTT-Economica.

Ramsay, J. O., & Silverman, B. W. (1997). *Functional Data Analysis*. New York: Springer.

Raskin, R. (2006). Facebook faces its future. *Young Consumers* [online] [Accessed 09 March 2009].

Raymond Mataloni. (2008). U.S. multinational companies: operations in 2006. *Survey of Current Business.* U.S. Department of Commerce, (November), 26-47.

RedCLARA. (2009). RedCLARA project homepage. Retrieved December 2009 from http://www.redclara.net

Reding, V. (2006a). The Review 2006 of EU Telecom rules: Strengthening Competition and Completing the Internal Market (Speech/06/422)

Reding, V. (2006b). From service Competition to Infrastructure Competition: the Policy Options Now on the Table (Speech/06/697)

Rens, A. (2008). What does an ECNS licence amount to anyway? Retrieved November 7, 2008 from http://www.shuttleworthfoundation.org/our-w ork/blogs/what-does-ecns-licence-amount-anyway

Results, I. C. T. (2009). Supercomputing gets its own superhero. Retrieved December 2009 from http://cordis.europa.eu/ictresults/

Review & Outlook. (2009, September 22). Bad News for Broadband. *Wall Street Journal* (p. A24).

Rey, P., & Tirole, J. (2007). A Primer in Foreclosure in Armstrong. In M. & R. H. Porter (Eds.), *Handbook of Industrial Organization* (Vol. 3, pp. 2145-222). Amsterdam: North-Holland.

Richardson, D., Ramirez, R., & Haq, M. (2000). *Grameen Phone (GP) village phone project: final report*. Canada: C.I.D.A. Guelph Ontario.

Riordan, M. H. (2008). Competitive Effects of Vertical Integration. In Buccirossi, P. (Ed.), *Handbook of Antitrust Economics* (pp. 145–182). Cambridge, MA: The MIT Press.

Rodgers, S., & Thorson, E. (2000). The interactive advertising model: How users perceive and process online ads. *Journal of Interactive Advertising*, *1*(1).

Roger, H., & Rajesh, R. (2006). *Empowering the Poor: Information and Communications Technology for Governance and Poverty Reduction.* UNDP, Asia Pacific Development information Program, Bangkok, Thailand.

Rogers, E. M. (1995). *Diffusion of Innovations* (*Vol. 4*). New York: Free Press.

Roller, L., & Waverman, L. (2001). Telecommunications infrastructure and economic development: a simultaneous approach. *The American Economic Review*, *74*, 909–923. doi:10.1257/aer.91.4.909

Roller, L. H., & Waverman, L. (1996). The Impact of Telecommunications Infrastructure and Economic Development. In Howitt, P. (Ed.), *The Implication of knowledge based Growth of Micro-Economic Policies*. Calgary: University of Calgary Press.

Romer, P. M. (1990). Endogenous Technological Change. *The Journal of Political Economy*, *98*(5), 71–102. doi:10.1086/261725

Romer, P. M. (1989). Capital accumulation in the Theory of long-run Growth. In Barro, R. J. (Ed.), *Modern Business Cycle Theory*. Cambridge, MA: Harvard University Press.

Rubenstein., et al. (1973). *Television and social behavior* (Technical report of the Surgeon General's Advisory Committee, Vol. IV). US Department of Health, Education and Welfare, Washington, DC.

Runnacles, M. (2006). Branding in the internet bazaar-struggling for control in an uncontrollable world. *Market Leader*, 33.

Santiago, T. (2000, June 10). Singapore's e-commerce net strategy bears fruit. *Electronic Engineering Times*.

Saunders, R. J., Warford, J. J., & Wellenius, B. (1983). *Telecommunications and Economic Development*. Baltimore: Johns Hopkins University Press.

Saunders, J. (2006). Growing up digitally: change drivers in marketing. *Market Leader*, 33.

Sawyer, A., Allen, J. P., & Lee, H. (2003). Broadband and Mobile Opportunities: A socio-Technical Perspective. *Journal of Information Technology*, *18*(2), 121–136. doi:10.1080/0268396032000101171

Schatz, A. (2009a, September 26). AT&T Asks for Curbs on Google. *The Wall Street Journal* (p. A5).

Schatz, A. (2009b, September 30). Full Web Access is Pegged at $20 Billion. *Wall Street Journal* (p. A2).

Scheuer, A. (2006). Traditional paradigms for new services? The Commission Proposal for a 'Audiovisual Media Services Directive'. *Communications & Strategies*, *62*(2), 71–91.

Schramm, W., Lyle, J., & Parker, E. (1961). *Television in the Lives of our Children*. Stanford, CA: Stanford University Press.

Schwab, K., & Porter, M. (2007). *The Global Competitiveness Report 2007-2008*. Geneva, Switzerland: World Economic Forum.

Sein, M. K., & Harindranath, G. (2004). Conceptualizing the ICT artefact: Toward understanding the role of ICT in national development. *The Information Society*, *20*(1), 15–24. doi:10.1080/01972240490269942

Sein, M. K., & Harindranath, G. (2004). Toward Understanding the Role of ICT in National Development. *The Information Society*, *20*(1), 15–24. doi:10.1080/01972240490269942

Sen, A. (2000). Social exclusion: concept, application and scrutiny. In *Social Development Paper No. 1*. Manila: Asian Development Bank.

Servaes, J. (2003). The European information society: a wake-up call. In J. Servaes (Ed.), *The European information society: A reality check* (pp. 11-32). Bristol, Portland, Or: Intellect.

SevenOneMedia (2002). *Die Sinus-Milieus 2001*. Das neue Gesamtdeutsche Modell, Unterföhring: SevenOne-Media.

Shih, T.H. (1999, April 8). Singapore at home and abroad. *Business Times* (p. 2).

Shirky, C. (2000, July 25). The Toughest Virus of All. *Business 2.0* [online] [accessed on May 3, 2009]

Sicilia, M., Ruiz, S., & Reynolds, N. (2006). Attitude formation online. *Journal of Consumer Psychology* [online]. [Accessed 13 March 2009].

Sieradzki, D. L., & Maxwell, W. J. (2008). The FCC's Network Neutrality Ruling in the Comcast Case: Toward a Consensus with Europe? *Communications & Strategies*, *72*(4), 73.

Simons, D. (2002). Telkom lights ADSL, but is the price right? Broadband Markets. Retrieved September 3, 2006 from http://www.iec.org/events/2002/ dslwfeurope/newsletter/feature.html

Smith, J. (2006). Becoming a part of the community. *Admap Magazine*, October.

Smoreda, Z., & Thomas, F. (2001). *Social Networks and Residential ICT Adoption and Use.* Paper presented at the EURESCOM Summit, Heidelberg.

SNG. (2009). Measuring the impacts of broadband. *Strategic Networks Group*. Retrieved November 11, 2009, from http://www.govtech.com/gt/692922

Solomon, M. (2004). *Consumer Behavior: buying, having and being* (6th ed.). New Jersey: Pearson Prentice Hall.

Sotiropoulos, D. (1996). Civil society and the Greek state in the Third Hellenic Republic. In Lyrintzis, C., Nikolakopoulos, E., & Sotiropoulos, D. (Eds.), *Society and Politics* (pp. 113–138). Athens: Themelio. (in Greek)

Southwood, R. (2006). Satellite dominates cellular backhaul. Balancing-act Africa. Retrieved November 6, 2006 from http://www.mybroadband.co. za/nephp/?m=show&id=4799

Sparkes, V. M. (1983). Public perception of and reaction to multi-channel cable television service. *Journal of Broadcasting*, *27*(Spring), 163–175.

Spiller, P., & Cardilli, C. (1997). The frontier of telecommunications deregulation: small countries leading the pack. *The Journal of Economic Perspectives*, *11*(4), 127–138.

Srivastava, R. K., Fahey, L., & Christensen, K. (2001). The Resource-Based View and Marketing: the Role of Market-Based Assets in Gaining Competitive Advantage. *Journal of Management*, *27*, 777–802. doi:10.1177/014920630102700610

Standard and Poor's. (2009). *Telecommunications: Europe*. New York: Standard and Poor's Equity Research Services.

Stanton, L. (2004). Factors influencing the adoption of residential broadband connections to the Internet. In *Proceedings of the 37th Hawaii International Conference on System Sciences*.

Stenberg, P. L. (2006). Investment and household adoption of communication and information services across the United States. In Johansson, B., Karlsson, C., & Stough, R. (Eds.), *The emerging digital economy: Entrepreneurship, clusters and policy* (pp. 263–276). New York: Springer.

Stenberg, P. L., & Morehart, M. (2007). The internet in U.S. farm and rural businesses. In Johansson, I. (Ed.), *Entrepreneurship and Development-Local Processes and Global Patterns*. Sweden: University West.

Stenberg, P. L. (2008). *Rural broadband internet use and the rural economy: An quasi-experimental design analysis of the broadband effect on rural economies*. Paper presented at the Economic Research Service's Broadband in the Rural Economy workshop, Washington, DC.

Stenberg, P. L., & Morehart, M. (2006). *Determinants of internet use for rural and farm economic sectors*. Paper presented at the Telecommunication Policy Research Conference, Arlington, VA.

Steyer, A., Garcia-Bardidia, R., & Quester, P. (2006). Online discussion groups as social networks: An empirical investigation of word of mouth on the internet. *Journal of Interactive Advertising*, *6*(2).

Stifel Nicolaus (2006, March). Net Neutrality: Value Chain Tug of War. Equity Research, Industry Update, Telecom & Media Regulatory.

Stiroh, K. J. (2002). Are ICT Spillovers Driving the New Economy? *Review of Income and Wealth*, *1*(48), 33–57. doi:10.1111/1475-4991.00039

Strategic Headquarter, I. T. (2006). New IT reform strategy: realizing ubiquitous and universal network society where everyone can enjoy the benefits of IT. Retrieved October 7, 2009, from http://www.kantei.go.jp/foreign/policy/it/ITstrategy2006.pdf

Subramaniam, G., & Varadhan, R. (2009). *UseR!2009*. Automatic Numerical Differentiation of Noisy, Time-Ordered Data. In R.

Subramaniam, G., & Varadhan, R. (2008). Borrowing Strength in Time Series Data Mining. In *Proceedings of American Statistical Association Conference, JSM, 2008.*

Sundaram, D., Mitra, K., & Webster, C. (1998). Word of mouth communications: A motivational analysis. *Advances in Consumer Research* [online]. [Accessed 01 May 2009].

Szoka, B., & Theirer, A. (2009, October). Net Neutrality, Slippery Slopes and High-Tech Mutually Assured Destruction. *Progress Snapshot, The Progress and Freedom Foundation, 5*(11).

Tajiri, N., & Okazaki, T. (2006). *Policies for Broadband Diffusion in Developing Countries*. Working Paper, Waseda University, GITS, Japan.

Tan, C. S. (2006, September 1). Steady flows of FDIs into ASEAN. *The Star Online*. Retrieved January 28, 2007 from http://sso.msc.com.my/index.php?option= com_content&task=view&id=184&Itemid=57

Tarasiewicz, M. (2003). Taking Gender Into Account. *UN Chronicle*, *40*(4), 37.

Taylor, A., & Harper, R. (2002, April). *Age-old practices in the 'new world': a study of gift-giving between teenage mobile phone users*. Paper presented at Changing Our World, Changing Ourselves, the SIGCHI Conference on Human Factors in Computing Systems. Minneapolis, Minnesota.

Teece, D. J., Pisano, G., & Shuen, A. (1997). Dynamic Capabilities and Strategic Management. *Strategic Management Journal*, *18*, 509–533. doi:10.1002/(SICI)1097-0266(199708)18:7<509::AID-SMJ882>3.0.CO;2-Z

Telecommunications Carriers' Forum. (2008). *Report on the Telecommunications Service Obligation for local service*. 23 July 2008. Retrieved from http://www.tcf.org.nz/library/df101bd9 -0668-452d-9227-e9e3b319d95d.cmr

Telecompaper (2009). Tele2 to introduce VDSL2 in Netherlands. Retrieved November 11, 2009, from http://www.telecompaper.com/news /article.aspx?cid=688321

Telegeography (2009). Vodafone to launch VDSL pilot in Heilbronn. Retrieved November 11, 2009, from http://www.telegeography.com /cu/article.php?article_id=29221

Temin, P. with L. Galambos (1987). *The fall of the Bell System*. New York: Cambridge University Press.

The Rights of Bits. (2009, September 26). *The Economist* (pp. 78-79).

The, F. C. C's Heavy Hand (2009, September 28). *WashingtonPost.com*. Retrieved from http://www.washingtonpost.com/wp-dy n/content/artic le/2009/09/27/AR2009092703026.html.

Thomas, F. (1995). *Telefonieren in Deutschland. Organisatorische, Technische und Räumliche Entwicklung eines Grosstechnischen Systems* (*Vol. 21*). Frankfurt, New York: Campus.

Thomas, F., & Mante-Meijer, E. (2001, June). *Internet Haves and Have Nots in Europe.* Paper presented at the e-Usages, Paris.

Thompson, H., & Garbacz, C. (2007). Mobile, Fixed Line and Internet Service Effects on Global Productivity Efficiency. *Information Economics and Policy*, *19*, 189–214. doi:10.1016/j.infoecopol.2007.03.002

Time Warner Cable, Inc. (2009, July 21). Reply Comments of Time Warner Cable, Inc. GN Docket No. 09-51.

Torero, M. B. J. (2006). *Information and communication technologies for development and poverty reduction: the potential of telecommunications.* Baltimore: Johns Hopkins University Press.

Trompenaars, F. (1993). *Riding the Waves of Culture*. Antwerp, Amsterdam: Economist Books.

Tsao, J. C., & Sibley, S. D. (2004). Displacement and reinforcement effects of the Internet and other media as sources of advertising information. *Journal of Advertising Research*, (March): 126–142. doi:10.1017/S0021849904040073

Tsatsou, P. (2009). Reconceptualising Time and Space in the Era of Electronic Media and Communications. *PLATFORM: Journal of media and communication, 1*, 11-32.

U.S. General Accounting Office. (2001). *Characteristics and choices of internet users*. Washington, DC (GAO-01-345).

UNCTAD. (2008). *World Investment Report, 2008*. Geneva: United Nations.

UNDP (2004). *Bridging the gender digital divide*. UNDP/UNIFEM. UNDP regional centre for Europe and the CIS and UNIFEM Central and Eastern Europe: Solvak Republic.

UNDP. (2005). *Promoting ICT for human development in Asia: Realizing the Millennium Development Goals.* Regional Human Development Report. UNDP.

UNECA. (2007). *Project on Information Society (IS) indicators – Current situation and prospects*. Geneva: United Nations.

UNIFEM. (2005). *Support to NEPAD*. Retrieved August 18, 2009 from http://www.un.org/africa/osaa/reports/UN%20folder2006-UNIFEM.pdf

United States Bureau of Census. (2008). *Statistical abstract of the United States, 2008*. Washington, D.C.: Bureau of Census.

United States Department of Commerce. (2001, October 10). NTT Launches world's first 3G cellular service. *International Trade Administration, International Market Insights*, Department of Commerce.

Van Ark, B., & Inklaar, R. (2005). *Catching Up or Getting Stuck? Europe's Troubles to Exploit ICT's Productivity Potential*. Groningen Growth and Development Centre, University of Groningen.

Vapi, X. (2005). Telkom drops ADSL and data prices further. Retrieved June 2, 2005 from http://www.telkom.co.za/common/aboutus/ mediacentre/pressrelease/articles/article_789b.html

Vaughan-Nichols, S. J. (2004). Achieving Wireless Broadband with WiMAX. *Computer*, *37*(6), 10–13. doi:10.1109/MC.2004.1266286

Verizon Communications, Inc. v. Law Offices of Curtis Trinko, LLP, 124 S. Ct. 872 (2004).

Vialle, P. (1998). *Stratégies des opérateurs de télécoms*. Paris: Editions Hermès.

Vialle, P. (Ed.). (2007). *Mutation des STICS. Acteurs, Ressources, Activités*. Paris: Hermes Science Publishing.

Vilpponen, A., Winter, S. & Sundqvist, S. (2006). Electronic WOM in Online Environments: Exploring Referral Network Structure and Adoption Behaviour. *Journal of Interactive Advertising* [online]. [Accessed 14 March 2009].

Voulgaris, Y., & Sotiropoulos, D. (2002). *Information Society, Sociology and Technology*. Athens: Operational Programme for the Information Society. (in Greek)

Walsham, G. (2005). Development, global futures and IS research: a polemic. *The Journal of Strategic Information Systems*, *14*, 5–15. doi:10.1016/j.jsis.2004.11.009

Walsham, G., & Sahay, S. (2006). Research on information systems in developing countries: current landscape and future prospects. *Information Technology for Development*, *12*(1), 7–24. doi:10.1002/itdj.20020

Walsham, G. (2002). Interpretive case studies in IS research: Nature and method. In Avison, D. E., & Myers, M. D. (Eds.), *Qualitative research in information systems: a reader* (pp. 101–113). London: SAGE.

Waverman, L., Meschi, M., & Fuss, M. (2005). The impact of telecoms on economic growth in developing countries in Africa: The Impact of Mobile Phones. *Vodafone Policy Paper Series*, *2*(March), 10–24.

Weiss, W. (1968). The impact of mass media. In Lindzey, G., & Aronson, E. (Eds.), *Handbook of Social Psychology* (*Vol. 5*, pp. 77–195). Reading, MA: Addison-Wesley.

Welker, C. (2002). The paradigm of viral communication. *Information Services and Use* [online] [Accessed 11 March 2009]

Wikipedia (2009). Neotel. Retrieved November 5, 2009 from http://en.wikipedia.org/wiki/Neotel

Wikipedia. (2009). Facebook [online]. Retrieved October 2, 2009 from http://en.wikipedia.org/wiki/Facebook

Wildman, S. (1992). *Information Technology, Productivity and Trade Implications*. Working Paper No 521, Columbia Institute for Tele-Information.

Williams, T. (1986). *The Impact of Television*. New York: Academic Press.

Williamson, O. (2000). New institutional economics: taking stock, looking ahead. *Journal of Economic Literature*, *38*, 595–613.

Williamson, O. E. (1985). *The Economic Institutions of Capitalism*. New York: Free Press.

Wilson, A. (1994). *Wire and wireless: a history of telecommunications in New Zealand, 1860-1987*. Palmerston North, New Zealand: Dunmore Press.

Wolfe, J. (2006). Space race. The Advertiser, August, 36-39 [online]. Retrieved March 11, 2009 from http://www.ana.net

World Economic Forum. (2007). The Africa Competitiveness Report 2007. Retrieved June 13, 2007 from http://www.weforum.org/en/initiatives/gcp/Africa %20Competitiveness%20Report/PastReports/index.htm

World Economic Forum. (2008). The Global Competitiveness Report 2008-2009. Geneva, Switzerland.

WSIS. (2005) *Tunis Agenda: for The Information Society.* Retrieved from http://www.itu.int/wsis/d ocs2/tunis/off/6rev1.html Accessed date: 18.09.2009

WSIS. (2008) *WSIS follow-up Report 2008*. Retrieved from http://www.unctad.org/e n/docs/none20081_en.pdf Accessed date: 18.09.2009

Xu, Y. (2006). The regulatory environment for future mobile multimedia services: The case of Hong Kong SAR and China. *ITU New Initiatives Workshop on the Regulatory Environment for Future Mobile Multimedia Services*. Geneva: International Telecommunication Union.

Yun, K., Lee, H., & Lim, So-Hye (2002). *The Growth of Broadband Internet Connections in South Korea: Contributing Factors.* Working paper, Asia-Pacific Research Center, Stanford University. Retrieved from http://APARC.stanford.edu

Zhu, K. (2001). Internet-based Distribution of Digital Videos: the Economic Impacts of Digitization of the Motion Picture Industry. *Electronic Markets*, *11*(4), 273–280. doi:10.1080/101967801753405562

About the Contributors

Yogesh K. Dwivedi is a senior lecturer at the School of Business and Economics at Swansea University in the UK. He was awarded his M.Sc. and Ph.D. by Brunel University in the UK, receiving a Highly Commended award for his doctoral work jointly by the European Foundation for Management and Development and Emerald Publishing Ltd. He is a qualified practitioner in Project Management Methodology PRINCE2. His research focuses on the adoption, diffusion and impact of ICT in organisation and society. He has authored a research monograph on consumer adoption and usage of broadband and has co-edited seven books and handbooks on various IS themes. Dr Dwivedi has written almost 100 refereed papers in international conferences (such as AMCIS, ECIS, PACIS and IFIP WG8.6) and leading journals including Communications of the ACM, the DATA BASE for Advances in Information Systems, the Information Systems Journal, Information Systems Frontiers, International Journal of Production Research, the European Journal of Information Systems, the Journal of the Operational Research Society, Journal of Computer Information Systems, Journal of Enterprise Information Management and Journal of Information Technology. He is the Senior Editor of the DATA BASE for Advances in Information Systems, Managing Editor of Journal of Electronic Commerce Research, Assistant Editor of Transforming Government: People, Process and Policy, member of the editorial board/review board of a number of other journals and as well as co-chairs of track and mini-tracks to international conferences. He has also co-edited special issues in leading journals that include: the DATA BASE for Advances in Information Systems, the Information Systems Journal, Information Systems Frontiers, Journal of Enterprise Information Management, Electronic Government, An International Journal and International Journal of Electronic Government Research. He is a member of the Association of Information Systems, IFIP WG8.6 and the Global Institute of Flexible Systems Management, New Delhi. He can be contacted at ykdwivedi@gmail.com.

* * *

Syeda Rownak Afza is a senior lecturer in BRAC Business school, BRAC University, Dhaka, Bangladesh. She has completed her MBA from University of Dhaka. Her research interests include Service Marketing, Women empowerment and ICT4D.

Md. Mahfuz Ashraf is a Lecturer in Department of Management Information Systems, University of Dhaka. His research interests include IT/IS adoption and implementation in rural-urban settings, impact study of ICT for development (ICT4D) and the digital divide.

Justin Beneke is a full-time marketing lecturer and programme convenor in the School of Management Studies at the University of Cape Town. His academic interests lie in the fields of electronic marketing, retailing management and higher education marketing.

Christos Bouras obtained his Diploma and PhD from the Department Of Computer Engineering and Informatics of Patras University (Greece). He is currently a Professor in the above department. Also he is a scientific advisor of Research Unit 6 in Research Academic Computer Technology Institute (CTI), Patras, Greece. His research interests include Analysis of Performance of Networking and Computer Systems, Computer Networks and Protocols, Telematics and New Services, QoS and Pricing for Networks and Services, e-Learning Networked Virtual Environments and WWW Issues. He has extended professional experience in Design and Analysis of Networks, Protocols, Telematics and New Services. He has published 300 papers in various well-known refereed conferences and journals. He is a co-author of 8 books in Greek. He has been a PC member and referee in various international journals and conferences. He has participated in R&D projects such as RACE, ESPRIT, TELEMATICS, EDUCATIONAL MULTIMEDIA, ISPO, EMPLOYMENT, ADAPT, STRIDE, EUROFORM, IST, GROWTH and others. Also he is member of experts in the Greek Research and Technology Network (GRNET), Advisory Committee Member to the World Wide Web Consortium (W3C), IEEE - CS Technical Committee on Learning Technologies, IEEE ComSoc Radio Communications Committee, IASTED Technical Committee on Education WG6.4 Internet Applications Engineering of IFIP, ACM, IEEE, EDEN, AACE, New York Academy of Sciences and Technical Chamber of Greece.

Nabyla Daidj is Associate Professor in Strategy at TELECOM Business School and Director of the Management, Marketing and Strategy department. Her teaching and research interests are corporate strategy, inter-organizational relationships (strategic alliances) and conglomerates' performance and corporate governance in a context of coopetition. She has also developed an expertise in the Media and Electronic Games industry. She has published in 2008 a book about cooperation, games theory and strategic management. Currently, she is studying the sources of value creation in international firms.

Qiuyan Fan is a lecturer at University of Western Sydney, Australia. Dr Fan has a particular research interest in the issues surrounding egovernment, e-commerce/e-business and broadband diffusion. Over the last few years she has published a number of referred Journal articles, book chapter and refereed conference papers in the areas above. Her doctorate research looked at how Australian government policies influence Internet access. She has been supervising and co-supervising honours and PhD students.

Elizabeth Fife is the Associate Dir. for Industry Studies at the Institute for Communication Technology Management (CTM), Marshall School of Business, University of Southern California and has a teaching appointment in the Viterbi School of Engineering. She has over 9 years research experience studying international telecommunications markets with expertise in the wireless industry and media/content business, developmental economics, and ICT adoption and innovation. Teaching interests include technical and professional communications for engineers and industry professionals; evolution of telecom/entertainment/digital markets.

Hidenori Fuke is Professor at Faculty of Global Media Studies, Komazawa University in Tokyo, Japan since April 2007. He has previously experienced regulatory matters as a senior executive at Nippon Telegraph and Telephone Corporation and done extensive research on regulatory policy as a board of director at InfoCom Research, Inc,. He was a member of Faculty of Informatics, Kansai University in Osaka from 2000 to 2007.He received B.A. in Economics from Tokyo University, and M.Litt. from Glasgow University in Scotland. He received Ph.D. in International Public Policy from Osaka University.

Roya Gholami is a lecturer in Operations and Information Management Group, Aston Business School, Birmingham, UK. Her current research interests are IT business value, Green IT, IT adoption and diffusion, IT and Development. She has published in IEEE Transactions on Engineering Management, Information & Management, Journal of Global Information Management, Information Resource Management Journal, World Economy, Technovation and Electronic Journal of Information Systems in Developing Countries.

Apostolos Gkamas obtained his Diploma, Master Degree and Ph.D from the Computer Engineering and Informatics Department of Patras University (Greece). He is currently Scientific cooperator at the Research Unit 6 of the Research Academic Computer Technology Institute, Patras, Greece, visiting lecturer in Hellenic Open University in the Informatics Curriculum and Visiting Assistant Professor in the Informatics Department of University of Ioannina. His research interests include Computer Networks, Telematics, Multimedia transmission and Cross Layer Design. More particular he is engaged in transmission of multimedia data over networks and multicast congestion control. He has published more than 60 papers in international Journals and well-known refereed conferences. He is also co-author of three books (one with subject Multimedia and Computer Networks one with subject Special Network Issues and one with subject IPv6). He has participated in various R&D project (in both EU and national) such as IST, FP6, FP7, Intereg eLearning, PENED, EPEAEK, Information Society.

Helena Grunfeld is a research scholar at the Centre of Strategic Economic Studies, Victoria University, Melbourne, Australia. Her research interests include evaluation of how ICT can contribute to development, focussing on its impacts on capabilities, empowerment, and sustainability.

Janice A. Hauge- Dr. Hauge is an associate professor of economics at the University of North Texas, and is also a senior research associate with the Public Utility Research Center at the University of Florida. Dr. Hauge's current research topics include competition policy in telecommunications and broadband, network studies, and regulatory analyses of both telecommunications and energy sectors. Recent work includes analyses of broadband Internet adoption in Europe and the US, analyses of broadband demand policies and programs, and the empirical studies of the effects of government policies on broadband adoption. Currently Dr. Hauge teaches strategic behavior and applied microeconomic theory courses at the undergraduate and graduate levels.

Bronwyn Howell is General Manager of the New Zealand Institute for the Study of Competition and Regulation and a faculty member of Victoria Management School, Victoria University of Wellington. In 2007, she was Visiting Research Scientist at the Networking Laboratory, Helsinki University of Technology. Bronwyn has a background as both an academic researching telecommunications, ICT and health markets, and as a practitioner, holding managerial positions in the ICT and health sectors in both New Zealand and multinational firms. Her research interests relate to the ownership, governance and management of firms operating in regulated industries.

Wen-yi Hsu is a part-time assistant professor of the Department of Radio and TV at National Chengchi University in Taiwan. She served as a senior reporter, anchorwoman and chief editor of the news department at the Broadcasting Corporation of China from June 1989 to December 2004. She was elected as an Honorary Member of the Phi Tau Phi Scholastic Honor Society of the Republic of China in 2009. Her research interests include radio theory and practice, digital broadcasting, radio news reporting and writing, radio in-depth reporting, radio programming & production, the ecology of electronic media, and new communications technology.

Vikas Jain holds a Ph.D. in Information Systems and Decision Systems from The George Washington University, Washington DC. He completed his masters in information systems from IIT, New Delhi, India and a Bachelor of Technology degree in Electronics Engineering from Delhi Institute of Technology, Delhi, India. He has worked for nearly 12 years in IT industry in India in different areas ranging from software development, IT market research and consulting, and e-learning. He has worked on projects for clients like IBM, Microsoft, HP, Prudential Insurance among many others. Currently, Dr. Jain is an assistant professor of information and technology management at the University of Tampa, Tampa, FL and conducts research in the areas of enterprise resource planning systems, electronic government, and IT business value. He has served as guest editor for number of journals including Electronic Government: An International Journal, Journal of Global Information Technology Management (JGITM), and International Journal of Organization and End User Computing. He also serves as an editorial review board member for International Journal of Electronic Government Research.

Mark A. Jamison - Dr. Mark Jamison is the director of the Public Utility Research Center at the University of Florida, which provides international training and research on business and government policy, focusing primarily on utilities and network industries. Dr. Jamison's current research topics include leadership and institutional development in regulation, competition and subsidies in telecommunications, and regulation for next generation networks. He has conducted education programs in Asia, Africa, Europe, the Caribbean, and North, South, and Central America. He is also a research associate with the UF Center for Public Policy Research and with Cambridge Leadership Associates. Dr. Jamison has served as chairperson of the National Association of Regulatory Utility Commissioners (NARUC) Staff Subcommittee on Communications, chairperson of the State Staff for the Federal/State Joint Conference on Open Network Architecture, and member of the State Staff for the Federal/State Joint Board on Separations.

Marijn Janssen is an Associate Professor at the Information and Communication Technology section and Director of the interdisciplinary SEPAM Master program of the Faculty of Technology, Policy and Management at Delft University of Technology. He has conducted and managed several simulation and gaming projects, and published over 150 refereed publications and serves on several editorial boards and conferences in the area of e-government. For more information, visit www.tbm.tudelft.nl/marijnj.

Yu-li Liu is a professor of the Department of Radio and TV at National Chengchi University in Taiwan. She served as one of the Commissioners of the National Communications Commission (NCC) from February 2006 to February 2008. She has received grants from the U.S. Fulbright Foundation and Taiwan's National Science Council for research projects in the areas of convergence, broadband networks, and digital TV in Taiwan. Her research interests include digital broadcasting, broadband communications, convergence, telecommunications law and regulation, new communications technology, and media management. She has published numerous books such as Multi-channel TV and Audience, Cable TV Management and Programming Strategy, Cable TV Programming and Policy in China, Radio and TV, Telecommunications, etc.

Bushra Tahseen Malik is a research in Brainstorm Bangladesh. He is currently working in SIRCA jointly funded by IDRC and NTU, Singapore under an impact study of ICT for development in Bangladesh.

Mircea I. Marcu - Dr. Marcu is a Postdoctoral Research Associate at the University of Florida Institute for Child Health Policy. He received his Ph.D. from the University of Florida in 2009, and M.A.s from Central European University (2001) and Babeş-Bolyai University (1999). Dr. Marcu's areas of expertise are Industrial Organization, Microeconomics, and Econometrics. His current research interests include the economics of two-sided markets, competition policy, regulation of public utilities, health care reform, and the estimation of health care costs and utilization. His research on telecommunications has been recognized at the 32nd and 35th Telecommunications Policy Research Conference, where he was awarded the second and first prize in the graduate student competition. He serves as a referee/reviewer for Utilities Policy.

John B. Meisel received a B.A. degree from St. Lawrence University in 1973 and a Ph.D. in economics from Boston College in 1977. He is a Professor of Economics and Finance at Southern Illinois University Edwardsville, where he has been employed since 1977. He served as staff member in the Office of Plans and Policy at the Federal Communications Commission while on sabbatical from the university. He has also worked as a consultant with the Illinois Commerce Commission and the St. Louis-based consulting firm, The Resource Group. His research interests are in the fields of industrial organization, antitrust and regulation. He has published over 20 articles in peer-reviewed economics journals, including The Review of Economics and Statistics. His current research interest involves the public policy implications of the Internet on market structure, specifically effects on the music, movies and television markets.

Mitchell Morehart is a senior agricultural economist at the Economic Research Service. He conducts research and develops current information and longer-term perspectives about the economic well-being of farm households, the financial structure and performance of farm businesses, and the economic contribution and performance of the farm sector of the U.S. economy. He has authored or co-authored of more than 80 ERS bulletins and journal articles that report research results in agricultural business, finance, and management.

Navonil Mustafee is a Lecturer in the School of Business and Economics at Swansea University. Prior to this, he was a Research Fellow in Brunel University and Warwick Business School. His research interests are in e-Infrastructures and Grid Computing, Information Systems, Operations Research and Healthcare Simulation. He completed his PhD in Information Systems and Computing from Brunel University in 2007. He is a member of the drafting group of the COTS Simulation Package Interoperability Product Development Group (CSPI-PDG) under the Simulation Interoperability Standards Organization. His email address is navonil.mustafee@gmail.com.

Banani Nandi is a Principal Technical Staff Member with AT&T Shannon Laboratories (formerly AT&T Bell Laboratories). She received her Ph.D. degree from New York University and her Ph.D. dissertation was awarded the Otto Ehrlich Memorial Award. Before joining AT&T, she taught at Calcutta University in India, Columbia University, New York University, and Rutgers University in U.S.A. Dr. Nandi's fields of specialization include International Trade and Econometrics. At present, her research focuses on the impact of telecommunications/information technologies on productivity and economic growth.

John C. Navin is a Professor of Economics and Finance at Southern Illinois University Edwardsville. He received his B.S. in Business Administration from Bowling Green State University in 1987. He received his Ph.D. from Michigan State University in 1992. Dr. Navin's areas of specialization include public finance, labor and health economics. He has published articles in such journals as Health Economics, Economic Inquiry, Public Finance Review and Growth and Change. Dr. Navin is also a partner in the economic consulting firm RSN Economic Group, Inc.

Anastasia Papazafeiropoulou is a lecturer in the Information Systems and Computing department at Brunel University, UK. Her PhD was awarded by Brunel University and she also holds a first degree in Informatics and a M.Sc. in Information Systems both from the Athens University of Economics and Business, Greece. She has been involved in a number of European and UK funded research projects with emphasis on electronic commerce and small and medium size enterprises (SMEs). She teaches information systems management and business integration at the postgraduate level. She also supervises researchers in the field of technology adoption by organisations with special interest in developing countries. She has 10 years of research experience on the study of diffusion and adoption of electronic commerce, broadband Internet, Enterprise Recourse Planning Systems (ERPs), Customer Relationship Management systems (CRMs), IP-telephony and mobile TV.

Francis Pereira is Director of Industry Research at the Institute for Communications, Technology and Management (CTM) and Lecturer in the Information & Operations Department, Marshall School of Business, University of Southern California. He received his Ph.D. in Political Economy and Public Policy from the University of Southern California, and teaches courses in economics, statistics and electronic commerce. For the past fifteen years, his research has focused on key business issues in telecommunications field, particularly telehealth, and adoption rates of E-commerce applications especially in the small and medium size enterprises, and business models in the new multimedia environment and the effects of emerging technologies on these models. Francis has authored several book chapters and has published papers in journals that include the Journal of Communications Networks and the Journal of the Institution of British Telecommunication Engineers.

Peter L. Stenberg is a senior economist at the Economic Research Service. He received his PhD in economics from the University of Minnesota and has conducted research with respect to North American, African, and Eastern European economies covering topics in economic development policy and human and physical infrastructure. He has written over 100 articles on telecommunications, economic impact of universities, science and technology policy, defense restructuring, high tech industry, and other economic development and policy subjects.

Ganesh K Subramaniam is a statistician and principal member of technical staff at AT&T Laboratories, Florham Park NJ. He joined AT&T in 1992 and since has worked on many statistical problems related to consumer behavior, network data, speech recognition, and financial modeling. His current research interest includes time series analysis, functional data analysis and computational statistics related to large data. He received his PhD in Statistics from the University of Leeds, UK.

Timothy Sullivan is an Instructor of Economics and Finance at Southern Illinois University Edwardsville, where he has been employed since 1995. He received B.S. Degrees in Economics and Computer Science at the University of Missouri – St. Louis in 1988 and a Ph.D. in Economics at the University of Maryland in 1995. Dr. Sullivan's research interests include statistical analysis, economic impact analysis and the Internet's effect on media industries. His research has been published in peer-reviewed journals, including Economic Development Quarterly, The Southern Journal of Business and Economics and info. He is a partner with the economic consulting firm RSN Economic Group, Inc.

Simon J. E. Taylor is the Founder and Chair of the COTS Simulation Package Interoperability Product Development Group (CSPI-PDG) under the Simulation Interoperability Standards Organization. He is the co-founding Editor-in-Chief of the UK Operational Research Society's (ORS) Journal of Simulation and the Simulation Workshop series. He was Chair of ACM's Special Interest Group on Simulation (SIGSIM) (2005-2008). He is a Reader in the School of Information Systems, Computing and Mathematics at Brunel and has published over 100 articles in modelling and simulation. His recent work has focused on the development of standards for distributed simulation in industry. His email address is simon.taylor@brunel.ac.uk.

Panayiota Tsatsou (BA, MA, MSc, PhD) is a lecturer in Media and Communication at Swansea University, United Kingdom. Her doctoral research at the London School of Economics and Political Science explored the role of everyday culture and decision-making in shaping digital divides and was supported by the Hellenic Republic State Scholarships Foundation. Previously her postgraduate work was supported by the N.A.T.O Science Fellowship Program and other fellowship programmes. She is involved in European research projects on Information and Communication Technologies (ICTs), with an emphasis on regulation and policy creation, as well as the role of ordinary people as users and actors in the Information Society. Her research aims to develop innovative and evidence-based solutions to issues arising in the Information Society through the submission of papers and reports to national and EU policy and regulation authorities. Along these lines, she publishes in the areas of digital divides, policy and regulation for the Information Society children and new media, cyberbullying, new media and gender.

Thrasyvoulos Tsiatsos obtained his diploma, his master's degree and his PhD from the computer engineering and informatics department of Patras University (Greece). He is currently lecturer in the Department of Informatics of Aristotle University of Thessaloniki as well as research member at Research Unit 6 of Research Academic Computer Technology Institute. His research interests include computer networks, telematics, networked virtual environments, multimedia and hypermedia.

Pierre Vialle is Professor at TELECOM Business School and Director of the MSc in ICT Business Management program. His teaching and research interests are inter-organizational relationships and networks, innovation, and management of high-technology firms. He has developed an expertise in the Information Communications Technology Industry and has published several books, chapters and articles. Currently he is studying the impact of convergence on the strategies of firms, and innovation in successful catching-up companies.

Index

F

G

H

I

Breinigsville, PA USA
10 December 2010
251061BV00002B/1-80/P

9 781609 600112